NEW PHENOLOGY

NEW PHENOLOGY

Elements of Mathematical Forecasting in Ecology

ALEXANDER S. PODOLSKY

A Wiley-Interscience Publication

JOHN WILEY & SONS

New York   Chichester   Brisbane   Toronto   Singapore

Library of Congress Cataloging in Publication Data:

Podolsky, Alexander S.
  New phenology.

  "A Wiley-Interscience publication."
  Bibliography: p.
  Includes index.
  1. Phenology. 2. Ecology—Mathematics. I. Title.
QH544.P677 1983   574.5   83-16723
ISBN 0-471-86451-X

Printed in the United States of America
10 9 8 7 6 5 4 3 2 1

Preface

This book is intended for use by a very wide range of practical workers, scientists, and students: agronomists, agrometeorologists, testers of crop varieties and crop breeders, entomologists and phytopathologists, medical and veterinary parasitologists, epidemiologists and epizootologists, cattle breeders, bee keepers, staff in quarantine establishments, and people working in agricultural aviation, as well as people dealing with "satellite phenology" and with medicinal plants.

The new method of phenological forecasts and bioclimatic estimations ("experimental and mathematical heat phenology") was published by me for the first time in April 1957. During the succeeding period of more than 25 years, the method was continuously developed by myself and my coworkers.

The first general description of the method was published in 1967 in the book *News in Phenological Forecasting: Mathematical Forecasting in Ecology*, in Moscow by Kolos Publishing House.

The second edition of the book, supplemented and revised, was published by the same publishing house in 1974 under the title *Phenological Forecasting: Mathematical Forecasting in Ecology*.

The present edition of *New Phenology* is the third, but this revision appears in English. The third edition is qualitatively and quantitatively different from the previous editions; it represents a total of many years' work by myself and by my coworkers, although the method is in fact still at a dynamic stage of development.

Nevertheless, it is appropriate to ask: can anything still remain unknown in such an ancient field of knowledge as phenology? (Let us remember that the first phenological calendar was compiled 2500 years ago in China; that the longest period of phenological observations of cherry trees in gardens of the Mikado is up to 1200 years; that the temperature totals method has existed now for 250 years—since Reaumur.)

Can anything new be found in a science where "there is not the smallest space to tread" (a picturesque expression used by a well-wisher during a time when the proposed method was publicly criticized)?

v

My answer is: such new knowledge can be found, and must be found. As you know, Rutherford named his investigations about converting one chemical element into another "new alchemy," although alchemy is "as old as the world is" and seemed to have been forgotten. Lord Thomson believed that it was impossible to discover anything new in physics, but Planck came along and founded quantum mechanics in contrast to the previous Newtonian mechanics.

I well understand that phenology and quantum physics are not comparable either in their content or in their level. But our scientific truth also had to go through three stages before entering into the fourth stage—the recognition of novelty. Rutherford expressed this concept well when he said that each scientific truth goes through three stages: at first it is considered absurd; then it is thought to have something to it; at last, it is said to have been known for a long time.

But we were not afraid of critics, since we were constantly supported by facts, and by the motto: not a single new step in theory without a verification of the truth of the previous step in practice.

As to the slight differences in the titles of the first, second, and third editions, it seems to me that phenology and phenological forecasts are synonomous to a high degree. Phenology, phenological observations, and phenomena in the past and in the present were always of interest to people from the point of view of extrapolations into the future.

It is true that phenology also includes a comparatively small (for the present) and subordinate section about observational methods. However, in this book I shall not touch on these methods, with a few exceptions. For instance, one such exception is some ideas about "satellite phenology" at the end of the book.

I want to express my deep gratitude to my post-graduate students and colleagues for their unbiased faith in the new method. I wish to express sincere gratitude to my wife Bertha Podolsky for inspiring me day by day, and also for her major role in translating the manuscript into English. I also wish to thank that extremely kindhearted person, Ann Riffin, as well as Helen Strumpf for her help in translating this book.

Readers' comments and suggestions are welcome. Please send them to the following address: Abarbanel str., 38/35, Beer-Sheva, Israel.

ALEXANDER S. PODOLSKY

Beer-Sheva, Israel
January 1984

Contents

Symbols and Abbreviations

Symbol	Signification
CBEZ	Central Black-Earth Zone in Russia (TzChO—in Russian)
ESU	European Section of the USSR
I_m	Imago
$L_{1,2,3,4}$	Larvae (caterpillars) of 1st, 2nd, 3rd, and 4th ages, respectively (on the graphs)
NTV	Network of centers for testing crop varieties in the USSR
O_v	Egg
P	Pupa
PTN	Phenotemperature nomogram
t_{av}	Average temperature over a certain period
VASHNIL	The All-Union Academy of Agricultural Sciences (in the USSR)
VIR	The All-Union Institute of Plant-Growing (in the USSR)
VIZR	The All-Union Institute of Plant Protection (in the USSR)

Note: Dates on the lines of average temperatures in the nets of PTNs: arabic numerals denote the days of the months, and roman numerals denote the months. For example, 5 IV is April 5. The same is true for the tables, and sometimes for formulas and the text. In the nets for Israel, all dates are designated by arabic numerals, for example, 5.04.

NEW PHENOLOGY

CHAPTER ONE

A New Method for Making Phenological Forecasts and Bioclimatic Estimations

The world's largest group of living organisms constitutes plants and cold-blooded creatures (poikilotherms).

Cold-blooded creatures include insects (up to 1.5 million species), ticks, worms, protozoa, an so on. Many are either crop pests or the bearers and causes of diseases in man and in livestock.

These organisms have specific periods of development, which are dependent basically on the sun's heat. Data about temperature can be used to make various calculations about the developmental dates of plants and cold-blooded organisms.

This book presents a method for making such calculations that is more scientific, more accurate, and quicker than other methods. This method also enables us to solve a number of problems in various areas of ecology that cannot be solved in any other way.

Briefly, this experimental-mathematical method involves a simple graphical solution of a system composed of two families of relatively complex empirical equations. One family deals with the heat resources of the geographical region. The other family deals with the heat requirements of the organisms in question. Where necessary, these requirements are determined under differing sets of nontemperature factors, for example, humidity, light, agricultural techniques, and so on. The method of the graphic solution presented is direct in that it does not use artificial temperature totals or thresholds.

The equation expressing the organism's requirements (or its reaction to the heat factor) represents an empirical curve, relating the length, in days (n), of one particular stage of the organism's development to the mean temperature during this concrete period of time, t_{av} (the regular temperature, regardless of any effective temperature thresholds). This kind of curve will be referred to as *an organism's*

1

developmental temperature curve, a phenological curve, or simply *a phenocurve.*

The developmental curve, which is stable for a given type of organism (Figs. 1.1, 1.2, 1.4, 1.5, 1.17, 1.27, 1.41, 1.45, 1.46, 1.47) but varies greatly for different organisms, may be considered as the organism's biological passport. The curve is constructed on the basis of a relationship between observations of the length of each developmental stage and the ambient temperatures. In the case of plants, local phenological field observations must be obtained for different years from various meteorological or experimental agricultural crop centers. This data will then be related to the atmospheric temperature recorded at the local meteorological stations. After observing the temperature differences for different years (cold and warm) and between different points in a given region (northern, southern, highlands, valleys), together with differences in the sowing dates, it is possible to arrive at a knowledge of the necessary temperature range for the developmental stages of a given crop.

Table 1.1 presents an example of the data required.

The results of normal phenological observations taken at either experimental agricultural crop centers or meteorological stations are given in Table 1.1 in the lines "Sowing date" and "Mass sprouting date."* The line headed "*n*" gives the difference in days between the sowing and the mass sprouting dates. The following line gives the results of totaling the 24-hour average air temperatures over this entire period. For example, for Shaartuz in 1957 the total required is that of all the 24-hour average air temperatures over a 22-day period, from the sowing date (April 1) until the day before the actual mass sprouting date (April 22) inclusive (i.e., excluding from the calculation the actual mass sprouting date, April 23).

Thus, in calculations relating to the period of time between any two developmental stages in any crop or pest, the 24-hour average air temperature on the day when the first stage is reached is included in the total, whereas that of the day when the second stage is reached is not included; the purpose of this is to ensure the necessary degree of consistency in the methods used. The temperature recordings must be relevant to the point where the phenological observations are carried out. In practice it is not always possible to be absolutely dogmatic about this, but at any rate the temperature data should be taken from the meteorological station that is nearest to the agricultural crop center where the phenological observations are made.

The line Σt in Table 1.1 gives the totals of the 24-hour average air

---

*In the USSR, "mass sprouting" is defined as the time when 50% of the plants are observed above the surface. The same percentage is referred to with other mass stages of development in cotton and other crops.

temperatures for the period of time between developmental stages. As can be seen, these totals are obtained without reference to any effective* or active† temperature thresholds, and are only necessary for calculating the ordinary average temperature t_{av} during the interstage period of n days. For example, in the case of Shaartuz (Tadzhikistan) in 1957, the average temperature during the period from the date when cotton variety 5904-I was sown (April 1) until its mass sprouting date (April 23) was 14.9°C. This was calculated by taking the total of the 24-hour average air temperatures throughout this period (April 1 to 22 inclusive) and dividing it by the number of days:

$$t_{av} = \frac{\Sigma t}{n} = \frac{327.6°\,C}{22} = 14.9°\,C \tag{1}$$

Point c with coordinates 14.9°C and 22 days can be found in Figure 1.1 (curve 1). The other points in Figure 1.1 were plotted in the same way: a (coordinates $t_{av} = 22.0°$C and $n = 13$ days); b ($t_{av} = 20.0°$C and $n = 13$ days); d ($t_{av} = 16.2°$C and $n = 11$ days); and so on, all of which relate to different years and different locations in Tadzhikistan.

The result of these calculations is a cluster of points in the lower part of Figure 1.1, forming a fairly regular band. If some points clearly fall outside the area where the majority fall, then they are left out of the succeeding analysis.

This cluster of points is used as the basis for drawing the most probable line relating the length of the cotton's interdevelopmental stage‡ (n) and the average temperature during the period (t_{av}). Strictly speaking, this line, which is known as a regression curve, should be calculated on the basis of the points' coordinates using well-known methods of mathematical statistics. For example, if we divide curve 4 in Figure 1.2 into two segments—from 18.0° to 23.6°C and from 23.7° to 27.3°C—then the curvilinear regression equations for these two segments will be, respectively:

$$n = \frac{1}{0.0016t - 0.009} \tag{2}$$

$$n = \frac{1}{0.0022t - 0.024} \tag{3}$$

*Effective temperature: any temperature that is higher than the lower threshold temperature required for the development of a given organism, minus this lower threshold.
†Active temperature: any temperature higher than 10°C.
‡"Interdevelopmental stage" means the length of time (n days) between any two moments in the development of an organism. For example, between sowing and sprouting, or between sprouting and budding, and so on.

TABLE 1.1. Data collected for constructing a phenological developmental curve for cotton variety 5904-I from sowing until mass sprouting; observations were made by experimental agricultural crop centers and meteorological stations in Tadzhikistan (USSR).

Factor	1955	1956	1957	1958	1959	1960	1961
Phenological observations are from the Shaartuz agricultural crop center; air temperatures were recorded by the "Shaartuz" meteorological station.							
Sowing date	24 IV	9 VI	1 IV	9 IV	17 IV	26 IV	14 IV
Mass sprouting date	7 V	22 IV	23 IV	24 IV	26 IV	14 V	25 IV
Length of time between developmental stages in n days	**13**	**13**	**22**	**15**	**9**	**18**	**11**
Total of 24-hour average temperatures Σt for n days (in °C)	286.5	260.5	327.6	297.5	195.6	359.9	177.7
Average temperature for whole period $t_{av} = \Sigma t / n$ (in °C)	**22.0**	**20.0**	**14.9**	**19.8**	**21.7**	**20.0**	**16.2**
Designation of corresponding point in Fig. 1.1 (on curve 1)	a	b	c	Not designated by letters (ND)			d
Phenological observations from the Leninabad agricultural crop center; air temperatures were recorded by the "Leninabad" meteorological station.							
Sowing date	17 IV	14 IV	13 IV	18 IV	26 IV	25 IV	23 IV
Mass sprouting date	6 V	24 IV	25 IV	4 IV	5 V	12 V	5 V

Length of time between developmental stages in n days	**19**	**10**	**12**	**16**	**9**	**17**	**12**
Total of 24-hour average temperatures Σt for n days (in °C)	335.3	173.5	186.4	281.1	163.1	294.9	245.3
Average temperature for whole period $t_{av} = \Sigma t/n$ (in °C)	**17.6**	**17.4**	**15.5**	**17.6**	**18.1**	**17.3**	**20.4**
Designation of corresponding point in Fig. 1.1 (on curve 1)	e	ND	f	ND	ND	q	ND

Phenological observations from the "Chapaev" meteorological station; air temperatures were recorded by the "Kurgan-Tyube" meteorological station.

Sowing date	—	4 IV	26 III	—	16 IV	I IV	—
Mass sprouting date	—	16 IV	20 IV	—	4 V	28 IV	—
Length of time between developmental stages in n days	—	**12**	**25**	—	**18**	**27**	—
Total of 24-hour average temperatures Σt for n days (in °C)	—	202.0	329.6	—	360.2	376.1	—
Average temperature for whole period $t_{av} = \Sigma t/n$ (in °C)	—	**16.8**	**13.2**	—	**20.0**	**13.9**	—
Designation of corresponding point in Fig. 1.1 (on curve 1)	—	ND	ND	ND	ND	h	—

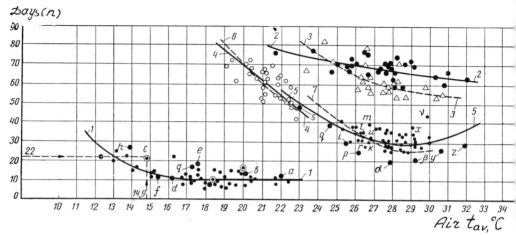

FIGURE 1.1. The phenological curves of a number of organisms. Curve 1: fine-fiber cotton *Gossypium barbadense* variety 5904-I from sowing to mass sprouting (plotted on the basis of phenological observations made by NV in Tadzhikistan from 1955-1961 inclusive, as well as on the basis of phenological and temperature recordings from the Tadzhikistan's network of meteorological stations from 1953 to 1960); curve 2: cotton variety 5904-I from mass flowering to mass boll unfolding (phenological observations from NTV, 1955-1961); curve 3: the same as curve 2, but with phenological and temperature recordings from the Tadzhikistan's meteorological network (in the Hydrometeorological Division's method, the seed boll is considered open as soon as the fiber begins to be visible). Curves for the other developmental stages in cotton variety 5904-I can be found in Fig. 1.31); curve 4: cotton bollworm (*Chloridea obsoleta F*) from that time in spring when the 24-hour average air temperature exceeds 15° C (i.e., the time when this species of pests awakens after the winter) until the beginning of the first mass egg laying (from field observations made during 1948-1961, according to the Tadzhikistan's entomological station records, and according to the corresponding temperature recordings from meteorological stations); curve 5: the same as curve 4, but between the beginning of the first and second, second and third, and third and fourth mass egg layings; curves 6 and 7: cotton bollworm, from observations in Azerbaidzhan (after A. R. Mustafaev; points are not illustrated).

However, since straight-line regression is simpler, we can first alter the quantity n to $1/n$. This gives us two straight-line regression equations $1/n = \varphi(t)$, which can then be easily transferred back to the curvilinear form $n = \psi(t)$.

After a little practice it is possible in most cases to draw the curve quite quickly without any need for these involved calculations. In order to do this, the band of points is divided into two or three parts of approximately equal length, and in each of these parts a test curve is drawn in pencil, in such a manner that the number of points above the curve approximately equals the number below it. Then the separate parts of the curve are joined into a single smooth curve, which is again checked by counting whether the number of points above the curve approximately equals the number below it, not including those points which lie on the line. In the case of curve 1, Figure 1.1, for example, there are 23 points above the curve, 24 below it, and 8 points lying on the curve.

The same method—that is, the geometrical alignment of a row of points—is used for constructing the dotted straight lines on Figures 1.8 and 1.9. As can be seen, these dotted lines, drawn "by hand," are located near the unbroken lines, which are drawn on the basis of regression equations.

$$n = -1.44 \, t_{av} + 59.4 \text{ (Fig. 1.8)} \tag{4}$$

and

$$n = -1.78 \, t_{av} + 60.6 \text{ (Fig. 1.9)} \tag{5}$$

In some instances there may be doubt as to the accuracy of the curve. In this case the accepted criterion for testing a curve's accuracy should be calculated for each separate segment of the curve as follows:

$$\frac{|r|}{|\sigma_r|} \geqslant 3 \tag{6}$$

where σ_r = the standard deviation in the correlation coefficient;
r = the correlation coefficient between the temperature (t_{av}) and the interdevelopmental stage (n).

Therefore, curve 1 in Figure 1.1 is the phenological developmental curve for cotton variety 5904-I during the period from sowing to sprouting.

Lines 4 and 5 (Figure 1.1) relate to the development of cold-blooded animal organisms. These lines are plotted in practically the same way as phenological curves for plants; phenological field observations for different years (cold and warm) are used, as well as different areas of incidence (warmer and colder areas) as reported by the Tadzhikistan Plant Protection Department, and these are related to air temperature recordings taken at the meteorological stations nearest to these areas (Table 1.2).

As can be seen, the method for dealing with data concerning cold-blooded organisms differs little from the method used for plants. First, the length of time elapsing between two consecutive stages (e.g., the number of days between the first and second or between the second and third egg layings) in the organism's life is determined by field observations. Second, using data from the nearest meteorological station, all the 24-hour average air temperatures for this period (ordinary temperatures, reckoned from 0° C, regardless of any tempera-

TABLE 1.2. Data collected for constructing a phenological developmental curve for cotton bollworm (*Chloridea obsoleta F*) between the beginning of 1st and 2nd, 2nd and 3rd, 3rd and 4th mass egg layings (beginning of mass egg laying is taken—25% of this species has laid eggs); phenological recordings are from stations of the Records Department Tadzhikistan Agricultural Ministry, related to air temperatures that were recorded at the local meteorological stations.

Factor	1957	1958	1959	1960	1961
Phenological observations from the Lenin Crop Pest Recording Station; air temperatures were recorded by the "Dushanbe" meteorological station in the Lenin district of Tadzhikistan.					
Beginning of 1st mass egg laying	27 VI	17 VI	22 VI	20 VI	10 VI
Beginning of 2nd mass egg laying	27 VII	17 VII	27 VII	29 VII	10 VII
Length of time between 1st and 2nd egg laying (in n days)	30	30	35	39	30
Total of 24-hour average temperatures Σt for n days (in °C)	764.7	800.5	923.6	1041.1	772.3
Average temperature for whole period $t_{av} = \Sigma t/n$ (in °C)	25.5	26.7	26.4	26.7	25.7
Designation of corresponding point in Fig. 1.1 (curve 5)	i	k	l	m	p
Beginning of 3rd mass egg laying	—	25 VIII	27 VIII	15 IX	12 VIII
n_1 (in days) between 2nd and 3rd egg laying	—	39	31	48	33
Total of average 24-hour temperatures Σt for n_1 days (in °C)	—	959.5	814.0	1102.1	872.3
Average temperature for whole period $t_{av} = \Sigma t/n_1$ (in °C)	—	24.6	26.3	23.0	26.4
Designation of corresponding point in Fig. 1.1 (curve 5)	—	q	r	s	u

Phenological observations from the Kulyab Crop Pest Recording Station; air temperatures were recorded by the "Kulyab" meteorological station

	12 VI	13 VI	10 VI	22 VI	5 VI
Beginning of first mass egg laying	12 VI	13 VI	10 VI	22 VI	5 VI
Beginning of 2nd mass egg laying	10 VII	10 VII	5 VII	4 VIII	3 VII
Length of time between 1st and 2nd egg layings (in n days)	28	27	25	43	18
Total of 24-hour average temperatures Σt for n days (in °C)	778.9	799.7	718.8	1285.0	776.1
Average temperature for whole period $t_{av} = \Sigma t/n$ (in °C)	27.8	29.6	28.8	29.9	27.7
Designation of corresponding point in Fig. 1.1 (curve 5)	Not designated (ND)			ν	ND
Beginning of 3rd mass egg laying	17 VIII	5 VIII	4 VIII	2 IX	1 VIII
n_1 (in days) between 2nd and 3rd egg laying	38	16	30	29	29
Total of 24-hour average temperatures Σt for n_1 days (in °C)	1107.5	794.9	899.2	841.6	926.1
Average temperature for whole period $t_{av} = \Sigma t/n_1$ (in °C)	29.1	30.6	30.0	29.0	31.9
Designation of corresponding point in Fig. 1.1 (curve 5)	x	y	ND	ND	z
Beginning of 4th mass egg laying	—	25 VIII	—	—	22 VIII
n_2 (in days) between 3rd and 4th egg layings	—	20	—	—	21
Total of 24-hour average temperatures Σt over n_2 days (in °C)	—	556.4	—	—	613.0
Average temperature for whole period $t_{av} = \Sigma t/n_2$ (in °C)	—	27.8	—	—	29.2
Designation of corresponding point in Fig. 1.1 (curve 5)	—	α	—	—	β

ture thresholds) are totaled, as in the above example for plants. Dividing the total of the temperature by the number of days (n) gives the average temperature (t_{av}) for that developmental period. It should be noted that the data for identical periods in different generations (e.g., between the first and second, second and third, or third and fourth egg layings) can be dealt with in exactly the same way for each generation, in a single table, and can then be transferred onto the same graph, since there is no particular reason for supposing that the biology of a cold-blooded organism's development in the first generation will be different from that in, say, the third (provided, of course, that neither of these generations includes a diapause). Different generations of the organisms, developing under different temperature regimes, merely develop faster or slower (i.e., n decreases or increases), but the sequence of development is not altered (of course, in different segments of the phenocurve the temperature thresholds and totals relate to different generations, but this is automatically calculated in the phenocurve and, later, in the nomogram). This means that research under field conditions is possible in a considerably wide range of temperatures, since, obviously, the first generation of the cotton bollworm (*Chloridea obsoleta F.*) develops at lower temperatures than the third.

The results of these calculations are then transferred onto one graph (Fig. 1.1, curve 5). For example, in the Lenin district in 1957 a 30-day period between egg layings (Table 1.2) was observed when the average temperature for this period was 25.5° C. Thus a point with coordinates t_{av} = 25.5° C and n = 30 days is transferred to Figure 1.1 and designated by the letter i. Points $k, l, m, p, q, r, s, u, \ldots, v, x, y, \ldots, z, \alpha,$ and β are plotted in the same way, as well as many other points, referring to different years, different places, and different generations, the data for which is not shown in Table 1.2. The result is the cluster of points seen in Figure 1.1, distributed in a regular pattern along curve 5.

We continue in the same way as with the phenocurve for the cotton: we draw curve 5 through this cluster of points in such a way that the number of points above the line roughly equals the number below it. Above curve 5 there are 24 points, below it there are 24, and on the line there are 7. This is the procedure for constructing a phenological temperature developmental curve (curve 5) for the cotton bollworm (*Chloridea obsoleta F*) between the beginning of the first and second, second and third, and third and fourth mass egg layings.

What remains is to construct a developmental curve for the cotton bollworm from its spring awakening* to the beginning of the first mass egg laying, since the species hibernates and awakes in the pupa phase;

---

*The resumption of development of insects after winter rest (i.e., after diapause).

the period between the pupa and the first mass egg laying is not identical to the periods between the first and second and between the second and third egg layings (whereas the periods between the egg layings of successive generations are identical). The construction of this last curve requires field data giving the time when the cotton bollworm awakens in the spring and the time of its first egg laying, in different years and in different areas.

Unfortunately, direct observations of this pest's awakening are rare. Most frequently, the time of its awakening is reckoned to be when the 24-hour average temperature rises regularly above this species' lower temperature threshold (+15°C). The problem lies in defining exactly when this threshold is exceeded. We shall take it that the threshold is exceeded on the date after which the 24-hour average temperature either does not fall below +15°C or falls below this level only rarely. Thus, by knowing the date of awakening and the date when the first eggs are laid, we can calculate the difference in days between them (n) and continue as above.

As can be seen (Fig. 1.1), curve 4 covers a relatively small range of temperatures, even though the data used for plotting it were taken from several different areas over a number of years. The same kind of picture is seen in the case of cold-blooded organisms that have only one generation. Cases in which there are several generations, which develop at different times of the year and under very different temperature conditions, are comparable with experiments in which plants are sown at different times of the year, when the researcher intentionally conducts the sowing under different temperature conditions. It is not possible to conduct this kind of experiment in the field with cold-blooded organisms that have only one generation. However, in the laboratory it is possible to obtain information about the development of cold-blooded organisms under different temperature conditions quite quickly, with the use of controlled temperature chambers. From this point of view curves derived from controlled temperature chambers are more convenient (although not better) than curves derived from field conditions. In some cases controlled temperature chamber curves can also be successfully used for constructing phenocurves. (This will be discussed in more depth later.)

With regard to plants, it should be noted that sowing at different times of the year is a very useful aid in the construction of temperature curves for crop development.

For example, we sowed cotton every 5 to 15 days from March to August in Dushanbe (in the extreme south of the USSR). The results are shown in Figures 1.2 and 1.3. In these figures the crops sown in spring are represented by the points at relatively low temperatures during the initial developmental stages, and at relatively high temperatures during subsequent developmental stages; and the crops sown in summer, vice-versa. Practice has proved that, having reliable observations, we can construct a phenological curve on the basis of two- or three-year observations even with the use of data for only one region.

We also carried out field experiments with crops sown at different times of the year, over a two-year period, in central regions of the European part of the USSR (Podolsky, A. S., 1978). Nevertheless, as can be seen from Figures 1.7, 1.8, and 1.9, the curves appear quite distinctly.

It is interesting and important that the encircled points and triangles in these figures relate to the uniquely hot year of 1972, which increased the effect of the experiments with sowing at different times of the year. It is no mere chance that the encircled symbols are located in the high-temperature section of the lines.

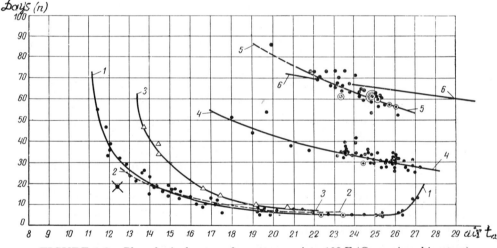

FIGURE 1.2. Phenological curves for cotton variety 108-F (*Gossypium hirsutum*). These curves, apart from curves 2 and 6, are constructed on the basis of results obtained from author's experiments with sowing at different times of the year, which were carried out from March to August, from 1947-1954, on plots of land at the Dushanbe agrometeorological station in Tadzhikistan (observations carried out by the Dushanbe station on fields belonging to a collective farm were also used). Curve 1: from sowing to mass sprouting, using normal agricultural methods; curve 2: the same as curve 1, from observations in Azerbaidzhan (after A. R. Mustafaev; points are not illustrated); curve 3: from sowing to mass sprouting, with soil crust and insufficient soil moisture; curve 4: from mass budding to mass flowering (for the period from sprouting to budding see Fig. 1.3); curve 5: from mass flowering to mass unfolding of the first seed bolls (according to the Hydrometeorological Division's method); curve 6: from mass flowering to mass unfolding of seed bolls (according to the method used by the NTV; the criterion for this is that the boll should have a complete opening).

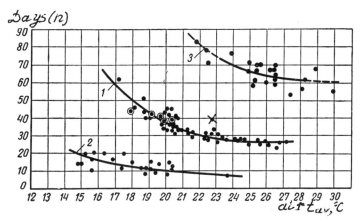

FIGURE 1.3. Phenological curves. Curve 1: cotton variety 108-F from mass sprouting to mass budding; curve 2: cotton variety 149-F from sowing to mass sprouting; curve 3: cotton variety 149-F from mass flowering to mass boll unfolding (for the period from sprouting to flowering see Fig. 1.31). Curve 1 is based on the data from author's experiments with sowing at different times of a year and on observations of collective farm fields in the Dushanbe region; curves 2 and 3 are based on phenological recordings from different agricultural centers in Tadzhikistan from 1956 to 1961.

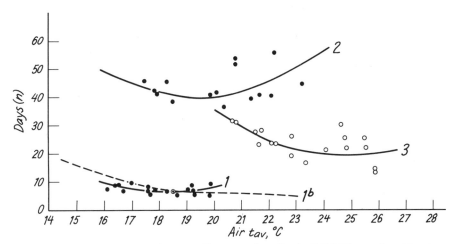

FIGURE 1.4. Phenological curves. Cotton variety *Acala S.J.-2* (*Gossypium hirsutum L.*): curve 1: from planting to germination (50%); curve 2: from germination to budding (1st bud); curve 3: from budding to flowering (1st flower). Cotton variety 108-F (*G. hirsutum*): curve 1[b]: from planting to sprouting (50%). Phenological curves were constructed by A. Podolsky as follows: curves 1-3: on the basis of data provided by A. Marani (different regions of Israel, 1978); curve 1[b]: on the basis of data provided by A. Podolsky (USSR, see Fig. 1.2).

13

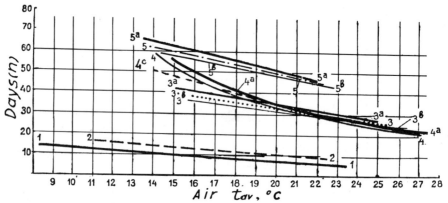

FIGURE 1.5. Phenological curves for spring wheat in the steppe zone of Kazakhstan. Curve 1: from sowing to mass sprouting for varieties Saratovskaya 29, Bezenchukskaya 98, and Kharkovskaya 46; curve 2: from mass sprouting to the beginning of tillering for varieties Saratovskaya 29, Bezenchukskaya 98, and Kharkovskaya 46; curve 3: from the beginning of tillering to mass heading for variety Saratovskaya 29; curve 3a: the same as curve 3, for variety Bezenchukskaya; curve 3b: the same as curve 3, for variety Kharkovskaya 46; curve 4: from full heading to full wax ripeness for varieties Saratovskaya 29 and Bezenchukskaya 98; curve 4a: the same as curve 4, for variety Kharkovskaya 46; curve 5: from sowing to full heading for variety Saratovskaya 29; curve 5a: the same as curve 5, for variety Bezenchukskaya 98; curve 5b: the same as curve 5, for variety Kharkovskaya 46 (from V. K. Adzhbenov). For comparison: curve 4^c: spring wheat variety Kharkovskaya 46 from heading to wax ripeness from observations made in the CBEZ of the USSR, after A. S. Podolsky and O. S. Yermakov. (At temperatures higher than 19°, line 4^c coincides completely with 4^a). Line 4^c is transferred from Fig. 1.6.

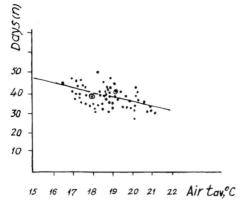

FIGURE 1.6. The phenological line for spring wheat variety Kharkovskaya 46 from heading to wax ripeness; from observations made in the CBEZ of the USSR. The equation of the line is $n = 82.98 - 2.38\, t_{av}$ (after O. S. Yermakov and A. S. Podolsky).

14

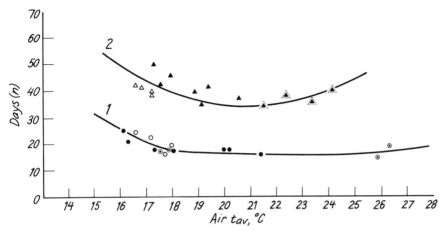

FIGURE 1.7. Phenological curves for spring wheat variety Kharkovskaya 46. Curve 1: from heading to milky ripeness; curve 2: from heading to wax ripeness. For the most part, field experiments are presented here, with sowing at different times of the year in Kursk during 1971 and 1972.

In contrast, the outlined points and triangles are located in the low-temperature part of the lines, since these symbols refer to the cold (as compared with the norm) year of 1973. (Several of the observations represented by these symbols on the Kursk graphs in Figures 1.7 and 1.9 were carried out in the fields of a collective farm in Livni, in the Orlov district, without regard for any special experiments. Observations were made by a student, V. M. Dolgikh, in 1973. The Orlov district is in the neighborhood of Kursk).

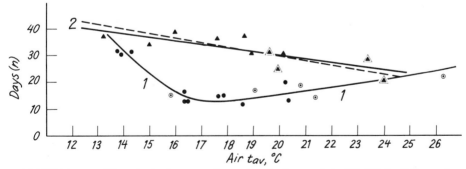

FIGURE 1.8. Phenological curves for spring barley variety "Valtitzky." Curve 1: from sprouting to stalk shooting; curve 2: from heading to wax ripeness (the solid straight line is calculated after the regression equation; the dotted line is plotted on the basis of visual judgment). The results are from field experiments with sowing at different times of the year in Kursk in 1971 and 1972.

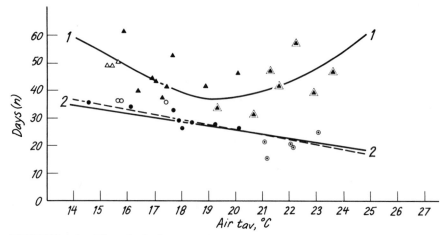

FIGURE 1.9. Phenological curves for oat variety Lgovsky 1026. Curve 1: from sprouting to the appearance of panicles; curve 2: from the appearance of panicles to wax ripeness (the solid straight line is calculated after the regression equation; the dotted line is plotted on the basis of visual judgment). For the most part field experiments are presented here, with sowing at different times of the year in Kursk during 1971 and 1972.

The phenological lines obtained from the experiments with sowing at different times of the year, during a two-year period in one location, are identical to the phenolines obtained from observations in a number of places during a number of years. This can be seen by comparing line 2 in Figure 1.9 with the corresponding line in Figure 1.11. However, sometimes there are differences.

One advantage of experimenting with sowing at different times of the year is that it makes it possible to find the temperature optimum.

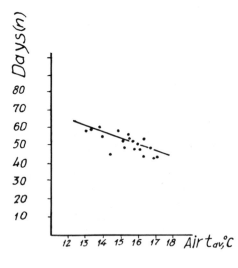

FIGURE 1.10. The phenological line for oats variety Lgovsky-1026 from sprouting to the appearance of panicles. Regular observations were made in the Kursk region, USSR. The equation of the line is $n = 108.23 - 3.61\ t_{av}$ (after O. S. Yermakov and A. S. Podolsky).

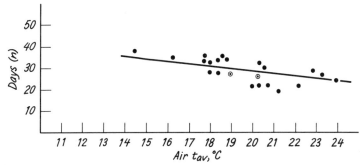

FIGURE 1.11. The phenological line for oats variety Lgovsky-1026 from the appearance of panicles to wax ripeness, according to data obtained from regular observations, which were carried out by the Hydrometeorological Division and by the NTV in the CBEZ of the USSR (after O. S. Yermakov and A. S. Podolsky).

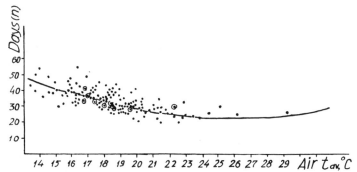

FIGURE 1.12. The phenological curve for oat variety Pobeda from the appearance of panicles to wax ripeness, according to data for the ESU (after O. S. Yermakov and A. S. Podolsky).

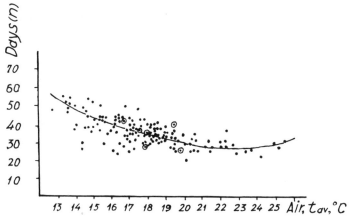

FIGURE 1.13. The phenological curve for spring wheat variety Lutescens-62 from heading to wax ripeness, according to data for the ESU (after O. S. Yermakov and A. S. Podolsky).

17

From this point of view let us compare line 1 in Figure 1.9 with the similar line in Figure 1.10. The first line is based on the experiments with crops sown at different times of the year, including crops sown extraordinarily late, when the period from sprouting to appearance of panicles occurs at optimal temperatures in the conditions of Kursk; the second line (Fig. 1.10) is based on the observations of crops sown ordinarily every year, as is the practice of farms. As can be seen, the line in Figure 1.10 bears a resemblance only to the below-optimal branch of curve 1 in Figure 1.9. However, there are no data obtained from ordinary practice for the developmental rate at above-optimal temperatures. A large amount of data is required to find in ordinary practice a below-optimal branch of the phenocurve (Figs. 1.12 and 1.13). Of course, in the homeland of one or another plant it is much easier to find the temperature optimum, because all developmental stages occur there, usually at optimal temperatures.

A great amount of data, often unused, has been collected in the records of experimental stations, institutes, meteorological stations, and so on, in the course of many decades of experimental work. This data can be used, for example, to provide "passports" for different plants, different varieties, and for cold-blooded organisms. Through the plotting of phenological developmental curves, a catalog or atlas of such curves can be compiled. This is an essential step in studying the biological, genetic, and ecological characteristics of these plants and cold-blooded organisms. The phenological curves shown in Figures 1.1 to 1.48 illustrate this. As can be seen, it is possible to draw such curves for a wide variety of plants and cold-blooded organisms. Many of these curves are needed for constructing nomograms for Tadzhikistan, Turkemistan, the Far East, and suburban Moscow; other curves will be useful for readers in other regions and countries. For the sake of economy of space, and for the purpose of showing all the curves and their constituent points in a clearly visible manner, the logical order of the curves in the figures has been sacrificed.

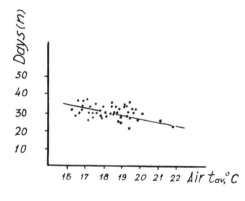

FIGURE 1.14. The phenological line for spring barley variety Valtitzky from heading to wax ripeness, according to observations made in the CBEZ of the USSR. The equation of the line $n = 54.2 - 1.26\ t_{av}$ (after O. S. Yermakov and A. S. Podolsky).

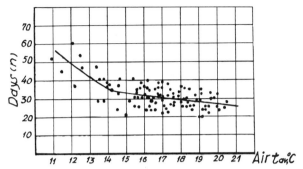

FIGURE 1.15. The phenological curve of barley variety Viner from heading to wax ripeness, according to data obtained by meteorological stations in Bryansk, Kalinin, and Vologda districts of the USSR (after O. S. Yermakov and A. S. Podolsky).

The developmental curve, which gives the equation for the species' heat requirements, demonstrates how the length of the period between developmental stages varies in relation to the average temperature over this period. Therefore, the region's heat resources, in turn, must be presented so as to show the average air temperature over any given period, at different times of year.

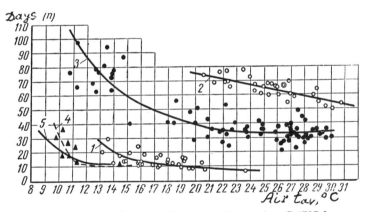

FIGURE 1.16. Phenological curves. Curve 1: cotton variety S-4727 from sowing to mass sprouting; curve 2: cotton variety S-4727 from mass flowering to mass boll unfolding (for the period from sprouting to flowering see Fig. 1.17); curve 3: Lucerne variety Tashkent 721 (irrigated) between spring growth or successive harvest times and the beginning of the following flowering; curve 4: mulberry trees from the swelling of the buds to the opening out of the first leaves (phenological and temperature recordings are from the Iolatan agrometeorological station, Turkmenistan, from 1950 to 1959, after G. M. Ostapenko, 1962); curve 5: the same as curve 4 (after A. R. Mustafaev, Azerbaidzhan, 1967). Curves 1, 2, and 3 are based on phenological recordings from Tadzhikistan's NTV: cotton, from 1955 to 1961; lucerne, from 1954 to 1958. The points and triangles (relating to different recordings) refer to Tadzhikistan and Turkmenistan.

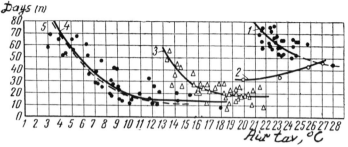

FIGURE 1.17. Phenological curves. Curve 1: cotton variety S-4727 from mass sprouting to mass flowering; curve 2: potatoes (irrigated) from sprouting to the moment when the tubers begin to form (after L. N. Babushkin, 1951); curve 3: barley varieties Khordzhau 18 and Persicum 64 sown in autumn and not irrigated, from full stalk shooting to full heading (the remaining phases are presented in Figs. 1.19, 1.31, and 1.32); curve 4: wheat variety Surkhak 5688 sown in autumn and not irrigated, from full sprouting to the beginning of tillering (the remaining phases are presented in Figs. 1.18 and 1.32); curve 5 (dashed line): wheat of Shark variety from full sprouting to the beginning of tillering (after A. R. Mustafaev, Azerbaidzhan, 1967). All the curves, except curves 2 and 5, are based on phenological recordings from the Tadzhikistan's NTV: cotton, from 1955 to 1961; barley and wheat, from 1949 to 1958. The points and triangles (relating to different recordings) refer to Tadzhikistan and Uzbekistan.

Let us take an annual temperature trend for a given climatic region and some starting date for this temperature trend. Then, taking different periods of time from this starting date, we shall obtain different average temperatures t_{av} over these different periods. The values t_{av} will change in relation to the increase of the periods of time (n) according to the law which can be expressed by a line of average temperatures in coordinates $t_{av} - n$ (see below). It is clear that this law is different for different starting dates, and that the lines of average temperatures will also be different, respectively. In other words, the heat resources equation is presented in the form of a line expressing the changes in the average temperature over a given period (t_{av}) in the annual temperature trend in relation to the length of these periods (n) as determined from a single starting date. This date serves to designate the line giving the average temperatures for particular periods. Different starting dates form together a family of intersecting lines that result in a chart of the average temperatures over different periods. Because this graphic presentation resembles a "net," it will be referred to as a *net of the region's heat resources.*

Such a net is based on the air temperature recordings at a given meteorological station over a period of many years. Figure 1.50 presents a net for the Yavan region of Tadzhikistan.

On this net, the line marked May 15 indicates that the 24-hour average air temperature on May 15, based on the annual average data,

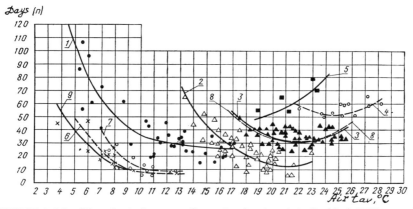

FIGURE 1.18. Phenological curves. Curve 1: wheat variety Surkhak 5688 sown in autumn and not irrigated, from the beginning of tillering to full stalk shooting; curve 2: the same as curve 1, from full stalk shooting to full heading; curve 3: the same as curve 1, from full heading to full wax ripeness; curve 4: fine-fiber cotton 5595-V from mass sprouting to mass flowering (the remaining phases are shown in Fig. 1.31); curve 5: soya of Easy-Cook variety irrigated from flowering to ripening (solid line with black squares); curve 6: early-ripening apricots from swelling of buds to flowering (Turkmenistan); curve 7: late-ripening apricots from swelling of buds to flowering (Turkmenistan); curve 8 (dashed line): wheat of Shark variety from full heading to full wax ripeness (Azerbaidzhan); curve 9: early-ripening apricots from swelling of buds to flowering (after A. R. Mustafaev, Azerbaidzhan, 1967). Curves 1 to 4 are based on phenological recordings from the Tadzhikistan's NTV: wheat, from 1949 to 1958; cotton, from 1958 to 1961. Curve 5 is based on recordings made by the Boz-Su agrometeorological station (near Tashkent) during experiments on sowing dates from 1930 to 1933 (after L. N. Babushkin, 1951); curves 6 and 7 are based on phenological and temperature recordings from the Iolotan agrometeorological station in Turkmenistan: early-ripening apricots, from 1940 to 1960 (with a break); late-ripening apricots, from 1948 to 1958 (after G. M. Ostapenko, 1962). The symbols (relating to different recordings) refer to Middle Asia.

is 20.2° C (see point A, where the line for May 15 leaves the horizontal axis on Figure 1.50; see also Figure 2.1, which shows the annual temperature trend). Over an 11-day period (i.e., from May 15 to 26), the average temperature will be 21.8° C (see Figure 1.50, on the line for May 15, the point B with coordinates $t_{av} = 21.8°$ C and $n = 11$ days; see also the corresponding point in Figure 2.1). Over a 21-day period (from May 15 to June 5 inclusive), the average temperature will be 23.0° C (see Figure 1.50, on the line for May 15, the point D with coordinates 23.0° C and 21 days). Further examples can also be cited.

Obviously, at the beginning of this line, which represents the average temepratures over various periods starting from May 15, t_{av} increases as n increases (since the temperatures increase in the first half of the summer); later, t_{av} decreases as n increases.

The same can be said about the lines starting from May 26, June 5,

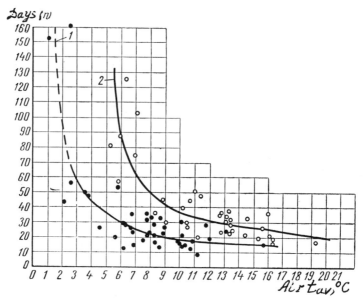

FIGURE 1.19.   Phenological curves for barley varieties Khorjau 18 and Persicum 64 sown in autumn and not irrigated. Curve 1: from full sprouting to the beginning of tillering; curve 2: from the beginning of tillering to full stalk shooting (for the period from stalk shooting to heading see Fig. 1.17; for the period from heading to wax ripeness see Fig. 1.31; for the period from sowing to sprouting see Fig. 1.32). The curves are based on phenological recordings from the Tadzhikistan's NTV from 1949 to 1958.

and so on. However, the lines starting from August 5, 15, 26, and so on, slope, on the whole, from right to left, indicating a decrease in t_{av} as n increases (since the temperatures decrease in the second half of the summer).

Therefore, if the annual temperature trend were rectilinear, we could say that *the lines representing average temperatures over certain*

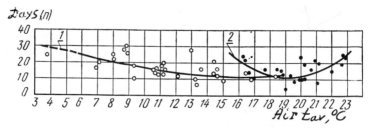

FIGURE 1.20.   Phenological curves for barley varieties Khorjau 18 and Jau-bapust sown in spring and not irrigated. Curve 1: from sowing to full sprouting; curve 2: from full stalk shooting to full heading (the remaining phases are presented in Figs. 1.21 and 1.32). The curves are based on recordings from the Tadzhikistan's NTV from 1949 to 1958.

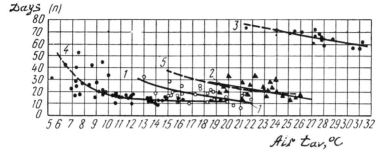

FIGURE 1.21. Phenological curves. Curve 1: barley varieties Khorjau 18 and Jau-bapust sown in spring and not irrigated, from the beginning of tillering to full stalk shooting; curve 2: barley, for the most part of the same varieties, from full heading to full wax ripeness (for the remaining phases see Figs. 1.20 and 1.32); curve 3: fine-fiber cotton 9123-I from mass flowering to mass boll unfolding (for the remaining phases see Fig. 1.31); curve 4: wheat variety Iroda 1006 sown in spring and not irrigated, from sowing to full sprouting (for the remaining phases see Figs. 1.22 and 1.32). The curves are based on phenological recordings by Tadzhikistan's NTV: barley and wheat, from 1949 to 1958; cotton, from 1958 to 1961. Curve 5: spring barley from heading to wax ripeness (data from A. A. Shigolev and A. E. Shklyar, in the European section of the USSR). Points and triangles (relating to different recordings) refer to Tadzhikistan.

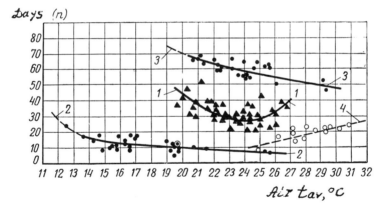

FIGURE 1.22. Phenological curves. Curve 1: spring wheat Iroda 1006 sown in spring and not irrigated, from full heading to full wax ripeness (for the remaining phases see Figs. 1.21 and 1.32); curve 2: maize varieties VIR 42 and Krasnodar 1/49 (data for both irrigated and nonirrigated crops) from sowing to full sprouting; curve 3: maize of the same varieties (data for both irrigated and nonirrigated crops) from full sprouting to full flowering; curve 4: maize of same varieties irrigated from full flowering of panicles to full milky ripeness (for final phase see Fig. 1.23). The curves are based on phenological recordings from Tadzhikistan's NTV: wheat, from 1949 to 1958; maize, from 1955 to 1958.

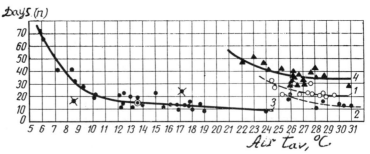

FIGURE 1.23. Phenological curves. Curve 1: maize varieties VIR 42 and Krasnodar 1/49 irrigated from milky to full (commercial) ripeness; curve 2: the same varieties in the same growth period, without irrigation; curve 3: sunflower variety Saratov 169 (data for both irrigated and nonirrigated crops) from sowing to full sprouting; curve 4: sunflower variety Saratov 169 (data for both irrigated and nonirrigated crops) from full flowering to full ripeness (for the remaining phases of sunflower see Fig. 1.31). The curves are based on phenological recordings from Tadzhikistan's NTV: maize, from 1955 to 1958; sunflowers, from 1948 to 1958.

periods are, generally speaking, consecutive parts of the same annual temperature trend, starting from different dates (where the usual designation of the axes used for drawing the annual temperature trend is reversed: i.e., the horizontal axis shows not time, but temperature, and the vertical axis shows time, not in dates, but in numbers of days elasped from an initial date). These ideas will be developed in Chapter 10 in relation to more recent ideas.

Graphic interpolation can be made if the abovementioned points A, B, D, and so on are joined by straight line segments. These can be used to determine the average temperature over any given period of time

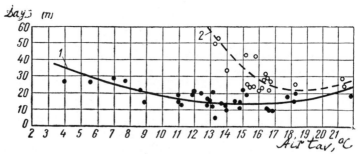

FIGURE 1.24. Phenological curves for crown flax (oil-producing) variety Hissar 474. Curve 1: from sowing to full sprouting [data from both nonirrigated (primarily) and irrigated crops]; curve 2: from full sprouting to full formation of clusters in nonirrigated crops (for the remaining phases see Fig. 1.31). The curves are based on phenological recordings from the Tadzhikistan's NTV, from 1949 to 1958.

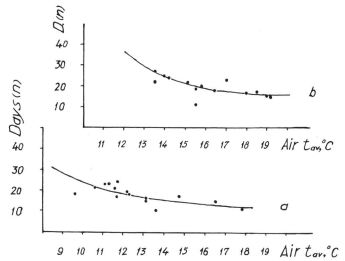

FIGURE 1.25. Phenological curves for lily-of-the-valley (*Convallaria majalis L*), a raw material for cardiological medicines, from the Kursk district, USSR: curve a: from the beginning of vegetation to the beginning of flowering; curve b: from the beginning of flowering to the end of flowering (after A. S. Podolsky, L. V. Shatunova, and V. D. Tokareva).

starting from a given initial date. In the same way, it is possible to make interpolations between the different lines on the net, and thus to select any desired starting date. For example, t_{av} for the 23-day period n starting from 15 May will be 23.2° C. To find this temperature we need only look for a point on the vertical axis (Fig. 1.50) that represents 23 days, and draw a horizontal line from this point to the line marked May 15. Moving down to the temperature axis from the point where this line (i.e., May 15) is intersected by our horizontal line, we find t_{av} = 23.2° C. The average temperature over a 23-day period starting from May 20 will be 24.2° C, because the imaginary line for May 20 will lie almost centrally between the lines designated May 15 and 26.

As an intermediary step, prior to drawing this net of the region's heat resources, a table should be compiled on the basis of the multiannual 10-day average air temperatures. Let us take as an example the calculation needed for drawing a line showing the average temperatures over periods starting on May 15, according to data from the Yavan meteorological station. We know that the multiannual average air temperature in Yavan is 20.2° C for the second 10 days of May; 23.4° C for the third 10 days of May; 25.5° C for the first 10 days of June; 26.7° C for the second 10 days of June; and so on. It is clear that the multiannual observations applied to a period as short as 10 days will give a figure for the average temperature that closely corresponds

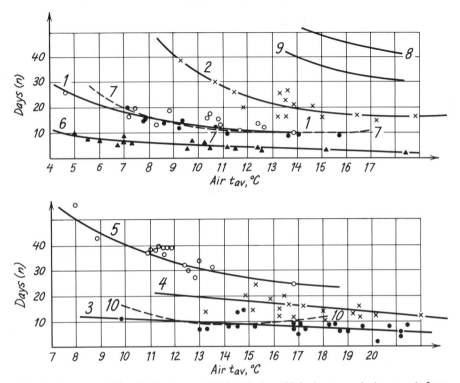

FIGURE 1.26. Phenological curves. For valerian (*Valeriana rossica*): curve 1: from the beginning of vegetation to budding (the solid curve through the white points); curve 2: from budding to the beginning of flowering; curve 3: from the beginning of flowering to mass flowering; curve 4: from mass flowering to the end of flowering; curve 5: from the beginning of vegetation to the end of flowering. For adonis (*Adonis vernalis*): curve 6: from the beginning of vegetation to the beginning of flowering; curve 7: from the beginning of flowering to mass flowering (the dashed curve through black points) (from V. A. Ryabov). For red bilberry (*Vaccinium vitisidae*): curve 8: from mass flowering to the beginning of ripening. For raspberry (*Rubus*): curve 9: from the beginning of flowering to the beginning of ripening. For bog bilberry (*Vaccinium uliginosum*): curve 10: from the beginning of flowering to mass flowering (from A. F. Cherkassov). All curves were constructed according to the Podolsky method and have been systematized by the latter.

to the midpoint of the 10-day period (provided that this 10-day period does not occur precisely at the peak of the annual temperature graph). Therefore, we can assume that the multiannual temperatures for May–June are distributed as follows: 15 May—20.2°C, 26 May—23.4°C, 5 June—25.5°C, 15 June—26.7°C, and so on. To calculate the true average temperature from May 15 to 26, it would be necessary to total all the multiannual 24-hour average temperatures for that 11-day period and to divide the total by 11.

Virtually the same result can be obtained much more simply by

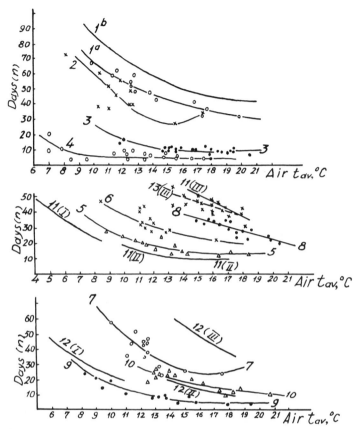

FIGURE 1.27. Phenological curves. *Medicago falcata:* curve 1[a]: from the beginning of vegetation to budding; curve 1[b]: from the beginning of vegetation to the beginning of flowering; curve 2: *Thymus marshallianus* from the beginning of vegetation to budding; curve 3: *Salvia pratensis* from the beginning of flowering to mass flowering; curve 4: *Hyacinthella leucophaea* from the beginning of flowering to mass flowering; curve 5: *Senecio czernjaevii* from budding to the beginning of flowering; curve 6: *Tragopogon orientale* from the beginning of vegetation to budding; curve 7: *Centaurea scabiosa* from the beginning of vegetation to budding; curve 8: *Centaurea scabiosa* from the budding to beginning of flowering; curve 9: *Orobus lacteus* from budding to the beginning of flowering; curve 10: *Poa angustifolia* from budding to the beginning of flowering (from V. A. Ryabov for the Central Black Earth Reserve in Russia, forest-steppe). For bilberry (*Vaccinium myrtillus*): curve 11 (I): from the time when the average 24-hour air temperature steadily exceeds 0°C to the opening of buds; curve 11 (II): from the opening of buds to flowering; curve 11 (III): from flowering to ripening of fruits. For bog bilberry (*Vaccinium uliginosum*): curves 12 (I), (II), and (III): developmental stages, respectively, to the stages mentioned previously (from V. B. Gedikh for Byelorussia, sphagnum bogs and bilberry pinery). For bilberry (*Vaccinium myrtillus*): curve 13 (III): from mass flowering to the beginning of ripening of fruits (the dashed line through crosses) (from A. F. Cherkasov for the Kostroma district, in the northern part of the European section of the USSR, dark-coniferous and larch forests and sparse growth of trees). All curves were constructed according to the Podolsky method and have been systematized by the latter.

27

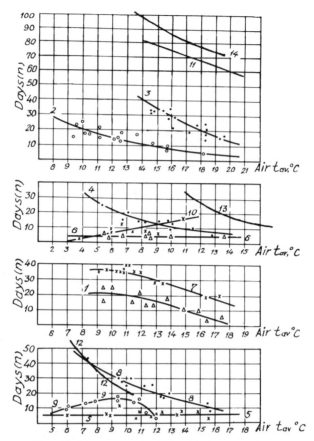

FIGURE 1.28. Phenological curves. English oak (*Quercus robur L*): curve 1: from the beginning of vegetation to the beginning of flowering; curve 2: from budding to the beginning of flowering. Euonymus (*Euonymus verrucosus Scop*): curve 3: duration of flowering. Drooping birch (*Betula pendula Roth*): curve 4: from the beginning of vegetation to the beginning of flowering; curve 5: duration of flowering; curve 6: from full coloration of foliage to the mass fall of the leaves. Mountain ash (*Sorbus aucuparia L.*): curve 7: from the beginning of vegetation to the beginning of flowering. Crab apple (*Malus silvestris Mill*): curve 8: from the beginning of vegetation to the beginning of flowering; curve 9: from full coloration of leaves to mass fall of leaves. Common buckthorn (*Rhamnus catharica L.*): curve 10: from full coloration of leaves to the mass fall of leaves (from V. A. Ryabov for the Central Black Earth Reserve in Russia, forest-steppe). Mountain ash (*Sorbus aucuparia L*): curve 11: from mass flowering to the beginning of ripening of fruits (from A. F. Cherkasov for the Kostroma district in the USSR, forests and sparse growth of trees). Cranberry *Oxycoccus Hill*: curve 12: from the time when the average 24-hour air temperature steadily exceeds 0° C to the opening of buds; curve 13: from the opening of buds to flowering; curve 14: from flowering to ripening of fruits (from V. Gedikh for Byelorussia). All curves were constructed according to the Podolsky method and have been systematized by the latter.

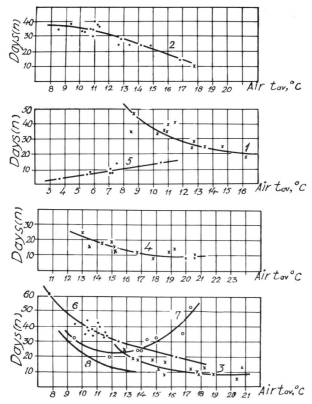

FIGURE 1.29. Phenological curves. Common buckthorn (*Rhamnus catharica L.*): curve 1: from the beginning of vegetation to the beginning of flowering; curve 2: from budding to the beginning of flowering; curve 3: duration of flowering. Alder buckthorn (*Frangula alnus Mill*): curve 4: duration of flowering; curve 5: from full coloration of leaves to the mass fall of leaves. Euonymus (*Euonymus verrucosus Scop.*): curve 6: from the beginning of vegetation to the beginning of flowering; curve 7: from mass ripening of fruits to full coloration of leaves (from V. A. Ryabov, forest-steppe). Mulberry (*Morus L.*): curve 8: from the swelling of buds to the unfolding of the first leaves (from A. R. Mustafayev, Azerbaidzhan). All curves were constructed according to the Podolsky method and have been systematized by the latter.

adding together and dividing by 2 the temperatures for the first and last days of the same period:

$$t_{\text{av, 11}}^{\text{May 15-May 26}} = \frac{20.2°\text{C} + 23.4°\text{C}}{2} = \frac{43.6°\text{C}}{2} = 21.8°\text{C}$$

In the same way, we obtain the average temperature for the 21-day period from May 15 to June 5 by totaling and dividing by 3 the

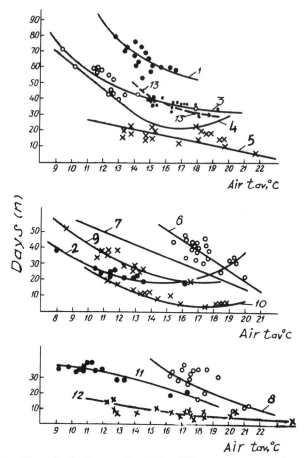

FIGURE 1.30. Phenological curves for wild melliferous plants and for a cultured one. Cornflower (*Centaurea scabiosa*): curve 1: from the beginning of vegetation to the beginning of flowering. Cornflower (*Centaurea sumensis*): curve 2: from budding to the beginning of flowering. *Echium russicum*: curve 3: from the beginning of vegetation to the beginning of flowering. Pea (*Vicia tenuifolia*): curve 4: from the beginning of vegetation to the beginning of flowering. *Senecio Czernjaevii*: curve 5: duration of flowering. Mountain clover (*Trifolium montanum*): curve 6: duration of flowering. Cow clover (*Trifolium pratense*): curve 7: from the beginning of vegetation to the beginning of flowering. Bluebell (*Campanula persicifolia*): curve 8: duration of flowering. Alder buckthorn (*Frangula alnus*): curve 9: from the beginning of vegetation to the beginning of flowering. Crab apple (*malus silvestris*): curve 10: duration of flowering. Mountain ash (*Sorbus aucuparia*): curve 11: from the beginning of vegetation to the beginning of flowering; curve 12: duration of flowering (from V. A. Ryabov). Common buckwheat (*Fagopyrum sagittatum*): curve 13: from sowing to the beginning of flowering (from T. M. Gavrilova). All curves were constructed according to the Podolsky method and have been systematized by the latter.

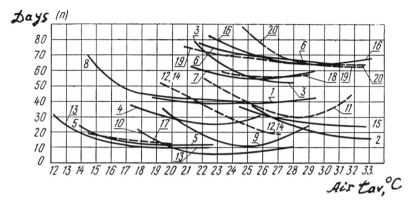

FIGURE 1.31. Curves for developmental stages omitted from Figures 1.1 to 1.3, and 1.16 to 1.24, plus the full developmental cycles of cotton varieties 504-V and 5769-V. Curve 1: cotton variety 5904-I from mass sprouting to mass budding; curve 2: cotton variety 5904-I from mass budding to mass flowering; curve 3: cotton variety 149-F from mass sprouting to mass flowering; curve 4: barley varieties Khorjau 18 and Persicum 64 sown in autumn, from full heading to full wax ripeness; curve 5: cotton 5595-V and 9123-I from sowing to mass sprouting; curve 6: cotton 5595-V from mass flowering to mass boll unfolding; curve 7: cotton 9123-I from mass sprouting to mass flowering; curve 8: sunflower variety Saratov 169 from full sprouting to the formation of the heads; curve 9: sunflower variety Saratov 169 from the formation of the heads to full flowering; curve 10: crown flax (oil-producing) variety Hissar 474 from the full formation of the clusters to full flowering (data for both nonirrigated and irrigated crops); curve 11: crown flax variety Hissar 474 from full flowering to commercial ripeness (irrigated); curve 12: crown flax variety Hissar 474 from full flowering to commercial ripeness (not-irrigated); curve 13: cotton variety 504-V from sowing to mass sprouting; curve 14: cotton variety 504-V from mass sprouting to mass budding; curve 15: cotton variety 504-V from mass budding to mass flowering; curve 16: cotton variety 504-V from mass flowering to mass boll unfolding; curve 17: cotton variety 5769-V from sowing to mass sprouting; curve 18: cotton 5769-V from mass sprouting to mass flowering; curve 19: cotton 5769-V from mass flowering to mass boll unfolding; curve 20: cotton variety 2I3 from mass flowering to mass boll unfolding. The curves are based on recordings from Tadzhikistan's NTV. (In the case of cotton varieties 504-V, 5904-I, and 2I3, phenological recordings from the meteorological network were also used.)

temperatures for the beginning, middle, and end of this period:

$$t_{av,\ 21}^{May\ 15\text{-June}\ 5} = \frac{20.2°\,C\ (May\ 15) + 23.4°\,C\ (May\ 26)}{3}$$

$$\frac{+\ 25.5°\,C\ (June\ 5)}{3} = \frac{69.1°\,C}{3} = 23.0°\,C$$

Or more simply, we add the temperature for June 5 to the previously calculated total and divide by 3:

$$t_{av} = \frac{43.6°\,C + 25.5°\,C}{3} = \frac{69.1°\,C}{3} = 23.0°\,C$$

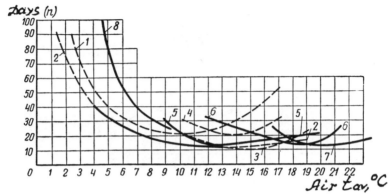

FIGURE 1.32. Curves for developmental stages omitted from Figures 1.1 to 1.3 and 1.16 to 1.24, plus a curve for peaches. Cereal crops without irrigation: curve 1: barley varieties Khorjau 18 and Persicum 64 sown in autumn, from sowing until full sprouting; curve 2: wheat variety Surkhak 5688 sown in autumn, from sowing until full sprouting; curve 3: barley variety Khorjau 18 sown in spring, from full sprouting to the beginning of tillering; curve 4: barley variety Jau-bapust sown in spring, from full sprouting to the beginning of tillering; curve 5: wheat variety Iroda 1006 sown in spring, from full sprouting until the beginning of tillering; curve 6: the same as curve 5, from the beginning of tillering to full stalk shooting; curve 7: the same as curve 5, from full stalk shooting to full heading; curve 8: peaches, from the time when the buds begin to swell (at the beginning of spring, soon after the time when the 24-hour average air temperature steadily exceeds 5°C) until the beginning of flowering (after Babushkin, Uzbekistan, 1951). The curves are based on data from Tadzhikistan's NTV, with the exception of curve 8.

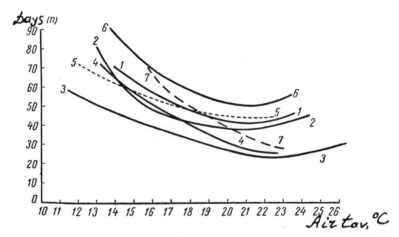

FIGURE 1.33. Phenological curves for various crops from sprouting to flowering at a latitude of 50-53° (15 hours of daylight per day in the growing season). Curve 1: oats variety Lokhovsky; curve 2: flax variety K-39; curve 3: peas variety Capital; curve 4: millet variety Saratov 748; curve 5: barley variety Viner; curve 6: sunflower variety A-41 Kruglik; curve 7: soya variety Krushulya. Obtained by A. S. Podolsky after isopleths provided by V. A. Smirnov.

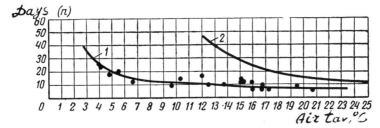

FIGURE 1.34. Phenological curves. Curve 1: geohelminth *Haemonchus contortus* from the egg to the third-stage (invasive) larva [constructed by the author on the basis of field observations made during two years (1957 and 1958) by I. F. Pustovoy (1963), in different seasons in Tadzhikistan]; curve 2: spider mite *Tetranychus urticae Koch.* between generations (after field observations by Yu. A. Piontkovskiy).

Similarly, the average temperature for the 31-day period determined from the same date, May 15, that is, until June 15, can be expressed as follows:

$$t_{av, 31}^{May\ 15\text{-}June\ 15} = \frac{20.2°C\ (May\ 15) + 23.4°C\ (May\ 26)}{4}$$

$$+ \frac{25.5°C\ (June\ 5) + 26.7°C\ (June\ 15)}{4} = \frac{95.8°C}{4} = 24.0°C$$

or

$$t_{av} = \frac{69.1°C +\ 26.7°C}{4} = \frac{95.8°C}{4} = 24.0°C$$

The average temperature for the "zero period" is the temperature for May 15 ($t_{av}^{May\ 15} = 20.2°C$).

The results of these calculations can be used to draw a line dated May 15. From this line it can be seen how the average temperatures for particular periods change in relation to the length of the periods

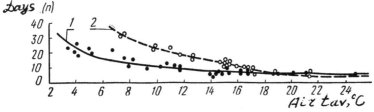

FIGURE 1.35. Phenological curves. Curve 1: the species of geohelminth *Bunostomum trigonocephalum* from the egg to the invasion stage of larva (after field observations by I. F. Pustovoi); curve 2: the same as curve 1, for the geohelminth *Chabertia ovina.*

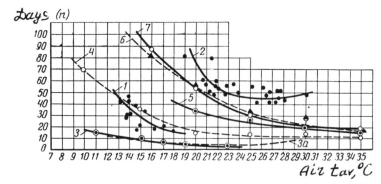

FIGURE 1.36. Phenological curves. Curve 1: piorcer (*Carpocapsa pomonella L*) from the time in spring when the 24-hour average temperature exceeds 10°C (when this species awakens) until the mass flight (50%) of butterflies of the wintering generation (constructed by the author on the basis of field recordings made by the Records Departments of entomological stations in Tadzhikistan from 1953 to 1961); curve 2: piorcer between the flight of butterflies of the wintering generation and that of the first generation, and between the flights of the first and second generations, using data from the same sources; curve 3: *Plasmopara viticola* (a fungus that causes mildew in vines) between outbreaks of the disease (the length of the incubation period is temperature-dependent), after data which D. D. Verderevskiy et al. (1961) obtained in Moldavia; curve 3[a]: the same as curve 3, after M. Drachovska, Czechoslovakia, 1962; curve 4: temperature-dependent duration of the period between the time when the females of the tick *Hyalomma detritum* leave the host's body (after feeding) and the time when they begin to lay eggs (after laboratory experiments by I. G. Galuzo); curve 5: duration of the egg-laying period in *H. detritum* (data from the same source); curve 6: development of the same ticks (with data from the same source) from the egg to the larva, with relative atmospheric humidity of 20-38%; curve 7: the same as curve 6, with relative atmospheric humidity of 50%. The transition from larva to nymph occurs after 8 to 12 days on the body of a warm-blooded host. A developmental curve for this tick from nymph to imago can be seen in Figure 1.38.

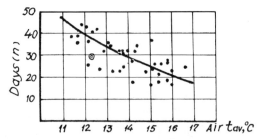

FIGURE 1.37. Phenological curve for piorcer (*Carpocapsa pomonella L.*) from the time in spring when the 24-hour average air temperature steadily exceeds +10°C to the beginning of the flight of the wintering generation according to observations made by Crop Pest Recording Stations in the Ukraine (after O. S. Yermakov and A. S. Pololsky).

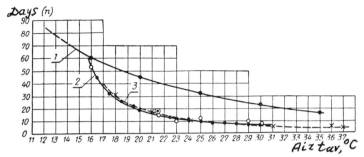

FIGURE 1.38. Temperature/developmental curves. Curve 1: *Hyalomma detritum* ticks from nymph to imago (after experiments made by I. G. Galuzo); curve 2: *Plasmodium vivax*, the pathogen in 3-day malaria, from gamont to sporozoite (after laboratory data provided by Nikolaev, black points; and after laboratory data provided by Roubad, Wenyon, Jansko, Grassi, and James, white points); curve 3 (dashed line through crosses): Yellow fever virus in the carrier from the moment it sucks infected blood until the moment it becomes a source of infection (constructed by the author using the results of observations reported by Sh. D. Moshkovsky et al., 1951).

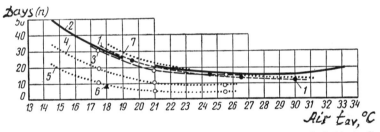

FIGURE 1.39. Phenological curves for malarial mosquitoes (*Culicidae*). Curve 1: *Anopheles superpictus Grassi* from the egg to the appearance of the wings (imago) (observations at 19.6°C by A. P. Gasanov in the field; at 24.5°C and 26.5°C by V. V. Almazova; at 25° and 30°C by A. V. Ulitcheva); curve 2: *A. maculipennis* from the egg to the appearance of the wings (after the data provided by Beklemishev and Shellford); curve 3: the same as curve 2 (after laboratory observations by Martini); curve 4: *A. maculipennis* from the egg to the fourth-stage larva (after Martini); curve 5: *A. maculipennis* from the egg to the second-stage larva (after Martini); 6 (one triangle): *A. superpictus* from the egg to the second-stage larva (after Gasanov); curve 7: *A. superpictus* from egg to egg (this curve was drawn on the basis of curve 1 with the addition of the length of the first gonotrophic cycle, taken from curve 2 in Fig. 1.40).

determined from a given starting date. We shall now proceed to make the same calculations for periods of time from the starting date of May 26:

$$t_{av,0}^{May\,26} = 23.4°\,C$$

$$t_{av,10}^{May\,26\text{-}June\,5} = \frac{23.4°\,C\,(May\,26) + 25.5°\,C\,(June\,5)}{2} = \frac{48.9°\,C}{2} = 24.5°\,C,$$

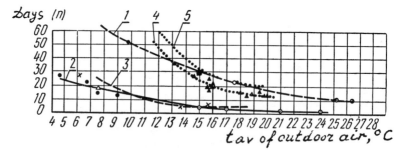

FIGURE 1.40. Age (without diapause) of old females of *Anopheles superpictus* that compose 2-3% of a population (1), and the length of a single gonotrophic period in the females (2), in relation to the average temperatures over these periods in a natural sitting [after data from E. S. Kalmikov, 1959 (white circles with a dot in the center); A. Ya. Storozheva (solid black circles); G. A. Pravikov and L. V. Popov (crosses); S. V. Pokrovskiy (white triangle); V. N. Beklemishev (black square); E. N. Pavlovskiy (plain white circle)]; curve 3: length of a single gonotrophic cycle in *A. maculipennis messeae* in relation to temperatures (after experimental data provided by M. F. Shlenova and A. Ya. Storozheva, with the addition of a small amount of time to allow for the search of food and for egg laying); curve 4: a phenological developmental curve of *A. maculipennis* in populations in the Moscow region, from the egg to the appearance of wings [observations made by N. K. Shipitzina in natural conditions during 1947, 1948, 1950, and 1951 (black triangles)]; curve 5: the development of *A. maculipennis* in populations in the Moscow region, from egg to egg (this curve is based on curve 4, with the addition of the duration of the first prolonged gonotrophic cycle, taken from curve 3).

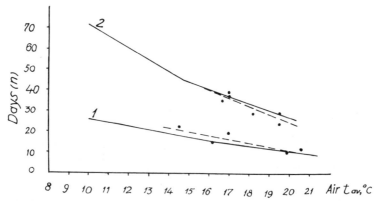

FIGURE 1.41. Phenological curves for the development of the Colorado potato beetle. Solid lines without points: observations for Pruzhany, Byelorussia, 1960-1969 (constructed on the basis of data provided by L. I. Arapova); dashed lines with points: observations for Kursk, Russia, 1974-1976 (constructed on the basis of data provided by V. I. Gudakova and, in part, by T. N. Shelukhina). Curve 1: from the beginning of egg laying to the beginning of the appearance of the second-age larvae; curve 2: from the beginning of the appearance of the second-age larvae to imago.

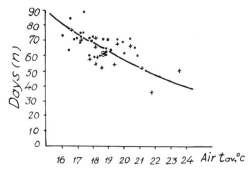

FIGURE 1.42. Turnip moth (*Agrotis segetum Schiff.*) from the beginning of flight of one generation to the beginning of flight of the next generation, in the Ukraine (points) and in the CBEZ (crosses) (from O. S. Yermakov and A. S. Podolsky).

$$t_{\text{av, 20}}^{\text{May 26-June 15}} = \frac{23.4°\,\text{C (May 26)} + 25.5°\,\text{C (June 5)} + 26.7°\,\text{C (June 15)}}{3}$$

$$= \frac{48.9°\,\text{C} + 26.7°\,\text{C}}{3} = \frac{75.6°\,\text{C}}{3} = 25.2°\,\text{C, etc.}$$

These calculations can be used to draw a line representing the average temperatures over given periods that all begin on May 26. Next, the line for June 5 is calculated. These calculations continue for the entire growing period, or any other period in which we are interested, or, finally, for the entire year. The calculations are shown in Table 1.3.

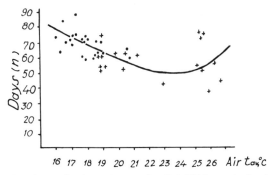

FIGURE 1.43. Turnip moth (*Agrotis segetum Schiff.*) from the beginning of flight of one generation to the beginning of flight of the next generation, in the Ukraine (points) and in Fergana, Uzbekistan (crosses) (from O. S. Yermakov and A. S. Podolsky).

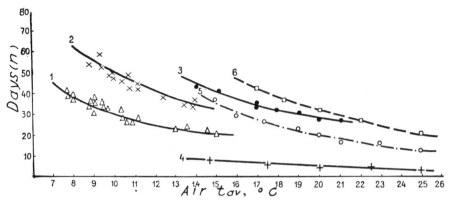

FIGURE 1.44. Phenological curves for the development of *Apamea anceps Schiff.-H. sordida Bkh* in the steppe zone of Kazakhstan. Curve 1: from the beginning of spring activity of wintering larvae (or from the time when the 24-hour average temperature steadily exceeds +5°) to the beginning of pupation; curve 2: from the beginning of spring activity of larvae to mass pupation; curve 3: from mass pupation to the mass flight of butterflies; curve 4: from the flight of butterflies from pupae to mass egg laying; curve 5: from mass egg laying to the mass appearance of the third-aged larvae; curve 6: from mass egg laying to the mass appearance of the fifth-age larvae. Curves are based on data from field observations over 1957-1969 and on data from ecological experiments. (From V. K. Adzhbenov on the basis of the Podolsky method.)

The next step is to make a graphic presentation of all the results that are contained in columns t_{av} and n in Table 1.3. Using millimeter graph paper, two axes are drawn, perpendicular to each other. The vertical axis is designated n days, and the horizontal axis is designated average temperature t_{av} [vertical axis (n) scale: 1 cm = 10 days; horizontal axis (t_{av}) scale: 1 cm = 1°C]. In all cases the scale of the graph of a region's heat resources should be the same as that of the graphs showing the organism's heat requirements, that is, the phenological curves: this becomes important when the latter are transferred onto the heat resources net. To proceed with the construction of the heat resources graph, on the basis of t_{av} and n, lines are drawn representing the average temperatures over given periods starting from May 15, May 26, June 5, June 15, and so on. Thus in Figure 1.50, the average temperature line designated May 15 is constructed as follows: in columns 6 and 7 of Table 1.3 it can be seen that when $t_{av} = 20.2°C$, period $n = 0$ days. We mark on Figure 1.50 a point A with coordinates 20.2°C and 0 days. This point will lie on the horizontal axis (since $n = 0$ days) between 20 and 21°C. In the second line of columns 6 and 7 in Table 1.3 it can be seen that when $t_{av} = 21.8°C$, period $n = 11$ days. We mark on Figure 1.50 a point B with coordinates 21.8°C and $n = 11$ days. This point lies where the vertical line rising from 21.8°C (on the horizontal axis) intersects the horizontal line that runs through the 11-day point on the vertical axis (this will, of course, be just above

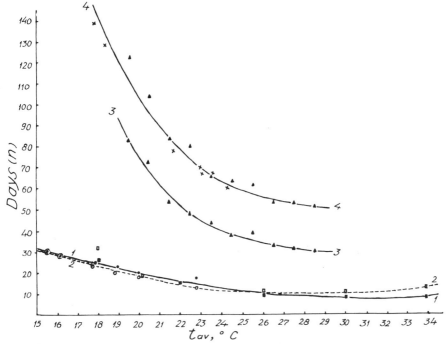

FIGURE 1.45. Phenological developmental curves for California red scale (*Aonidiella aurantii*). Curve 1: from first-age larva to second-age larva; curve 2: from second-age larva to third-age larva; curve 3: from the appearance of the female to the birth of the first-age larva of this female's offspring (Preoviposition period); curve 4: from first-age larva of one generation to first-age larva of the next generation (total development). Laboratory data: from Bliss, California, 1931 (black dots for L_1-L_2; white dots for L_2-L_3); from Hunger (black squares for L_1-L_2; white squares for L_2-L_3); from Bodenheimer, Israel, 1931 (triangles); from Jones, S. Rhodesia, 1935, on leaves (crosses).

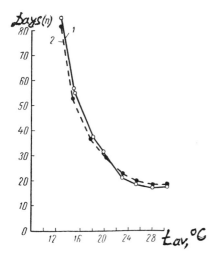

FIGURE 1.46. Phenological curves (the effect of temperature on the length of development) for various geographical populations of the caterpillar *Spilosoma menthastri*. Curve 1: Sukhumi (Caucasus) population; curve 2: Leningrad region population (after A. S. Danilevsky, 1961).

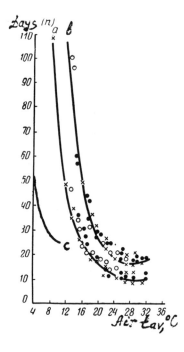

FIGURE 1.47. Phenological curves. Curve a: caterpillars, curve b: caterpillars and pupae of cabbage white butterfly (*Pieris brassicae*) in various geographical populations (including Palestina and USSR), according to data provided by various authors at different times (after A. S. Danilevsky, 1961); curve c: development of hawthorn tortricid (after V. A. Chekhonadskikh, 1973).

the horizontal line marked by the number 10 on the vertical axis). In the third line of columns 6 and 7 we see that $t_{av} = 23.0°C$ and $n = 21$ days. We plot point D according to these coordinates, and then we plot points E, F, L, M, I, O, . . . , P, Q, T, and V (e.g., we read the coordinates of point V in the last line of columns 6 and 7: $t_{av} = 23.4°C$ and $n = 184$ days). When we join all the points by straight line segments, we obtain a

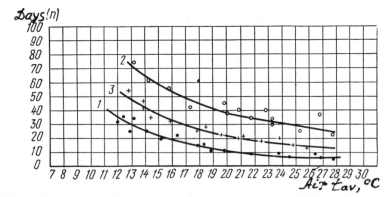

FIGURE 1.48. Phenological curves for fall webworm (*Hyphantria cunea D.*). Curve 1: from egg to larva; curve 2: from larva to pupa; curve 3: from pupa to butterfly. Data from various authors were compiled by A. Ye. Stadnikov. Curve 2 coincides entirely with a similar curve constructed by A. S. Podolsky and E. A. Sadomov on the basis of 53 observations in various years at various places and for different generations.

curved line, which we will designate May 15. This represents the average temperatures over periods of time that begin on May 15.

The line for May 26 is constructed in the same manner (compare Figure 1.50, points $W, Z, g, h, r, \gamma, \delta, \varepsilon, \ldots, \eta, \vartheta, \lambda, \mu$, and β, with the data found in Table 1.3, columns 10, 11, and 12). The lines for June 5, 15, and 25 . . . and August 5 are constructed in the same way (for August 5 compare points $\pi, \ldots, \rho, \sigma, \tau, \varphi, \psi, \omega$, a, b, c, d, e, and f with the data found in columns 32, 33, 34). The same applies to August 15 and 26, . . . , December 26, . . . , January 5, . . . , February 15 and 25, . . . , May 5.

The various lines intersect to form the heat resources net for a climatic region.

As can be seen, the lines of heat resources net are smooth, and the entire net is "beautiful," although it is based completely on empirical data. This follows from the fact that the temperatures are averaged to a high degree: first, multiannual data are used; then they are averaged over various periods $(10, 20, 30, \ldots, 80,$ etc. days). With this averaging, the deviations caused by the superposition of accessory factors (wind, precipitation) onto the key astronomical factor of the Sun-Earth system level each other out. It remains, let us say, a pure law. When the degree of averaging is less (at the beginning of the lines of average temperatures), the smoothness of the lines is slightly broken, as can be seen from Figure 1.50 and from many other similar figures.

It turns out that, provided nature is cleared of "accidental" deviations, it appears extremely harmonious. It is probably the task of science to find these harmonious laws in nature. The shape of the net resembles the outline of an equilateral trapezium. Its base line presents the multiannual amplitude, within which the average 10-day temperatures flucutate in a given climatic region. The base line in Figure 1.50 begins at a temperature of about $4°C$ and ends at about $31°C$. That means that the multiannual amplitude of average 10-day temperatures in Yavan is $31° - 4°C = 27°C$. The height of the trapezium is determined by the maximum value of n days appearing in the calculations in Table 1.3. In this table maximum n is 181 to 184 days and is almost identical for every month; therefore, the height of the trapezium is 181 to 184 days and is virtually identical at all points.

Figure 1.49 is a graphic representation of Table 1.3 in its entirety. In this figure there are two hatched areas, one is a diagonal band, and the other, at the upper right-hand side, is a triangle. These hatched areas represent the data that appear in Table 1.3, that is, calculations of the heat resources net. As the band becomes narrower, the area of the triangle in the upper corner increases, since an increasingly greater part of the maximum value of n days refers to the following year—after January 1. It can be seen that the lower the selected maximum value of n days, the narrower the diagonal band will be.

It is possible in this way to "cut off" the heat resources net at any

TABLE 1.3. Calculation of heat resources net; data collected by the meteorological

10-Day Period; Name of Month	Midpoint Date of 10-Day Period	Multiannual 10-Day Average Air Temperature (in °C) and Its Number ν (for computer)		Before 15 V	15 V, N = 1			Point in Fig. 1.50	26 V, N = 2			Point in Fig. 1.50	5 VI, N = 3		
			ν		Total	t_{av}	n		Total	t_{av}	n		Total	t_{av}	n
1	2	3,	ν	4	5	6	7	8	9	10	11	12	13	14	15
1st; Jan.	5 I	3.8	.												
2nd; Jan.	15 I	4.3	.												
3rd; Jan.	26 I	4.8	.												
1st; Feb.	5 II	4.6	.												
etc.	.	.													
2nd; May	15 V	20.2	1		20.2	20.2	0	A							
3rd; May	26 V	23.4	2		43.6	21.8	11	B	23.4	23.4	0	W			
1st; June	5 VI	25.5	3		69.1	23.0	21	D	48.9	24.5	10	Z	25.5	25.5	C
2nd; June	15 VI	26.7	4		95.8	24.0	31	E	75.6	25.2	20	g	52.2	26.1	10
3rd; June	25 VI	28.9	5		124.7	24.9	41	F	104.5	26.1	30	h	81.1	27.0	20
1st; July	5 VII	30.6	6		155.3	25.9	51	L	135.1	27.0	40	r	111.7	27.9	30
2nd; July	15 VII	30.4	7		185.7	26.5	61	M	165.5	27.6	50	γ	142.1	28.4	40
3rd; July	26 VII	30.9	8		216.6	27.1	72	I	196.4	28.1	61	δ	173.0	28.8	51
1st; Aug.	5 VIII	30.2	9		246.8	27.4	82	O	226.6	28.3	71	ε	203.2	29.0	61
etc.	.	.													
2nd; Oct.	15 X	16.1	.		409.9	25.6	153	P	389.7	26.0	142	η	366.3	26.2	132
3rd; Oct.	26 X	14.2	.		424.1	24.9	164	Q	403.9	25.2	153	ϑ	380.5	25.4	143
1st; Nov.	5 XI	11.8	.		435.9	24.2	174	T	415.7	24.4	163	λ	392.3	24.5	153
2nd; Nov.	15 XI	8.8	.		444.7	23.4	184	V	424.5	23.6	173	μ	401.1	23.6	163
3rd; Nov.	25 XI	7.3	.						431.8	22.7	183	β	408.4	22.7	173
1st; Dec.	5 XII	7.8	.										416.2	21.9	183
2nd; Dec.	15 XII	4.9	.												
3rd; Dec.	26 XII	4.0	.												

given level, which corresponds to the maximum required values of n days. This means that in the columns headed by letter n in Table 1.3 a maximum value of, for example, 102 to 103 days must be taken but not 183 to 184 days. Then the number of data in the other columns is reduced respectively. However, sometimes a net of this nature proves to be unnecessarily large: for regions with severe winters the left-hand, or winter, part of the net is usually not required. In the case of the intermediate and northerly latitudes, as well as valleys in highland regions, it is advisable to draw up a table of calculations starting at the beginning of the warm season. In the graphic representation of the table (Figure 1.49), the straight line AB, which divides the diagram into station in Yavan (Tadzhikistan)—(fragment).

Temperature Totals and Average Temperatures (in °C) for given Periods (n Days), as Reckoned from Given Starting Date

15 VI, N = 4			25 VI, N = 5			5 VII, N = 6			15 VII, N = 7			26 VII, N = 8			5 VIII, N = 9			Point in Fig. 1.50	etc.
Total	t_{av}	n	Total	t_{av}	n	Total	t_{av}	n	Total	t_{av}	n	Total	t_{av}	n	Total	t_{av}	n		
16	17	18	19	20	21	22	23	24	25	26	27	28	29	30	31	32	33	34	35
						347.8	18.3	184	317.2	17.6	174	286.8	16.9	163	255.9	16.0	153	c	...
									321.5	16.9	184	291.1	16.2	173	260.2	15.3	163	d	...
												295.9	15.6	184	265.0	14.7	174	e	...
															269.6	14.2	184	f	...
26.7	26.7	0																	
55.6	27.8	10	28.9	28.9	0														
86.2	28.7	20	59.5	29.8	10	30.6	30.6	0											
116.6	29.2	30	89.9	30.0	20	61.0	30.5	10	30.4	30.4	0								
147.5	29.5	41	120.8	30.2	31	91.9	30.6	21	61.3	30.6	11	30.9	30.9	0					
177.7	29.6	51	151.0	30.2	41	122.1	30.5	31	91.5	30.5	21	61.1	30.6	10	30.2	30.2	0	π	
340.8	26.2	122	314.1	26.2	112	285.2	25.9	102	254.6	25.5	92	224.2	24.9	81	193.3	24.2	71	ρ	
355.0	25.4	133	328.3	25.3	123	299.4	25.0	113	268.8	24.4	103	238.4	23.8	92	207.5	23.1	82	σ	
366.8	24.5	143	340.1	24.3	133	311.2	23.9	123	280.6	23.4	113	250.2	22.7	102	219.3	21.9	92	τ	
375.6	23.5	153	348.9	23.3	143	320.0	22.9	133	289.4	22.3	123	259.0	21.6	112	228.1	20.7	102	φ	
382.9	22.5	163	356.2	22.3	153	327.3	21.8	143	296.7	21.2	133	266.3	20.5	122	235.5	19.6	112	ψ	
390.7	21.7	173	364.0	21.4	163	335.1	20.9	153	304.5	20.3	143	274.1	19.6	132	243.2	18.7	122	ω	...
395.6	20.8	183	368.9	20.5	173	340.0	20.0	163	309.4	19.3	153	279.0	18.6	142	248.1	17.7	132	a	...
			372.9	19.6	184	344.0	19.1	174	313.4	18.4	164	283.0	17.7	153	252.1	16.8	143	b	...

two, shows that it is possible to begin calculations from April 15, and then continue toward the right.

Someone who has enough experience can calculate and construct a net in this manner in only a few hours, although the phenotemperature nomogram, once constructed, is relevant for a considerable number of years.

The nomogram may be completed even more quickly if the net is based not on 10-day average temperatures, but on monthly averages. However, this method is not recommended except in cases where no information about average temperatures for 10-day periods is available. The use of monthly data leads to substantial inaccuracies in the

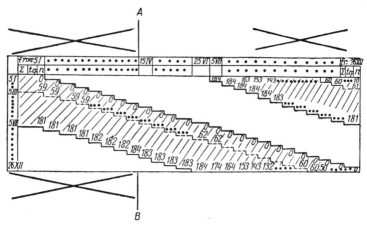

FIGURE 1.49. A graphic representation of Table 1.3 for calculating the heat resources net for a region.

construction of the average-temperature lines at the right and left extremities of the net. If this sytem is used, it is essential to interpolate 10-day lines between the monthly lines (see examples in Figs. 1.52, 1.53, and 1.57).

It is possible for a computer with a graphic output facility (plotter) to both compute and construct heat resources nets. For this purpose it is necessary to first work out an algorithm of Table 1.3, and then a program for a computer with a plotter.

Let us designate the number of each initial date by N, the numbers of average 10-day temperatures by ν, the average 10-day temperatures by $t_N \cdots t_\nu$, and the average temperature over a certain period by $t_{\mathrm{av},\nu}$. Then the algorithm will be represented by the following formula:

$$t_{\mathrm{av},\nu} = \frac{t_N + \cdots + t_\nu}{\nu - N + 1}$$

where $\nu \geqslant N$.

For example, on the basis of this formula we obtain for the initial date May 15 ($N = 1$) the following average temperature over the 11-day period starting from this date ($\nu = 2$):

$$t_{\mathrm{av},2} = \frac{20.2 + 23.4}{2 - 1 + 1} = 21.8°$$

For starting date June 5 ($N = 3$) over a 40-day period ($\nu = 7$):

$$t_{av,7} = \frac{25.5 + 26.7 + 28.9 + 30.6 + 30.4}{7 - 3 + 1} = 28.4$$

These same results are seen in the corresponding columns of Table 1.3.
   In relation to this algorithm, the following program has been worked out (Table 1.4).

TABLE 1.4.   The program for computer "CDC CYBER 170-720," NOC 1.4, and for PLOTTER PACKAGE of CALCOMP, for calculating and constructing the heat resources nets, as well as the humidity resources nets, of a given geographical region.

```
      PROGRAM DIAGRAM(INPUT,OUTPUT)
      DIMENSION X(74),Y(74)
      INTEGER DURATI,DUR,DAYS,STATION
      DIMENSION TEMPER(36),DATE(36),MONTH(12),TMEAN(72,9)
      DIMENSION DAYS(36),M(72,9)
      DATA MONTH/3HJAN,3HFEB,3HMAR,3HAPR,3HMAY,3HJUN,3HJUL,
    =            3HAUG,3HSEP,3HOCT,3HNOV,3HDEC/
      DATA DATE/5.01,15.01,26.01,5.02,15.02,24.02,5.03,15.03,26.03,
    =            5.04,15.04,25.04,5.05,15.05,26.05,5.06,15.06,25.06,
    =            5.07,15.07,26.07,5.08,15.08,26.08,5.09,15.09,25.09,
    =            5.10,15.10,26.10,5.11,15.11,25.11,5.12,15.12,26.12/
      DATA DAYS/2*10,11,2*10,8,2*10,11,3*10,2*10,11,3*10,2*10,11,
    =            2*10,11,3*10,2*10,11,3*10,2*10,11/
      A=1.
      PRINT 95
  95  FORMAT(/////)
      READ*,(TEMPER(I),I=1,36),STATION,LEVEL,DURATI,RIN,KON
      ARIN=A−RIN
      IF(LEVEL)4,5,6
   4  LEVEL=−LEVEL
      PRINT 96,LEVEL
  96  FORMAT(86X,≠(GROUND AT A DEPTH OF ≠,I3,≠ CM.)≠)
      GO TO 99
   5  PRINT 97
  97  FORMAT(86X,≠(GROUND SURFACE)≠)
      GO TO 99
   6  PRINT 98
  98  FORMAT(86X,≠(AIR)≠)
  99  PRINT 100,STATION
 100  FORMAT(*+*,39X*CALCULATION OF HEAT RESOURCES FOR *,A10///)
 101  FORMAT(1H ,135(1H*))
 102  FORMAT(2H *,2(10X≠*≠),12X≠*≠,98X≠*≠)
```

TABLE 1.4. *(Continued)*

```
103   FORMAT(2H *,≠ TEN-DAY  *  MIDDLE  *  MEAN  *≠,8X≠MEAN ≠,
      =≠CENTIGRADE TEMPERATURES FOR THE GIVEN BELOW PERIODS OF
        M DAYS ≠,
      =≠COUNTED OFF FROM≠,7X1H*)
104   FORMAT(2H *,≠ PERIODS *DAY OF THE*TEMPERATURE *≠,40X≠THE ≠,
      =≠FOLLOWING INITIAL DATES≠,31X≠*≠)
105   FORMAT(2H *,≠OF A MONTH*  PERIOD  *  OF MANY≠,3X100(1H*))
106   FORMAT(2H *,2(10X≠*≠),12X1H*,9(5X6H*     *))
107   FORMAT(2H *,2(10X1H*),≠ YEARS FOR   *≠,9(F5.2,6H*  M *))
108   FORMAT(2H *,2(10X1H*),≠ THE PERIOD *≠,9(5X6H*       *))
109   FORMAT(2H *,2XA4,I1,3X1H*,F7.2,3X1H*,F8.1,4X1H*,=(F5.1,1H*,I4,1H*)
      ==(5X6H*      *))
110   FORMAT(2H *,2XA4,I1,3X1H*,F7.2,3X1H*,F8.1,4X1H*,9(F5.1,1H*,I4,1H*)
      =)
111   FORMAT(2H *,2XA4,I1,3X1H*,F7.2,3X1H*,F8.1,4X1H*,=(5X6H*       *),
      ==(F5.1,1H*,I4,1H*))
112   FORMAT(1H ,135(1H*)/1H1)
      XLONG=25,+KON-RIN$YLONG=20
      CALL INITITAL(XLONG,YLONG,A,RIN)
      DUR=DURATI+1
      NDUR=DUR+2
      DO 1 I=1,4
      PRINT 101
      PRINT 102
      PRINT 103
      PRINT 102
      PRINT 104
      PRINT 102
      PRINT 105
      PRINT 106
      NLEFT=9*I-8
      NRIGHT=9*I
      PRINT 107,(DATE(N),N=NLEFT,NRIGHT)
      PRINT 106
      PRINT 108
      PRINT 106
      PRINT 101
      JLEFT=(I-1)*9+1
      JRIGHT=I*9+DURATI
      DO 2 K=1,9
      DO 2 J=JLEFT,JRIGHT
      LEFT=(I-1)*9+K
      M(J,K)=0
```

TABLE 1.4. *(Continued)*

```
      TMEAN(J,K)=TEMPER(LEFT)
      IF(J.LE.LEFT) GO TO 2
      LLEFT=LEFT+1
      DO 3 L=LLEFT,J
      LL=L
      IF(L.GT.36)LL=L-36
      TMEAN(J,K)=TMEAN(J,K)+TEMPER(LL)
  3   M(J,K)=M(J,K)+DAYS(LL)
      TMEAN(J,K)=TMEAN(J,K)/(J-LEFT+1)
  2   CONTINUE
      IX=JLEFT
      JDUR=JLEFT+DURATI
      DO 88 KK=1,9
      NNUM=9*(I-1)+KK
      DO 77 JJ=IX,JDUR
      JF=JJ-IX+1
      X(JF)=TMEAN(JJ,KK)
      Y(JF)=M(JJ,KK)
 77   CONTINUE
      IX=IX+1
      JDUR=JDUR+1
      FNJM1=DATE(NNUM)
      CALL LIN(FNUM1,X,Y,NDUR, ARIN,I)
 88   CONTINUE
      JRIGHT=DUR+8
      DO 24 J=1,JRIGHT
      PRINT 106
      N=9*(I-1)+J
      NMONTH=((I-1)*9+J+2)/3
      NTEN=(I-1)*9+J+3-3*NMONTH
      N1=N
      IF(N.GT.36)N1=N1-36
      IF(NMONTH.GT.12)NMONTH=NMONTH-12
      IF(J.GE.9)GO TO 11
      K1=9-J
      PRINT 109,MONTH(NMONTH),NTEN,DATE(N1),TEMPER(N1),J,
    =         ((TMEAN(N,K),M(N,K)),K=1,J),K1
      GO TO 24
 11   IF(J.GT.DUR) GO TO 12
      PRINT 110,MONTH(NMONTH),NTEN,DATE(N1),TEMPER(N1),
    =         ((TMEAN(N,K),M(N,K)),K=1,9)
      GO TO 24
 12   KL=J-DUR+1
```

TABLE 1.4. *(Continued)*

```
      K1=J-DUR
      K2=9-K1
      PRINT 111,MONTH(NMONTH),NTEN,DATE(N1),TEMPER(N1),K1,K2,
    =          ((TMEAN(N,K),M(N,K)),K=KL,9)
 24   CONTINUE
  1   PRINT 112
      CALL SYMBOL(A,23.,.7,30HPHENO-TEMPERATURE NOMOGRAM
        FOR,0.,30)
      CALL SYMBOL(A+18.5,23.0,.7,STATION,0.0,10)
      CALL PLOT(XLONG+4,0.0,-3)
      CALL ENDPLT
      STOP
      END
      SUBROUTINE INITIAL(XLONG,YLONG,A,RIN)
      CALL NAMPLT
      FACT=1./2.54
      CALL FACTOR(FACT)
      CALL PLOT(A,0.0,-3)
      YY=0.0
      DO 20 K=1,21
      XX=XLONG+1
      CALL PLOT(XX,YY,2)
      XX=1.0
      YY=YY+1
      CALL PLOT(XX,YY,3)
 20   CONTINUE
      CALL PLOT(A,0.0,3)
      XX=1.0
      KKK=XLONG+1.
      DO 30 K=1,KKK
      YY=20.0
      CALL PLOT(XX,YY,2)
      XX=XX+1.0
      YY=0.0
      CALL PLOT(XX,YY,3)
 30   CONTINUE
      Y0=-0.5
      X0=0.5
      NUM=RIN
      DO 60 I=1,KKK
      CALL NUMBER(X0,Y0,0.3,NUM,0.0,2HI3)
      NUM=NUM+1
      X0=X0+1.0
```

TABLE 1.4.   *(Continued)*

```
60  CONTINUE
    XO=0.2$YO=0.0$NUM=0
    DO 70 I=1,21
    CALL NUMBER(XO,YO,0.3,NUM,0.0,2HI3)
    NUM=NUM+10$YO=YO+1.0
70  CONTINUE
    CALL SYMBOL(A-1.0,21.0,0.6,4HDAYS,0.0,4)
    CALL SYMBOL(A+XLONG,0.5,0.6,4HTEMP,0.0,4)
    RETURN
    END
    SUBROUTINE LIN(FNUM1,X,Y,NDUR, ARIN,I)
    DIMENSION X(NDUR),Y(NDUR)
    RAD=180./3.1415625
    N=NDUR-2
    Y(N+1)=0.$Y(NDUR)=10.$X(N+1)=-ARIN$X(NDUR)=1.
    CALL LINE(X,Y,N,1,0,0)
    IF(I.EQ.3.OR.I.EQ.4) GO TO 55
    L=3$IND=1$GO TO 33
10  L=14$IND=3$GO TO 33
55  L=6$IND=2$GO TO 33
20  L=16$IND=3
33  XL=X(L+1)-X(L)
    XO=X(L)+ARIN$YO=Y(L)/10.
    TETA=RAD*ACOS(XL/(XL*XL+1)**0.5)
    CALL NUMBER(XO,YO,0.2,FNUM1,TETA,4HF5.2)
    GO TO (10,20,30),IND
30  RETURN
    END
10.14.05.UCLP, 0013,     0.510KL
```

The succession of commands for the computer is as follows:

1. Number of the account
2. Password
3. BAT
4. Construction of a data file:
    ITERM, MEMAPL. These designations must be typed on the recording terminal. When working with a luminous screen (terminal), FULL must be typed.
    EDIT. Then the chosen name of the file must be taken, for example, NAH (but not more than six characters).

5. A
6. < ....
    ....
    ....

    Here we typed the average 10-day temperatures in °C. In each
    line there are 12 numbers (together there are 36 numbers,
    corresponding to the number of 10-day periods in the year).
    Periods are used in decimals. Commas are used to separate
    numbers from each other. Then we type the following symbols:
    ≠ and the name of the meteorological station, for example,
    NAHARIJA (not more than 10 letters); ≠, 1 (for air); ∅ (for the
    soil surface) or −5 (minus 5 for a depth of 5 cm below soil
    surface); 20 (number of 10-day periods, which limits the net in
    the vertical).
        The beginning of coordinate net is set (the minimum average
    temperature over a 10-day period, rounded off to an integer,
    minus 3: e.g., if the minimum temperature is 12.1°C, then we
    type 9, because $12 - 3 = 9$). Then the end of the coordinate net
    is set (the maximum average temperature over a 10-day period,
    rounded off to an integer, plus 3, minus 25: e.g., if the maximum
    temperature is 25.8°, we type 4, since, $26 + 3 - 25 = 4$) <. We
    end with the symbol with which we began.

7. END
8. SAVE, ... Here we type the name of the data file, for
    example, NAH. All of this is a request to record the data onto
    the disk.
9. |GET, ... Here we type the name of the program, for example,
    HEATR. |FTN, I = HEATR, L = ∅. This is a call for the
    program at the beginning of the working session.
10. LGO, the file name (e.g., NAH), the name of the result (e.g.,
    TOZN; for each heat or humidity resources net a new name is
    required). All of this is a request to record the result onto the
    tape.
11. T
12. REWIND, the name of the result (TOZN). This is a request to
    rewind the tape back to the beginning of the result.
13. ROUTE, the name of the result (TOZN), DC = PR, ID = 1. This
    is the request to type the results in the table. If the table is not
    necessary, this command can be skipped and we begin to deal
    directly with the plotter.
14. ROUTE, PLOT, DC = PH, ID = 4. This is a request to transfer
    the results of the calculation onto the plotter.

If the work is continued with the next net, then the following applies:

15. |REWIND, ... Here we type again the name of the program, for example HEATR. This is a request to rewind the program back to the beginning.

16. |RETURN, LGO. This is a command to clean LGO.

17. EDIT, ... Here we type the name of the new data file for another station, for example, GEVA. Thus we come back to command 4 and continue as above.

When the work is finished, we have to clear the memory of the computer of unnecessary files with the help of the following combination:

|PURGE, ... Here we type the chosen names of the files, for example, the above mentioned NAH, GEVA.

The program for a computer and plotter, as given above, and the succession of the commands given are good for constructing both heat resources nets (on the basis of temperatures of the air, soil, or water) and humidity resources nets. We have to remember that the data file consists of fractional numbers, and therefore the relative air humidity should be represented by a numeral with a zero after the decimal point. For example, the relative humidity 41% should be represented by a numeral 41.0. When calculating the beginning and the end of the coordinate net, we have to take 30 and 250 instead of the constants 3 and 25 (see command 6).

The program for computer as given above is based on the FORTRAN-4 language. This is a most widely used language, suitable for many types of computers. For some computers minor alterations will be necessary. For example, line 1 of the above program:

PROGRAM DIAGRAM (INPUT, OUTPUT)

is replaced by the corresponding control statements. Symbol $ is excluded; the rest of the line after this symbol becomes a new line.

In commands FORMAT the asterisk symbol * is replaced by the quotes sign (").

The PLOTTER PACKAGE of CALCOMP was used by this author. If another package is used, commands for the plotter must be changed.

But, generally speaking, each reader, knowing the type of the computer and the type of plotter he is using, will know which kind of alterations must be made in the program, particularly if a specialist helps him.

The program given above was constructed with the help of mathematicians A. Galperin and J. Iomdin (Ben-Gurion University, Beer-

Sheva, Israel); the author wishes to express his sincere gratitude to them.

Calculation and Constructing of Autumn-Frosts Zone, as well as of Spring-Frosts Zone.

"Autumn-frosts" zone refers to the part of the heat resources net where any point lying on the autumn-summer average temperature lines corresponds to a time after the beginning of the first frosts in autumn. In Figure 1.50 this zone will be the entire area to the left of the boundary with the hatching. For example, point a, which lies on the line with starting date August 5, has an ordinate of 132 days (see dashed line with arrow); therefore, point a corresponds to August 5 + 132 days = December 15. Since the multi-annual average first-frost date in Yavan is November 17, this means that by December 15 there are heavy frosts. This explains why point a lies well within the frosts zone. On the other hand, point φ, which lies on the same line (August 5), corresponds to a prefrost period: the ordinate of point φ is 102 days, which means that this point corresponds to November 15 (August 5 + 102 days = November 15). November 15 is close to the date of the first frosts, but it lies in the prefrost zone.

The boundary between the prefrosts and frosts periods is marked on Figure 1.50 as follows: On the multiannual average, the first frosts in Yavan occur on November 17. Therefore, each of the summer-autumn average-temperature lines consists of two parts: prefrost and frost period. For example, the line designated November 15 has a two-day prefrost segment (November 17 − November 15 = 2 days), and the remaining segment corresponds to a period after the onset of the frosts. We look on the November 15 line for a point with an ordinate of 2 days, and we mark it k. The next line, November 5, consists of a 12-day prefrost segment lm (from November 5 to November 17), and the remaining segment refers to the frost period (after November 17). The next line, October 26, consists of a 22-day segment su (from October 26 to November 17), and the remaining segment refers to the frost period, and so forth. If we join points k, m, and u by straight line segments, we obtain a boundary to the left of which begins the autumn-frosts zone. This boundary, of course, cuts the horizontal axis at point i, which corresponds to the line for November 17, between the lines for November 15 and 25. If we were to draw a line for November 17, it would have a prefrost segment of zero length, that is, the entire line would refer to the frosts period. It is point i that has an ordinate of 0 days, since it lies on the horizontal axis.

In a similar manner we can also construct a spring-frosts zone. For example, the average multiannual date of the last spring frosts in the air of Yavan is March 12. Thus in Figure 1.50 the beginning of the spring-frosts zone will be at point J on the horizontal axis, between the average-temperature lines dated March 5 and 15. The line of average temperatures rising from the initial date of March 5 will consist of a seven-day frost section (March 5 + 7 days = March 12) and the remainder, the after-frost section—see point q. The line of average temperatures from the initial date of February 25 will consist of a 15-day frost section (February 25 + 15 days = March 12) and an after-frost section—see point G. The line of average temperatures from the initial date of February 15 will have a 25-day frost section (February 15 + 25 days = March 12), and the rest of this line will be the after-frost section—see point H, and so on. If we join points J, q, G, H, and so on with a line, then to the left of this line we obtain a hatched zone of spring frosts, in the lower left-hand corner of the PTN (this zone intersects, as can be seen, the autumn-frosts zone).

The principle of constructing frost zones turned out to be very fruitful. For instance, it is useful for calculating the time zone of highly critical temperatures, for calculating the time zone of the drying up of reservoirs (the habitat of malaria mosquitoes), for calculating the time zone of the amount of daylight critical for the development of plants or poikilotherms, and on the whole for taking into account the light factor, as well as the time boundaries crossed by a plant at any particular developmental stage. The latter is important for making combined forecasts regarding the plant/pest system. A detailed description of this idea will be found in the relevant sections of this book.

In the preceding pages we first explained how to construct phenological curves for various organisms, giving their heat requirements, and second, we described how to construct a heat resources net for a given region. The next step is to compare requirements and resources by superimposing the phenological curves onto the heat resources net. This is possible because both the net and the phenological curves have the same coordinates—t_{av} and n days—and are drawn on the same scale.

When we superimpose the biological curves onto the meterological net we obtain a phenotemperature nomogram (PTN), based on a new theory (see Chapter 2). Thus, for example, in Figure 1.50 a number of phenological curves from Figures 1.1 to 1.48 have been transferred onto the net.

As necessary, other biological curves from Figures 1.1 to 1.48 can be transferred onto tracing paper, and the tracing paper can be laid over Figure 1.50. Since tracing paper is reasonably transparent, it is possible to see the net clearly and thus to carry out calculations as though the growth curve had actually been drawn onto the net.

It is preferable not to clutter the heat resources net by the addition of

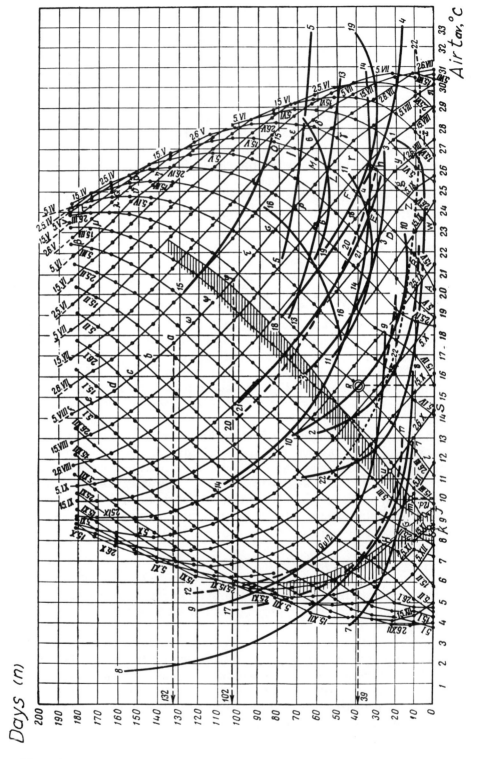

Days (n)

Air tav, °C

54

FIGURE 1.50. A phenotemperature nomogram (PTN) for solving combinations of equations, giving first, the temperature requirements of certain organisms, and second, the heat resources of Yavan, Tadzhikistan (height above sea level $h = 664$ m). Phenological curves for the following organisms have been superimposed onto the net: **Cotton:** curve 1: variety 108-F (*G. hirsutum*) from sowing to mass sprouting, using normal agricultural methods (this curve is taken from Fig. 1.2); curve 2: variety 108-F from sowing to mass sprouting, with soil crust and insufficient soil moisture (taken from Fig. 1.2); curve 3: variety 108-F from mass sprouting to mass budding (from Fig. 1.3); curve 4: variety 5904-I (*G. barbadense*) from mass budding to mass flowering (from Fig. 1.31); curve 5: variety 5904-I from mass flowering to mass boll unfolding (from phenological recordings at NTV; from Fig. 1); curve 6: variety 5595-V (*G. barbadense*) from mass sprouting to mass flowering. **Wheat** variety Surkhak 5688 sown in autumn (not irrigated): curve 7: from sowing to mass sprouting; curve 8: from mass sprouting to the beginning of tillering; curve 9: from the beginning of tillering to mass stalk shooting (for regions with cold winters); curve 10: from mass stalk shooting to mass heading; curve 11: from mass heading to waxy ripeness. **Barley** varieties Khorjau 18 and Persicum 64 sown in autumn (not irrigated); curve 12: from the beginning of tillering to mass stalk shooting (the middle section of curve 12 coincides with curve 9). **Maize** varieties VIR 42 and Krasnodar 1/49; curve 13: from full sprouting to full flowering of the panicles. **Lucerne** variety Tashkent 721 (irrigated): curve 14: between spring growth (which usually occurs when the 24-hour average temperature exceeds 5°C in the south) or the successive harvest and the beginning of the following flowering. **Soya** variety Easy-Cook (irrigated): curve 15: from sowing to flowering; curve 16: from flowering to ripening. **Peaches:** curve 17: from the time when the buds begin to swell (at the beginning of spring, soon after the average 24-hour air temperature exceeds 5°C) until flowering begins. Cotton bollworm (*Chloridea obsoleta F.*): curve 18: from spring date when the 24-hour average air temperature exceeds 15°C (i.e., the species awakens) until the beginning of the first mass egg laying; curve 19: between the beginning of the first and second, second and third, and third and fourth mass egg layings. **Ticks** *Hyalomma detritum* from egg to larva: curve 20: with a relative atmospheric humidity of 20-38%; curve 21: with a relative atmospheric humidity of 50%. **Spider mite** (*Tetranychus urticae Koch.*): curve 22: between generations (dotted line). The heat resources net was based on average data from 11 years of observations (1951–1961). Hatched band *ikmu*, and further up to the right, shows the usual beginning of autumn frosts in the air of Yavan. Hatched band *JqGH*, and further up to the left, shows the normal ending of spring frosts in the air of Yavan. For the dates on the graph, see the symbols and abbreviations.

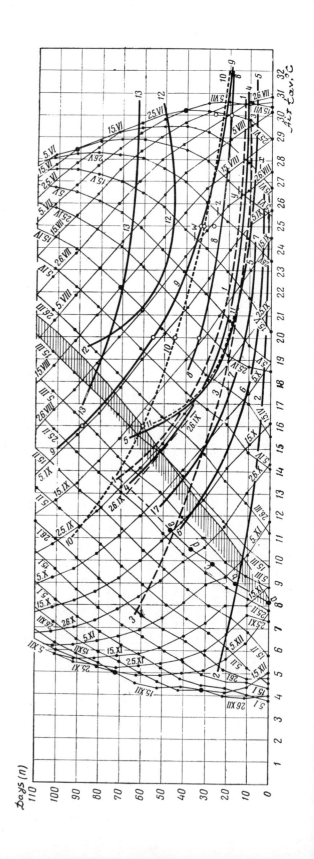

FIGURE 1.51. A PTN for solving combinations of equations, giving first, the heat requirements of parasites (as well as of cotton variety 5769-V), and second, the heat resources of Yavan (Tadzhikistan). Phenological curves for the following organisms have been superimposed onto the net: **malaria mosquito** (*Anopheles superpictus*): curve 1: phenological curve of the development from egg to egg, or between other like stages, of successive generations (i.e., dependence of the length of this period on the average temperature over this period); curve 2: lengths of a single gonotrophic cycle in relation to the temperature; curve 3: lifetimes (without diapause) of old females which compose 2-3% of a population; curve 4: the curve of development from egg to the appearance of wings (imago); curve 5: duration of sporogony of the 3-day malaria pathogen *Plasmodium vivax* in a mosquito's body; curve 6: development of *Anopheles maculipennis* from egg to larva of the second age; curve 7: development of *Anopheles maculipennis* from egg to larva of the fourth age (curves 1, 4, 6, and 7 are transferred from Fig. 1.39; curves 2 and 3 are transferred from Fig. 1.40; curve 5 is transferred from Fig. 1.38). **Ticks** *Hyalomma detritium*: curve 8: duration of egg layings in relation to the temperature; curve 9: the phenotemperature curve from egg to larva (complex: the section of low temperatures, for high relative air humidity; the section of high temperatures, for low relative air humidity); curve 10: curve of development from nymph to imago (curves 8 and 9 are transferred from Fig. 1.36; curve 10 is transferred from Fig. 1.38). Curve 11: duration of the development of **yellow fever virus** in the body of mosquito *Aedes* during the period from the sucking of infectious blood by the carrier to the moment when the carrier becomes infectious, in relation to average temperatures during this process (the curve is transferred from Fig. 1.38). From 22° C and higher, curve 11 coincides with curve 5. Curve 12: development of **piorcer** (*Carpocapsa pomonella L.*) between mass flights of butterflies of different generations (transferred from Fig. 1.36). Curve 13: development of **cotton** variety 5769-V from mass flowering to mass ripening (transferred from Fig. 1.31) (an example of a vegetable indicator for the developmental dates of poikilothermic animals, etc.). The hatched band shows the usual beginning of autumn frosts in the air of Yavan (November 17).

phenological curves. Only the most frequently required curves should be drawn directly onto the net, and then only if they are few in number.

A useful piece of equipment is a circular plastic plate on which heat resource nets are scribed. The plate is divided into sections, with each sector representing a different region. Attached to the lower plate is a transparent plastic overlay that contains various phenological developmental curves. This transparency is then allowed to rotate in order to evaluate each region (Podolsky, A. S., 1958).

Owing to the consistency of the basic biological characteristics of an individual species, and of their relationship to temperature, the same temperature/developmental curve for a particular organism can be used in more than one contiguous region. Each natural climatic region, on the other hand, has a unique pattern of heat resources and, therefore, a unique net. As is known, in a flat country the temperature observations made at a meteorological station can be relevant for a surrounding area 100 km in diameter, with an exactness of up to $\pm0.5°$. This exactness ensures sufficient accuracy for phenological calculations, ±2 to 3 days. Therefore, in a flat country one and the same net is effective over a diameter of 100 km.

However, it is not necessary for the purposes of the bioclimatic calculations to construct a net for each single meteorological station. It is sufficient to construct a net according to the data from a "base" meteorological station in a particular climatic region, and then to use this for contiguous regions, where the temperature deviations in comparison with the "base" station are known. For this purpose the required average-temperature lines should be shifted (either mentally or with a pencil) to correspond to the extent of the abovementioned deviation (for further details see Section 4.2). This approach may also be used for purposes of prognosis, and it extends the diameter of effective action of one net up to 200 km or more in a flat country.

In Figures 1.50 to 1.108 and 1.111 to 1.129 we give a number of PTNs and plain heat resources nets for various regions of the USSR (including Moldavia, Western Ukraine, Western Byelorussia, the Baltic Republics, and Kaliningrad—which is the former German province of Königsberg), as well as for Israel and Denmark, and for three environments: for air, and sometimes for soil and water. The majority of these figures will be useful for the further discussion of the problem.

In order to reduce the number of nets for a given country, and to choose correctly the "base" places for making calculations, it is necessary, first of all, to understand which are the main factors that cause climatic differences within a given country. For such a vast territory as the USSR, one of the principal factors causing the temperature differences in individual regions is the latitude of the

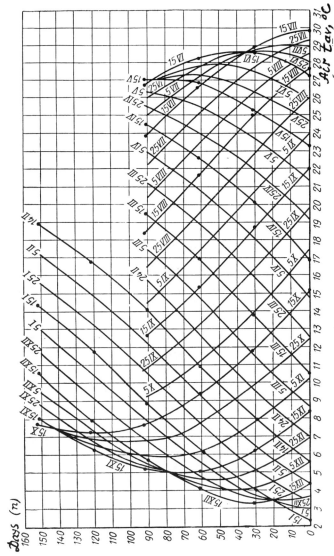

FIGURE 1.52. Heat resources net for a PTN for Kirovabad (Tadzhikistan; $h = 363$ m). This net was constructed using monthly average air temperatures recorded over a period of 10 years (1948-1957); lines representing average 10-day temperatures have been interpolated between the monthly lines.

59

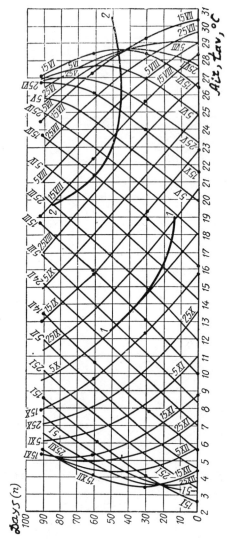

FIGURE 1.53. A PTN for Kulyab (Tadzhikistan; $h = 584$ m). Biological curves; curve 1: development of **piorcer** from the time when the 24-hour average air temperature steadily exceeds $10°C$ (i.e., the time when this species awakens) to the time when 50% of the butterflies of a wintering generation fly; curve 2: **piorcer** between mass (50%) flights of wintering butterflies and those of the first generation, and between flights of the first and second generations. The heat resources net was constructed using monthly average air temperatures recorded over a 10-year period (1948–1957); lines for 10-day average temperatures have been interpolated between the monthly lines; the phenological curves are taken from Figure 1.36.

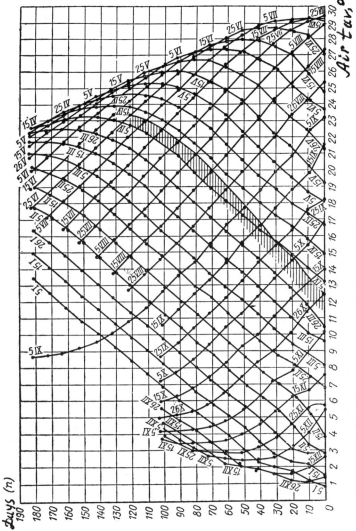

FIGURE 1.54. Heat resources net for Dangara (Tadzhikistan; $h = 660$ m) based on average 10-day air temperatures recorded over a period of 18 years (1941-1944 and 1947-1960). The hatching shows the usual beginning of the autumn frosts in the air in Dangara (October 24).

61

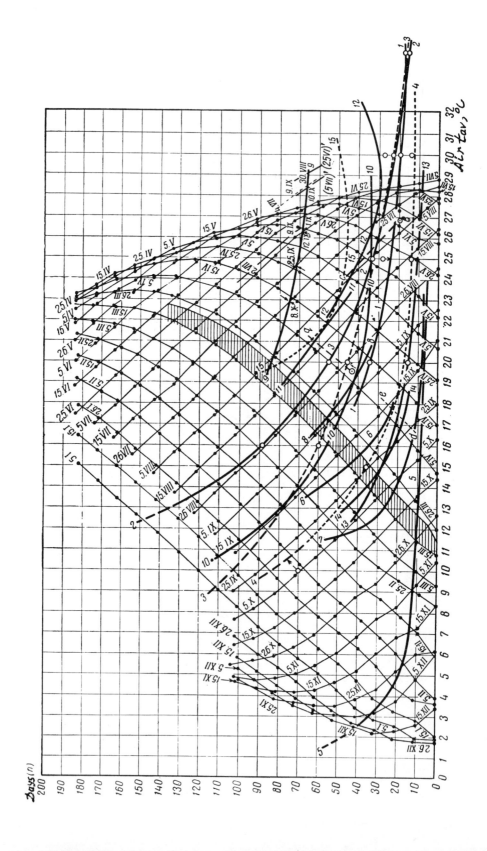

FIGURE 1.55. A PTN for Kurgan-Tyube (Tadzhikistan, Vakhsh Valley; $h = 427$ m). Biological curves: **Ticks** *Hyalomma detritum*: curve 1: duration of the egg-laying period in relation to temperature (the curve is taken from Fig. 1.36); curve 2: duration of development from egg to larva (for temperatures over 16°C, with an atmospheric humidity of 20-38%; for temperatures of 16°C and under, with a humidity of 50%) (the composite curve is derived from experimental curves 6 and 7 from Fig. 1.36); curve 3: development from nymph to imago (from Fig. 1.38); curve 4: time lapse between the time when the females leave the host's body until the beginning of egg-laying (from Fig. 1.36). Curve 5: development of **geohelminth** *Haemonchus contortus* from egg to third-stage larva (from Fig. 1.34). **Wheat** Surkhak 5688 sown in autumn (nonirrigated): curve 6: from mass stalk shooting to mass heading (from Fig. 1.18). **Cotton:** curve 7: 108-F from sowing to mass sprouting, using normal agricultural methods (from Fig. 1.2); curve 8: 108-F from mass sprouting to mass budding (from Fig. 1.3); curve 9: 5904-I from mass flowering to mass boll unfolding (according to observations made by NTV) (from Fig. 1.1). **Lucerne** variety Tashkent 721 (irrigated): curve 10: between spring growth or successive harvest and the beginning of the following flowering (from Fig. 1.16). **Cotton bollworm** (*Chloridea obsoleta F.*): curve 11: from the time when the 24-hour average temperature steadily exceeds 15°C until the first mass egg laying begins; curve 12: between the beginning of the first and second, and second and third, and third and fourth mass egg layings (curves 11 and 12 are taken from Fig. 1.1). **Spider mite** (*Tetranycus urticae Koch*): curve 13: between generations (from Fig. 1.34). **Piorcer** (*Carpocapsa pomonella L*): curve 14: from the time when the 24-hour average temperature steadily exceeds 10°C (i.e., species awakens) until the mass (50%) flight of butterflies of the wintering generation; curve 15: from the flight of one generation to the flight of the next (curves 14 and 15 are taken from Fig. 1.36). The hatching shows the normal beginning of the autumn frosts in the air of Kurgan-Tyube. The heat resources net is based on 10-day average air temperatures recorded over a 21-year period (1940-1960).

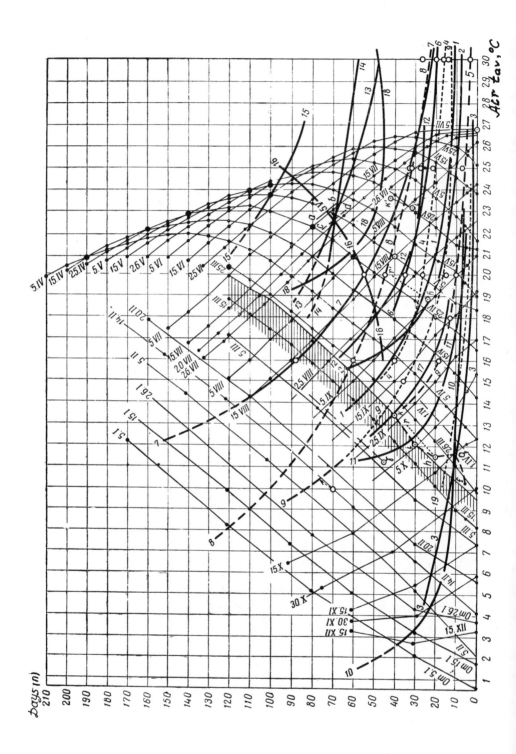

FIGURE 1.56. A PTN for Dushanbe (Tadzhikistan, Hissar Valley; h = 803 m). Biological curves: curve 1: development of **malarial mosquitoes** (*Anopheles superpictus*) from egg to wing appearance (from Fig. 1.39); curve 2: development of **malarial agents** (*Plasmodium vivax*) from gamonts to sporozoites; curve 3: length of a single gonotrophic cycle in *A. superpictus*. Development of *Musca domestica* after Harms: curve 4: from imago to imago or from egg to egg in consecutive generations (taking into account the conditions in Tadzhikistan); curve 5: from pupa to imago. **Ticks** *Hyalomma detritum*: curve 6: duration of egg laying; curve 7: from egg to larva (composite curve); curve 8: from nymph to imago (between temperatures of 17° and 28°C, this curve virtually coincides with the phenological curve for 108-F cotton from mass budding to mass flowering); curve 9: time lapse between the time when females leave the host's body and the time when egg layings begin. Curve 10: development of 108-F **geohelminth** *Haemonchus contortus* from egg to third-stage larva. Development of 108-F **cotton**: curve 11: from sowing to mass sprouting, using normal agricultural methods; curve 12: from mass sprouting to mass budding; between 17° and 28°C, the curve from mass budding to mass flowering virtually coincides with curve 8; curve 13: from mass flowering to mass unfolding of the first bolls (as assessed by the method of Hydrometerological Division); curve 14: the same as curve 13 (as assessed by the method of NTV). **Easy-Cook soya** (irrigated): curve 15: from sowing to flowering; curve 16: from flowering to ripening. **Piorcer** (*Carpocapsa pomonella L*): curve 17: from the time when the 24-hour average temperature steadily exceeds 10°C (i.e., the species awakens) until the mass (50%) flight of butterflies of the wintering generation; curve 18: between the mass flights of butterflies of wintering and first generations, and of first and second generations. Curve 19 (dotted line at the bottom of the nomogram): development of **fungus** *Plasmopara viticola* (cause of mildew in vines) between outbreaks of the disease (i.e., relationship between the length of the incubation period and the temperature) (from Fig. 1.36). The hatching shows the usual beginning of autumn frosts in the air of Dushanbe. The heat resources net is based on 10-day average temperatures of the outdoor air.

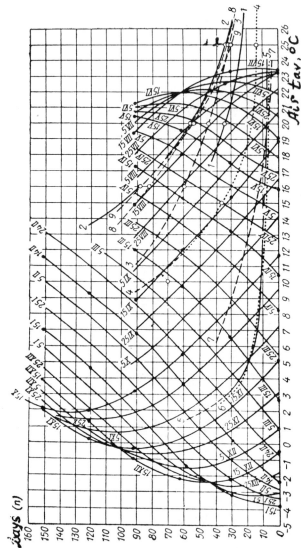

FIGURE 1.57. A PTN for Garm (Tadzhikistan highlands; $h = 1317$ m). Biological curves: curve 1: duration of egg layings of *Hyalomma detritum* ticks; curve 2: composite phenological curve (consisting of curves 8 and 9, "stuck together") for *H. detritum* from egg to larva; curve 3: development of *H. detritum* from nymph to imago; curve 4: time lapse between the time when the females of *H. detritum* leave the host's body and the time when egg laying begins. Curve 5: development of **geohelminth** *Haemonchus contortus* from egg to third-stage larva. Curve 6: development of **geohelminth** *Bunostomum trigonocephalum* from egg to the invasion stage of larva. Curve 7: the same as curve 6, but for the species *Chabertia ovina*. Curve 8: development of *H. detritum* from egg to larva, with a relative atmospheric humidity of 20-38%; curve 9: the same as curve 8, with a humidity of 50%. The heat resources net was based on monthly average air temperatures recorded over a 10-year period (1948–1957); 10-day average lines have been interpolated between the monthly ones.

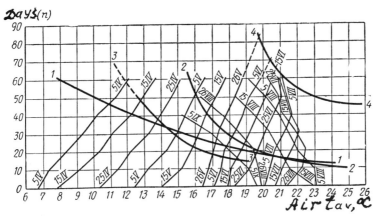

FIGURE 1.58. A PTN for Khorog (a high mountainous region in Tadzhikistan; $h = 2077$ m). Biological curves: curve 1: lifetime of old females (composing about 2-3% of the population) of the *Anopheles superpictus* species in relation to air temperature (the curve is taken from Fig. 1.40); curve 2: duration of sporogony of *Plasmodium vivax* in relation to temperature (from Fig. 1.38). Curve 3: piorcer (*Carpocapsa pomonella L.*) from spring awakening (when the 24-hour average temperature exceeds 10°C) until the mass flight of butterflies of the wintering generation; curve 4: the same as curve 3, but between generations (curve 3 and 4 are taken from Fig. 1.36). The heat resources net was calculated and drawn by parasitologists at the Institute of Medical Parasitology and Tropical Medicine (Moscow) on the basis of 10-day average temperatures of outdoor air recorded over a six-year period (1950-1955), under the guidance of the author.

locations. It is clear that for such a small country as Israel the notions "south-north" are very relative. Here the differentiation of climates is influenced more by the following factors: the height of the place above or below sea level, the proximity to the Mediterranean Sea, and, perhaps, wind. Figures 1.109 and 1.110 show that temperature differences are influenced in the greatest degree, to all appearances, by the height above or below sea level. Taking into consideration specifically this latter factor, a number of appropriate regions were chosen for constructing heat resources nets in Israel. These are hot regions and relatively cool regions, distributed evenly over the territory of Israel (Figures. 1.111-1.128 for air, soil, and water). In all of these figures average data over a number of years were used.

Unfortunately, at meteorological stations of the climatological type in Israel, there are no night observations (at the same time, recordings made by automatic recorders are usually not adjusted by observers; but even if they were adjusted, without night observations by a mercury thermometer such adjustments would not be reliable). The average 24-hour temperature is taken here as being half of the total of the temperatures shown by the maximum and minimum thermometers.

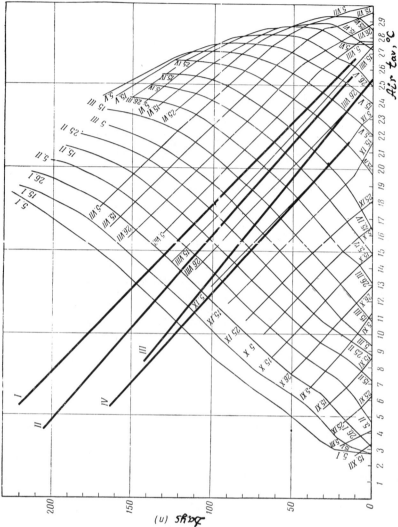

FIGURE 1.59. Ecological-temperature nomogram for Nizhne-Surkhandaryinsky physico-geographical region in Uzbekistan. Biological straight lines represent the relationship of survival dates of the oncospheres of beef tapeworm (*Taeniarhynchus saginatus*) to the air temperature. Curve I: in grass, in the shade; curve II: in grass, in the sun; curve III: on bare soil, in the shade; curve IV: on bare soil, in the sun.

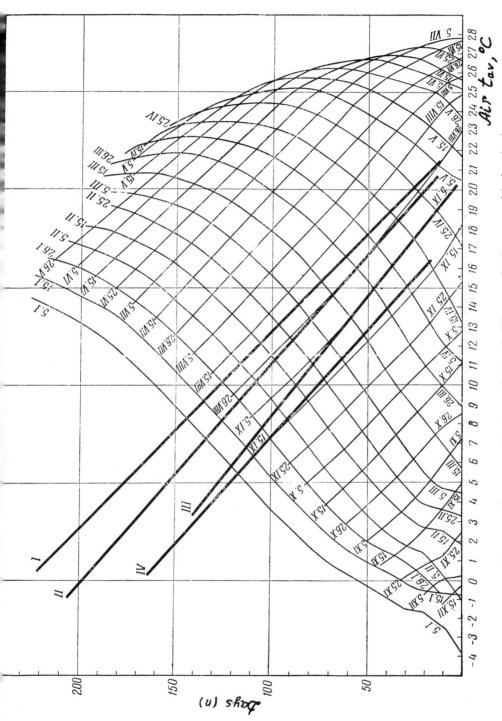

FIGURE 1.60. The same as Figure 1.59, but for the Khorezm physicogeographical region in Uzbekistan.

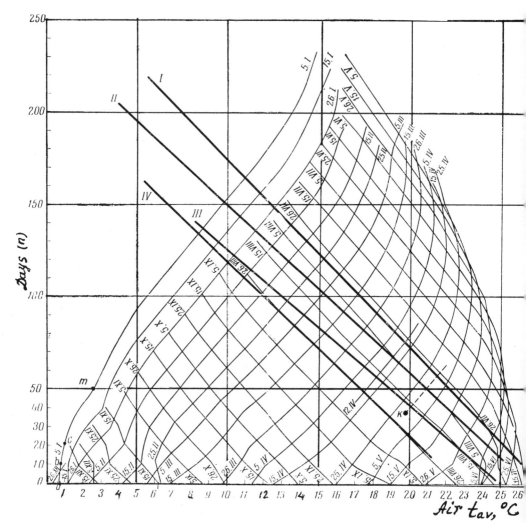

FIGURE 1.61. The same as Fig. 1.59, but for the Samarkand physicogeographical region in Uzbekistan.

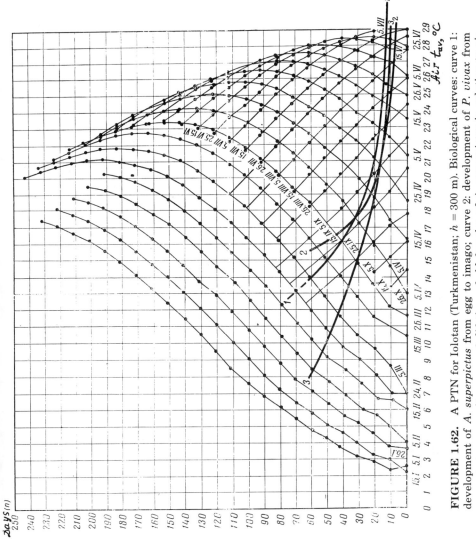

FIGURE 1.62. A PTN for Iolotan (Turkmenistan; $h = 300$ m). Biological curves: curve 1: development of *A. superpictus* from egg to imago; curve 2: development of *P. vivax* from gamonts to sporozoites; curve 3: lifetime of old females of *A. superpictus* in relation to air temperature. The heat resources net is constructed on the basis of multiannual recordings of 10-day average outdoor air temperatures. (From G. M. Ostapenko, 1962, and A. S. Podolsky).

71

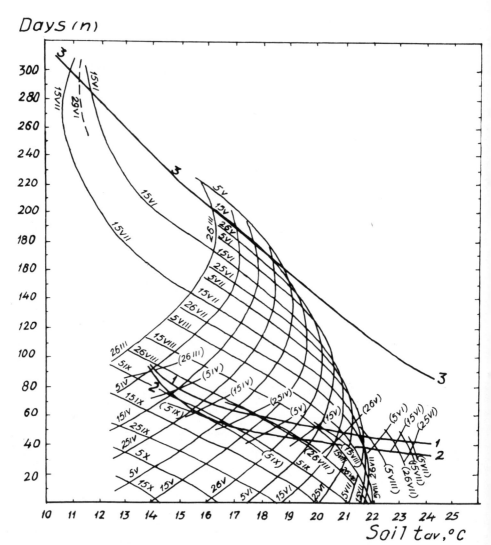

FIGURE 1.63. A PTN for the village of Bakanas in the Alma-Ata district in Kazakhstan. Biological curves: curve 1: fleas *Xenopsula gerbilli* (parasitizing on *Rhombomys optimus Licht*) from egg to imago at a relative humidity of 75-80% (from N. T. Kunitzkaya, 1977); curve 2: the same as curve 1, but at a relative humidity of 89-91%; curve 3: lifetime of souslik fleas and marmot fleas infected by plague (from I. G. Ioff, 1941, and A. S. Podolsky). The net is based on subsoil temperatures in 1970: the entire net is based on a depth of 60 cm, and the fragment of the net marked by the dates in brackets is based on a depth of 40 cm (from Kunitzkaya and Podolsky).

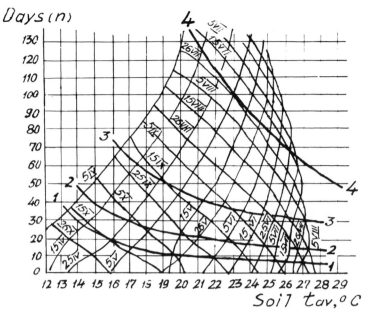

FIGURE 1.64. A PTN for the suburbs of the town of Shevchenko in Kazakhstan (seaboard of the Caspian Sea). Biological curves: curve 1: development of the eggs of fleas *X. skrjabini* parasitizing on *Rhombomys opimus Licht*, at a relative humidity of 75-80% (from B. M. Yakunin, 1979); curve 2: the same as curve 1, for the development of larvae; curve 3: the same as curve 1, for the development from egg to imago; curve 4: length of life of souslik fleas and marmot fleas infected with plague (from I. G. Ioff, 1941, and A. S. Podolsky). The net was based on the multiannual temperatures of the subsoil at a depth of 40 cm (from Yakunin, after Podolsky).

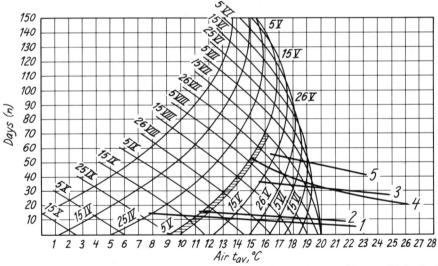

FIGURE 1.65. A PTN for Uritzk in the Kustanay district in Kazakhstan. Biological curves are for the development of spring wheat variety Saratov 29. Curve 1: from sowing to mass sprouting; curve 2: from mass sprouting to the beginning of tillering; curve 3: from the beginning of tillering to full heading; curve 4: from full heading to full wax ripeness; curve 5: from sowing to full heading. The curves are transferred from Figure 1.5. The heat resources net is calculated on the basis of multiannual air temperatures (from V. K. Adzhbenov after A. S. Podolsky).

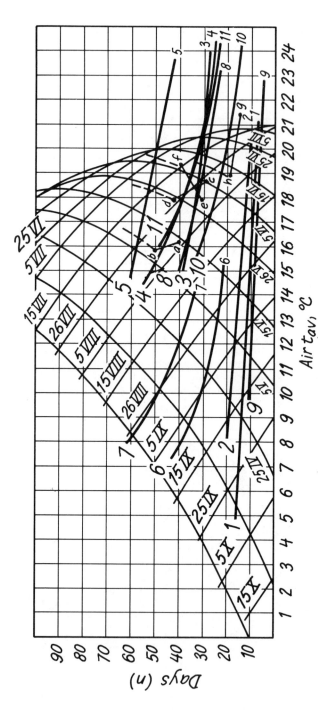

FIGURE 1.66. A PTN for Yesil in the Turgay district in northern Kazakhstan. Phenological curves: **spring wheat** variety Saratovskaya 29: curve 1: from sowing to mass sprouting; curve 2: from mass sprouting to the beginning of tillering; curve 3: from the beginning of tillering to full heading; curve 4: from full heading to full wax ripeness; curve 5: from sowing to full heading. **Cereal pest** (*Apamea anceps Schiff.-H. sordida Bkh*): curve 6: from the beginning of the spring activity of larvae (or from date when air temperatures exceed +5°) to the beginning of pupation; curve 7: from the beginning of the spring activity of larvae to mass pupation; curve 8: from mass pupation to the mass flight of butterflies; curve 9: from the flight of butterflies from pupae to mass egg laying; curve 10: from mass egg laying to the mass appearance of the third-age larvae; curve 11: from mass egg laying to the mass appearance of the fifth-age larvae. The phenocurves are taken from Figures 1.5 and 1.44 (from V. K. Adzhbenov after A. S. Podolsky).

74

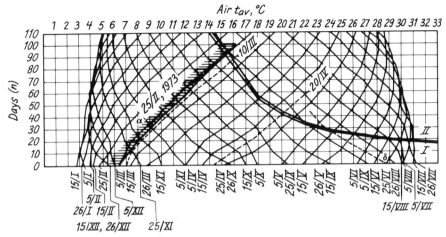

FIGURE 1.67. A PTN for the Apsheron peninsula in Azerbaidzhan. Phenological curves are for the development of geohelminths. Curve I: eggs of **ascarides** (*Ascaris lumbricoides*); curve II: eggs of whipworm (*T. trichiurus*) (from R. E. Chobanov, after Podolsky). The hatching shows the beginning of autumn frosts. The heat resources net is based on the multiannual soil temperatures at a depth of 5 cm.

Our observations in Russia showed that the root-mean-square deviation of average 24-hour temperatures obtained by this oversimplified method approximates ±2° in comparison with the real average 24-hour temperature (obtained from eight observations over a 24-hour period, by a mercury thermometer). Such a deviation takes place when we are dealing with soil temperatures; we may suppose that for air temperatures, which are more mobile, this deviation is bigger. However, when constructing heat resources nets we deal not with single average 24-hour temperatures, but with temperatures that are averaged over 10-day periods and over a number of years; therefore, the deviation of such multiannual average temperatures must not be very great. Nevertheless, we have to take into consideration the fact that the heat resources nets for Israel are based on the oversimplified average 24-hour temperature described above (in contrast to the nets for the USSR).*

There are some difficulties in deducing the average 24-hour temperature for soil, if there are no night observations (automatic recorders for

---

*When an investigator is interested in only the average 24-hour temperature, and this need not be very precise (and the tapes of the automatic recorders are not adjusted), he can manage with only a single observation: the morning observation with a maximum thermometer gives the maximum temperature for the preceding day; the observation with the minimum thermometer gives the minimum temperature for the preceding night. The inconvenience is that in a hot climate there may be evaporation of spirit into the free section of the capillary in the minimum thermometer.

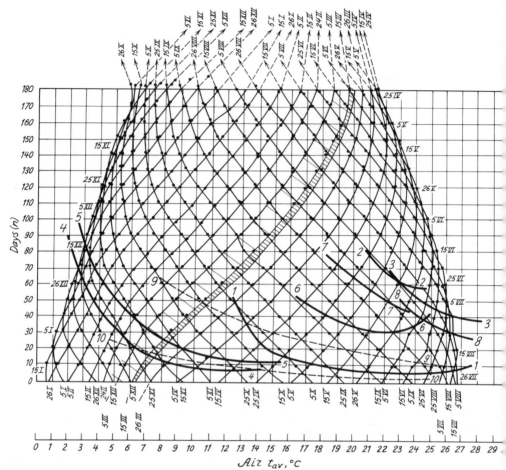

FIGURE 1.68. A PTN for Pushkino in Azerbaidzhan. Phenocurves: **Cotton** variety 2421u: curve 1: from sowing to mass sprouting; curve 2: from mass sprouting to flowering; curve 3: from flowering to boll unfolding. **Winter wheat** "Shark": curve 4: from sowing to mass sprouting; curve 5: from sprouting to tillering; curve 6: from heading to wax ripeness. **Cotton bollworm:** curve 7: from the time in spring when the average 24-hour temperatures exceed +15°C (i.e., from the moment when the pest awakens) to the beginning of the first mass egg laying; curve 8: between the beginning of the first and the second, the second and the third mass egg layings (from A. R. Mustafayev after A. S. Podolsky). Curve 9: age (without diapause) of old **mosquito** females (*Anopheles superpictus*), whose quantity in a population is 2-3%; curve 10: length of a single gonotrophic cycle of *A. superpictus* (curves 9 and 10, from Podolsky). The hatching shows the beginning of the autumn frosts zone.

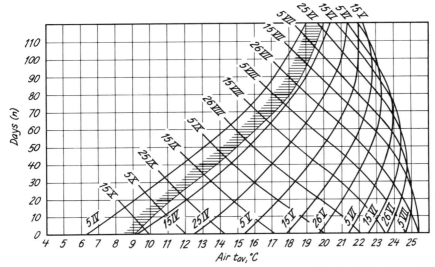

FIGURE 1.69. Heat resources net for Astrakhan, Russia. The hatching shows the normal beginning of autumn frosts.

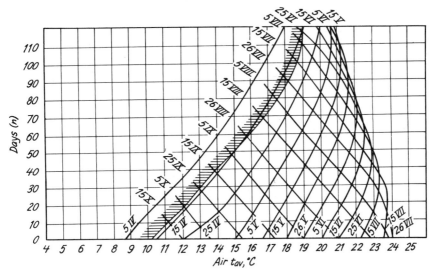

FIGURE 1.70. Heat resources net for Krasnodar, Russia. The hatching shows the normal beginning of autumn frosts.

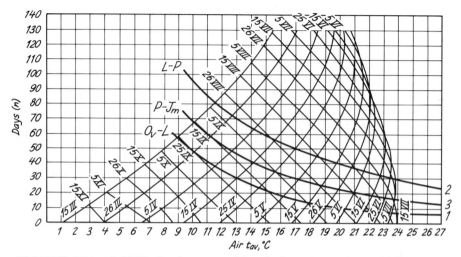

FIGURE 1.71. A PTN showing a graphic solution of equations which give the temperature requirements for the fall webworm (*Hyphantria cunea Dr*) and the heat resources for Eysk of the Krasnodar district in Russia.

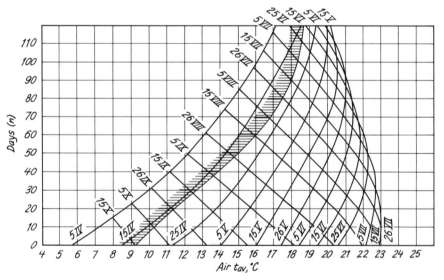

FIGURE 1.72. Net of heat resources for Rostov-na-Donu, Russia. The hatching shows the normal beginning of autumn frosts.

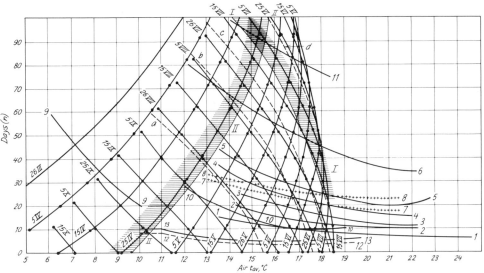

FIGURE 1.73. A PTN for Pruzhany in Byelorussia. Phenological curves: **Colorado potato bettle** (*Leptinotarsa decemlineata*): Curves 1 to 4: from egg to larva of the first age (1), second age (2), third age (3), and fourth age (4); curve 5: from egg to the time when larvae begin to go into the soil for pupation; curve 6: from egg to beginning of appearance of imago. **Potato:** curve 7: from planting to the appearance of shoots on sandy loam soil; curve 8: the same as curve 7, on loamy soil. **Cranberry:** curve 9: from the time when the average 24-hour air temperature exceeds 0° to the opening of buds; curve 10: from the opening of buds to flowering; curve 11: from flowering to ripening of fruits. Pathogen of **phytophthoroze** on tomatoes [*Phytophthora infestans (Mont.) de By*]: curve 12: on leaves; curve 13: on fruits. **Potato:** curves **a, b, c, d:** the beginning of the appearance of shoots, the beginning of budding, flowering, and withering of tops (see Section 2.3 for explanation). **Hatched bands:** curve I: the border of the period with a certain day length after which (as the days become progressively shorter) beetles do not mature for egg laying (see Section 2.3 for explanation); curve II: the normal beginning of autumn frosts (see Chapter 1 for explanation). (From L. I. Arapova and V. B. Gedikh after A. S. Podolsky.)

soil temperatures are available at only a few stations, and moreover they are not reliable without constant verification by direct observations). In addition to all this, maximum and minimum thermometers, as a rule, are not used when soil temperature is being taken (although primitive equipment for this purpose can be constructed). For these reasons, we deduced the average 24-hour temperature at a depth of 5 cm using observations made at 8 AM hours and 2 PM hours: the first observation gives a somewhat *higher* temperature than the minimum, but the second observation gives a somewhat *lower* temperature than the maximum (taking into account Fourier's laws about the spread of temperature oscillations). The average of these two temperatures probably approximates the average temperature over the 24 hours.

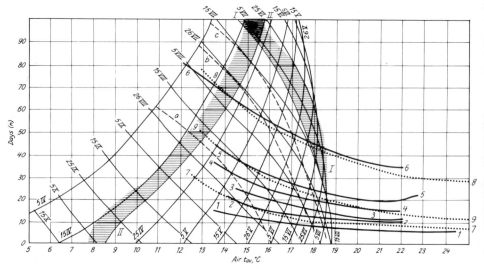

FIGURE 1.74. A PTN for Gomel in Byelorussia. Phenological curves: **Colorado potato beetle** (*Leptinotarsa decemlineata*): Curves 1 to 4: from eg to larva of the first age (1), second age (2), third age (3), and fourth age (4); curve 5: from egg to the time when larvae begin to go into the soil for pupation; curve 6: from egg to beginning of the appearance of imago. **Fall webworm** (*Hyphantria cunea D.*): curve 7; from the beginning of egg laying to the beginning of the appearance of larvae; curve 8: from the beginning of the appearance of larvae to the beginning of pupation: curve 9: from the beginning of pupation to the beginning of the flight of butterflies. Curves **a, b,** and **c:** the beginning of the appearance of shoots, the beginning of budding, and flowering (see Section 2.3 for explanation). Hatched bands: curve I: border of the period with a certain day length after which (as the days become progressively shorter) beetles do not mature for egg laying (see Section 2.3 for explanation); curve II: the normal beginning of autumn frosts (see Chapter 1 for explanation). (From A. S. Podolsky and L. I. Arapova.)

As has been proved many times by many scientists, the key factor for the development dates of plants and cold-blooded organisms (only for the dates, not for the yield's biomass or for the numbers of individuals in poikilotherms) is the temperature factor—provided, of course, that the nontemperature factors are not in a state lethal to the organism (more details about this will be found in succeeding chapters). However, in exceptional cases it may be found that the key factor for development dates is, for instance, the atmospheric humidity.

Among the various characteristics of atmospheric humidity, relative humidity is apparently the most important. First, the state of the organism depends more on the relative air humidity than, for instance, on the absolute humidity, since the intensity of the interchange of moisture between the organism and the environment depends more on the relative humidity. Second, the relative humidity is, to a certain degree, a complex indicator, which reflects both the state of humidity in

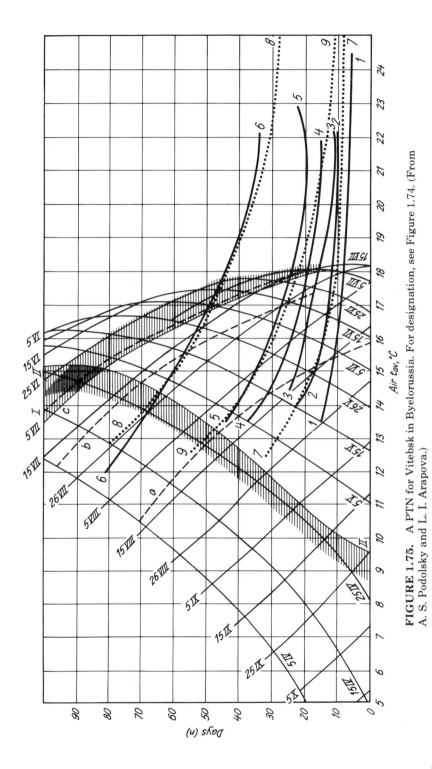

FIGURE 1.75. A PTN for Vitebsk in Byelorussia. For designation, see Figure 1.74. (From A. S. Podolsky and L. I. Arapova.)

81

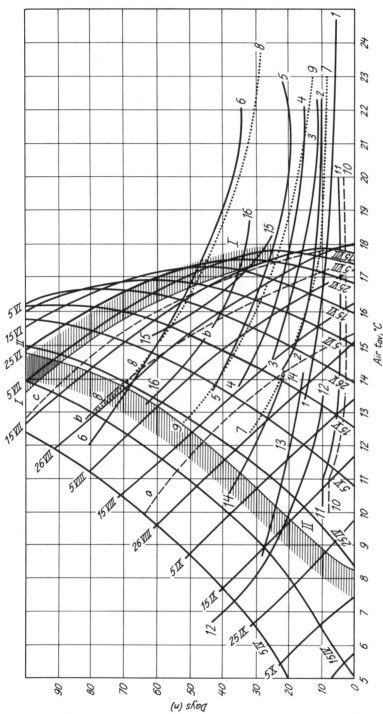

FIGURE 1.76. A PTN for Minsk in Byelorussia. Phenological curves for the pathogen of **phytophthoroze** in tomatoes [*Phytophthora infestans* (*Mont.*) *de By*]: curve 10: on leaves; curve 11: on fruit. **Cabbage pest** (*Hylemyia brassicae Bouch*): curve 12: from the date when the average 24-hour air temperature exceeds +5°C to the flight of imago; curve 13: from the same initial date (as in curve 12) to the beginning of egg laying; curve 14: from the date when the average 24-hour air temperature steadily exceeds +10°C to mass egg laying; curve 15: **cabbage pest** (*Hylemyia floralis*) from the appearance of imago to mass egg laying; curve 16: **cabbage moth** (*Barathra brassicae*) from the date of the appearance of larvae to the appearance of pupae. For other designations, see Figure 1.74. (From A. S. Podolsky, L. I. Arapova, and N. N.

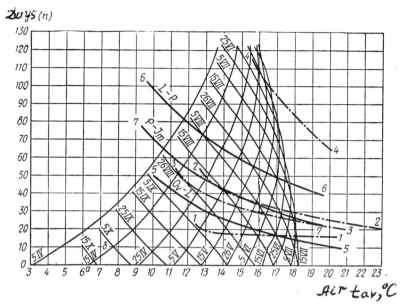

FIGURE 1.77. A PTN for Grodno in Byelorussia. Phenological curves: curve 1: spring barley from sprouting to stalk shooting; curve 2: spring barley from heading to wax ripeness. Curve 3: winter and spring wheat from stalk shooting to heading. Curve 4: millet from sowing to ripening. Curves 5 to 7: fall webworm (*Hyphantria cunea Dr*).

the air and the temperature. This will be clear if we remember the formula for relative air humidity:

$$R = \frac{e}{E} \cdot 100\% \tag{8}$$

where R is relative air humidity in percent; e is the water vapor pressure at a given moment in time (this is often called "absolute humidity"); E is maximum water vapor pressure at a given temperature, that is, saturation water vapor pressure.

The higher the temperature, the greater E is. However, when the temperature gets higher, e rises too, because of increasing evaporation. Usually e rises less than E does. That explains why, as temperature increases, the relative humidity R usually decreases, and vice versa. In particular, in the summer, R is lower than in winter.

This can be clearly seen, for example, from recordings in Eilat, Israel. Haifa shows the absolute antithesis: the maximum relative humidity occurs not in winter, as should usually occur, but in summer. It may be that the closeness of Haifa to the Mediterranean Sea, which is an unlimited source of humidity, causes such intensive evaporation

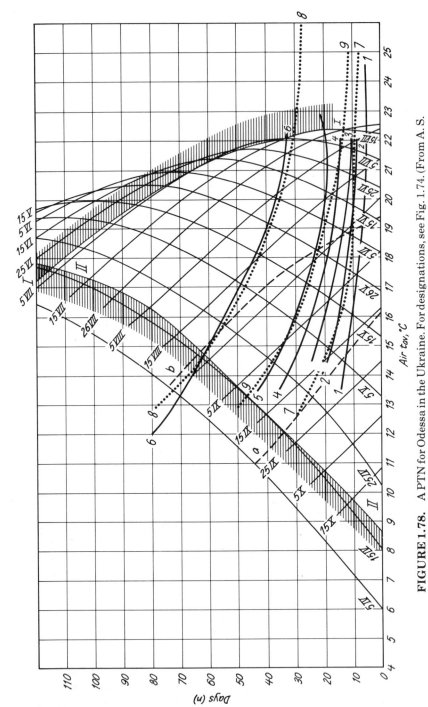

FIGURE 1.78. A PTN for Odessa in the Ukraine. For designations, see Fig. 1.74. (From A. S. Podolsky, L. I. Arapova, and E. A. Sadomov.)

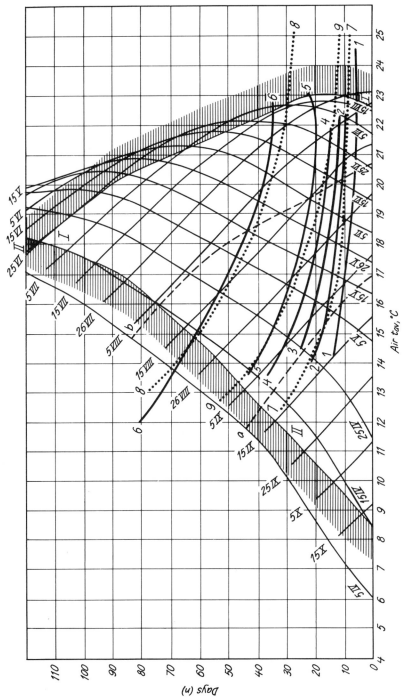

FIGURE 1.79. A PTN for Zaporojye in the Ukraine. For designations, see Figure 1.74. (From A. S. Podolsky and L. I. Arapova.)

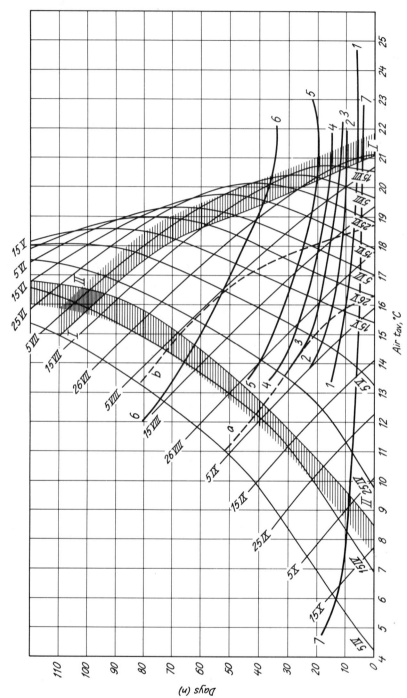

FIGURE 1.80. A PTN for Poltava in the Ukraine. Phenological curves: curve 7: length of the development of uredogeneration in brown rust (*Puccinia triticina*) of winter wheat. For other designations, see Figure 1.74. (From L. I. Arapova and K. P. Shashkova, after A. S. Podolsky.)

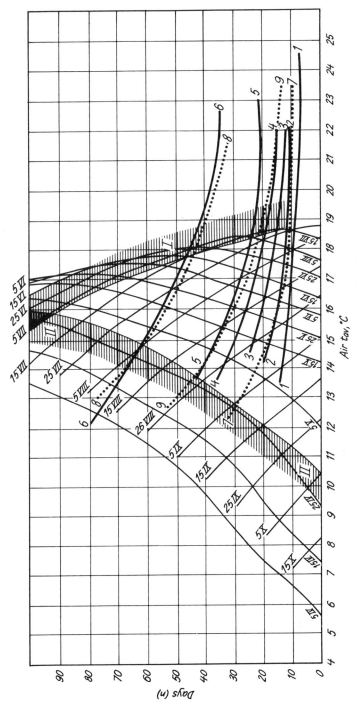

FIGURE 1.81. A PTN for Lvov in the West Ukraine. For designations, see Figure 1.74. (From A. S. Podolsky and L. I. Arapova.)

87

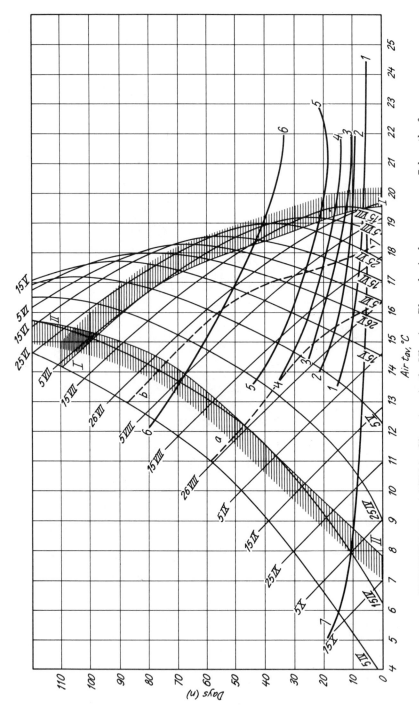

FIGURE 1.82. A PTN for Chernigov in the Ukraine. Phenological curves: curve 7: length of the development of uredogeneration in brown rust of winter wheat; for other designations, see Figure 1.74. (From L. I. Arapova and K. P. Shashkova, after A. S. Podolsky.)

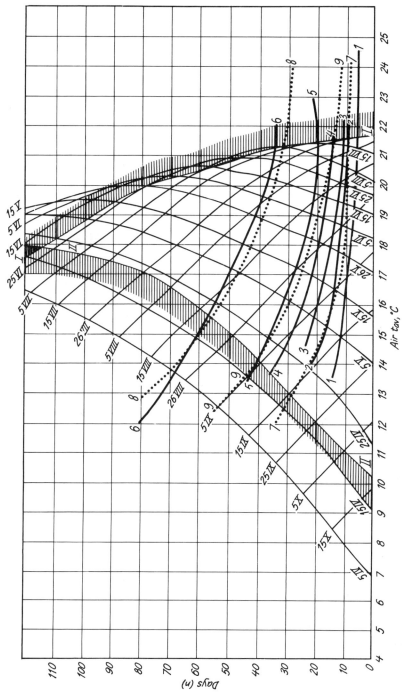

FIGURE 1.83. A PTN for Kishinev in Moldavia. For designations, see Figure 1.74. (From A. S. Podolsky and L. I. Arapova.)

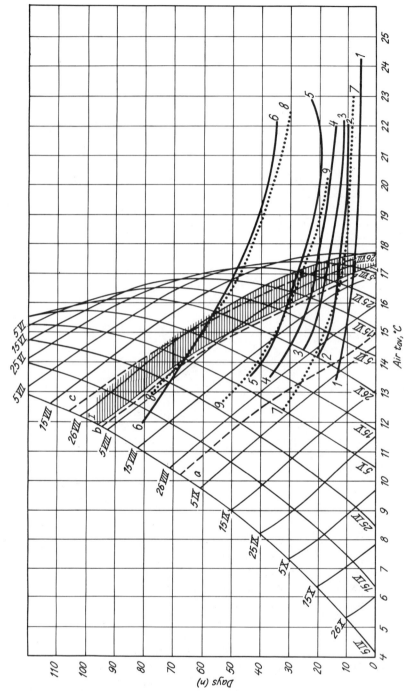

FIGURE 1.84. A PTN for Kaliningrad (formerly Konigsberg). For designations, see Figure 1.74. (From A. S. Podolsky and L. I. Arapova.)

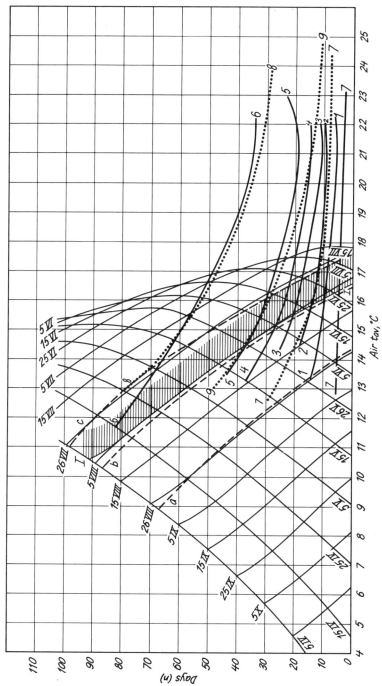

FIGURE 1.85. A PTN for Riga, Latvia. For designations, see Figure 1.74. (From A. S. Podolsky and L. I. Arapova.)

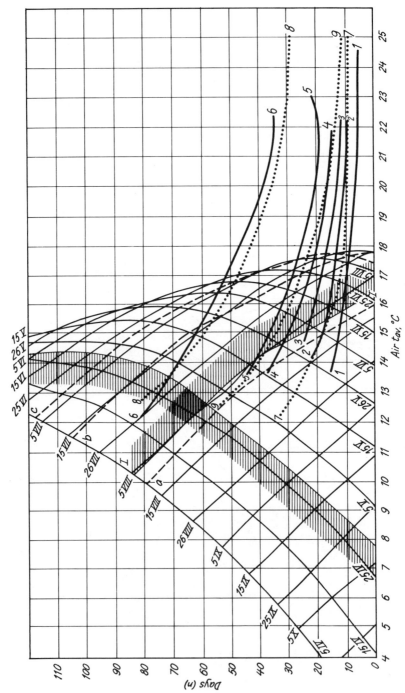

FIGURE 1.86. A PTN for Pskov, Russia. For designations, see Figure 1.74. (From A. S. Podolsky and L. I. Arapova.)

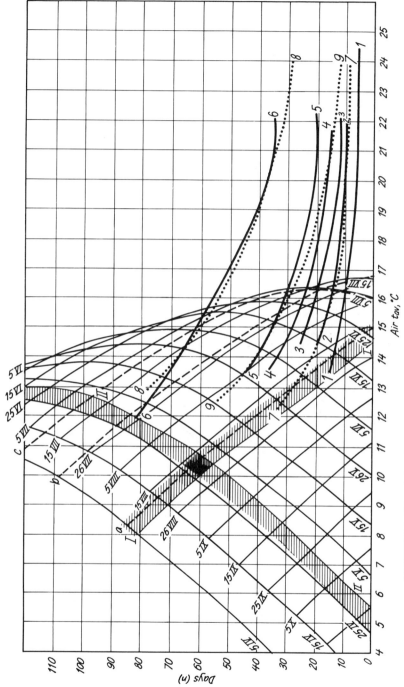

FIGURE 1.87. A PTN for Tallin, Estonia. For designations, see Figure 1.74. (From A. S. Podolsky and L. I. Arapova.)

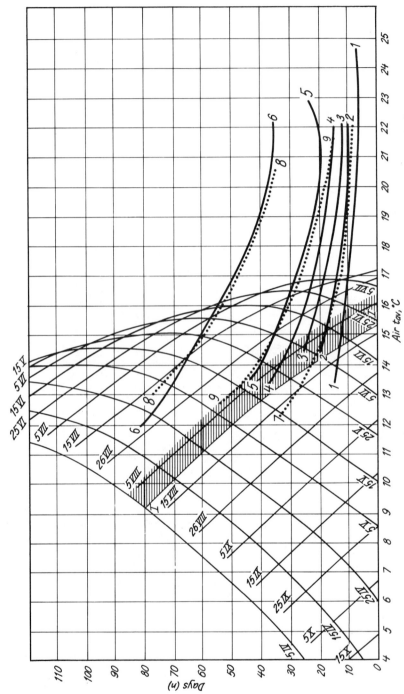

FIGURE 1.88. A PTN for Tartu, Estonia. For designations, see Figure 1.74. (From A. S. Podolsky and L. I. Arapova.)

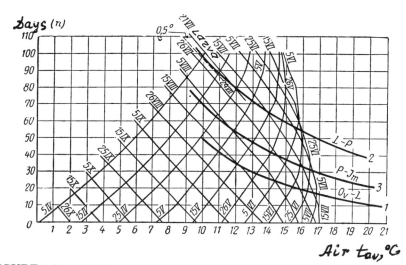

FIGURE 1.89. A PTN showing a graphic solution of equations which give temperature requirements of the fall webworm (*Hyphantria cunea Dr.*) and heat resources of Volkhov in the Leningrad district, latitude 59°56′.

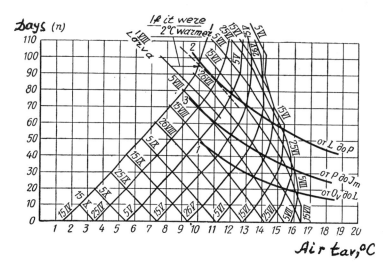

FIGURE 1.90. A PTN for Petrozavodsk (Karelia) with phenological curves for the fall webworm (*Hyphantria cunea Dr*).

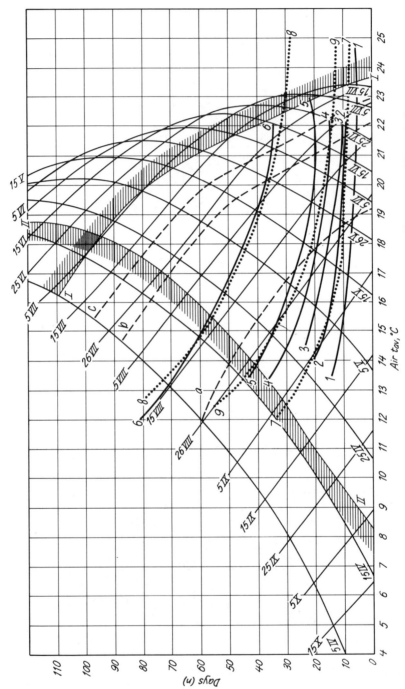

FIGURE 1.91. A PTN for Saratov, Russia. For designations, see Figure 1.74. (From A. S. Podolsky and L. I. Arapova.)

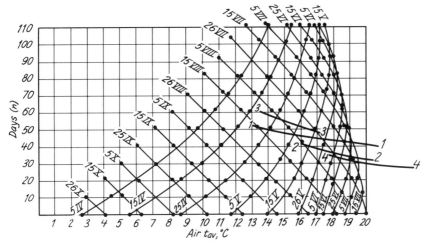

FIGURE 1.92. A PTN for Oboyan in the Kursk district, Russia. Developmental curves for spring wheat variety Lutescens 62: curve 1: from sprouting to heading; curve 2: from heading to wax ripeness. Developmental curves for oat variety Lokhovsky 1026: curve 3: from sprouting to the appearance of panicles; curve 4: from the appearance of panicles to wax ripeness. (From A. S. Podolsky and O. S. Yermakov.)

at high air temperatures that, in the formula given above, the numerator e rises quicker than the denominator E. This is not the case in Eilat: Eilat is located on the shore of the extremely narrow Gulf of Eilat, formed by the Red Sea, which is also narrow. Therefore, Eilat can be considered, in comparison to Haifa, as a "continental" location.

In conclusion, Beer-Sheva, because of its location, displays the intermediate state of the annual trend of relative air humidity. All of this can be clearly seen from Figure 1.130. It must be taken into consideration, however, that the lack of night observations required us to use morning or evening observations instead of the average 24-hour values for relative air humidity. (In Fig. 1.130 the 24-hour average is used for obtaining the monthly average.) All three—morning, evening, and 24-hour average—are more or less close to each other. Figure 130 gives the data for constructing humidity resources nets for Israel (Figs. 1.131-1.133).

As can be seen, the humidity resources net for Eilat (Fig. 1.131) is a mirror reflection of the temperature resources net for Eilat: the winter lines of average humidity over a certain period (R_{av}) are located on the right-hand side of the net, and the summer lines are on the left-hand side.

The humidity resources net for Haifa (Fig. 1.133) "follows" the temepratures: the summer lines for average humidity over a certain period are located on the right-hand side, and the winter lines are on the

Days (n)

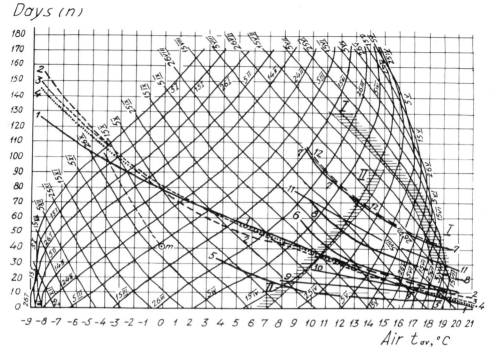

Air t_av, °C

FIGURE 1.93. Temperature-ecological nomogram for Kursk, Russia. Phenological curves for the relationship between the survival dates of **foot-and-mouth disease virus** type A_{22} in the shade and at air temperature: curve 1: on straw; curve 2: on wood; curve 3: on cotton cloth; curve 4: on paper. Curve 5: development of larvae of **nematode** *Dictyocaulus viviparus* to the third invasion stage. **Fall webworm** (*Hyphantria cunea Dr.*): curve 6: from Ov to L; curve 7: from L to P; curve 8: from P to Im. Curve 9: length of incubation of **apple scab** (*Fusicladium dendriticus*). **Colorado potato beetle** (*Leptinotarsa decemlineata*): curve 10: from the beginning of egg laying to the beginning of the appearance of the second-age larvae; curve 11: from the beginning of the appearance of the second-age larvae to the beginning of appearance of imago; curve 12: from the beginning of egg laying to the beginning of the appearance of imago. Hatched bands: curve I: the border of the period with day length below which Colorado potato beetles do not mature for egg laying; curve II; the normal beginning of autumn frosts in the air (October 9).

left-hand side. But since in the winter months in Haifa R changes little (Fig. 1.130), the winter lines are laid densely and almost vertically. Because of this circumstance, the net has a complicated outward appearance (Fig. 1.133). The net for Beer-Sheva (Fig. 1.132) looks especially complicated.

It remains to be mentioned that these humidity resources nets have been constructed for the first time.

A number of nets for Israel are not given here. In all, there are more

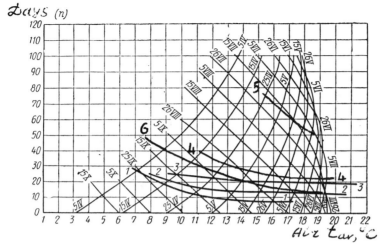

FIGURE 1.94. A PTN for Lgov (Kursk district) with phenological curves. **Winter rye** from sprouting to tillering, with different levels of soil moisture in the ploughed layer: curve 1: 35-55 mm (good moisture level); curve 2: 25 mm (moderate moisture); curve 3: 15 mm (low moisture). **Biohelminth** *Fasciola hepatica*: curve 4: from egg to mass appearance of miracidia in the environment; curve 5: from miracidia to cercariae in the body of mollusk. Curve 6: **Mollusk** *Galba truncatula*, from egg to mass appearance of young mollusks.

than 30. All of them (including the humidity nets) were calculated with the help of a computer and were constructed automatically by a plotter (with the exception of the net for Avdat).

Using these nomograms, it is relatively simple to solve ecological questions. Let us take, as an example, the question: on what date (on the multiannual average) does wheat of variety Surkhak 5688 form

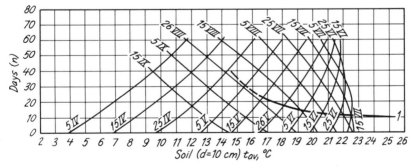

FIGURE 1.95. A PTN for Lgov (Kursk district) with phenological curve 1 for sugar-beet root aphid (*Pemphigus fuscicornis Koch.*), from larva to imago (from A. S. Podolsky and Yu. S. Vovchenko). The net is based on soil temperatures.

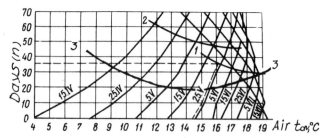

FIGURE 1.96. A PTN for Ryazan, Russia, with phenological curves for cultured and wild melliferous: curve 1: common buckwheat (*Fagopyrum sagittatum*) from sowing to the beginning of flowering; curve 2: phacelia from sowing to the beginning of flowering; curve 3: cornflower (*Centaurea sumensis*) from budding to the beginning of flowering. (From I. N. Gavrilova and A. S. Podolsky).

heads in Yavan, if the multiannual average date for mass stalk shooting is March 26?

In order to answer this question, we have to compare the temperature requirements for wheat, in the period from stalk shooting to heading, with Yavan's heat resources from March 26 until head time. Using our new method, we can make this comparison by solving two mathematical equations simultaneously: (1) the equation for this variety of wheat's heat requirements, which is shown on the PTN for Yavan (Fig.

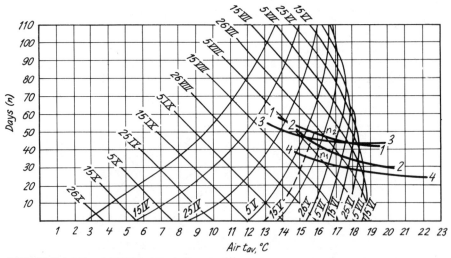

FIGURE 1.97. A PTN for Trubchevsk (Bryansk district, Russia) with phenological curves. Curve 1: oat variety Lokhovsky from sprouting to the appearance of panicles; curve 2: oat variety Lokhovsky from the appearance of panicles to wax ripeness. Curve 3: barley variety Viner from sprouting to heading; curve 4: barley variety Viner from heading to wax ripeness. (From A. S. Podolsky and O. S. Yermakov.)

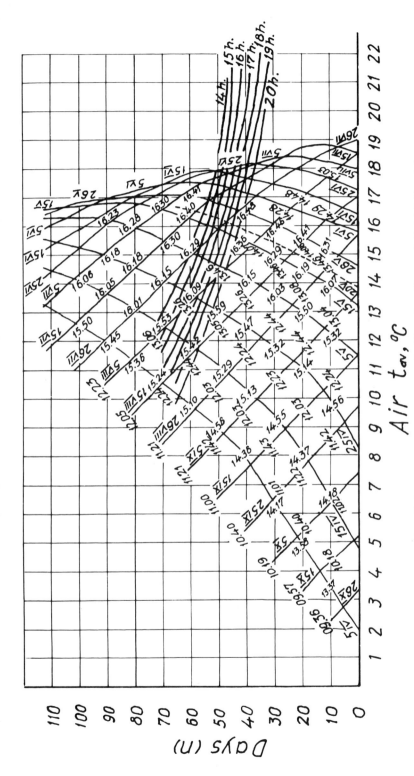

FIGURE 1.98. Phenolight temperature nomogram for Karachev (Bryansk district, Russia) with phenological curves for barley variety Viner from sprouting to heading at different lengths of the day. (From O. S. Yermakov and A. S. Podolsky; phenocurves were obtained on the basis of isopleths provided by V. A. Smirnov, 1960).

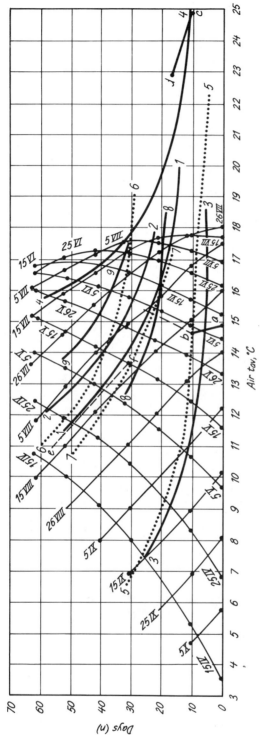

FIGURE 1.99. A PTN for the suburbs of Moscow (h = 150 m). Biological curves: curve 1: development from egg to the appearance of the wings of malarial mosquitos *Anopheles maculipennis* in populations in the suburbs of Moscow; curve 2: the same as curve 1, from egg to egg; curve 3: length of a single gonotrophic cycle of *A. maculipennis* in relation to temperatures (curves 1, 2, and 3 are from Fig. 1.40); curve 4: length of sporogony of malarial agent *Plasmodium vivax* (the curve is from Fig. 1.38); curve 5: length of incubation period in pathogen of wheat stem rust (*Puccinia graminis*) (after K. U. Stepanov 1940, 1972); curve 6: duration of development of spring barley variety Viner from heading to wax ripeness; curve 7: piorcer (*Carpocapsa pomonella L.*) from spring awakening to the beginning of flight of the wintering generation; curve 8: *Trichogramma cacoecia pallida Meyer* from egg to imago; curve 9: common buckwheat (*Fagopyrum sagittatum*) from sowing to the beginning of flowering.

102

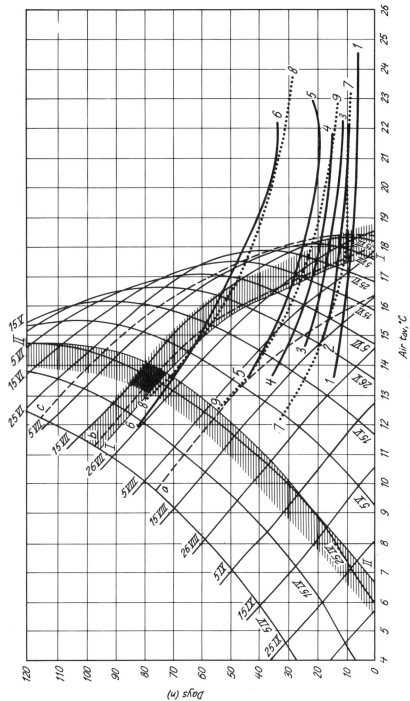

FIGURE 1.100. A PTN for Gorky, Russia. For designations, see Figure 1.74. (From A. S. Podolsky and L. I. Arapova.)

103

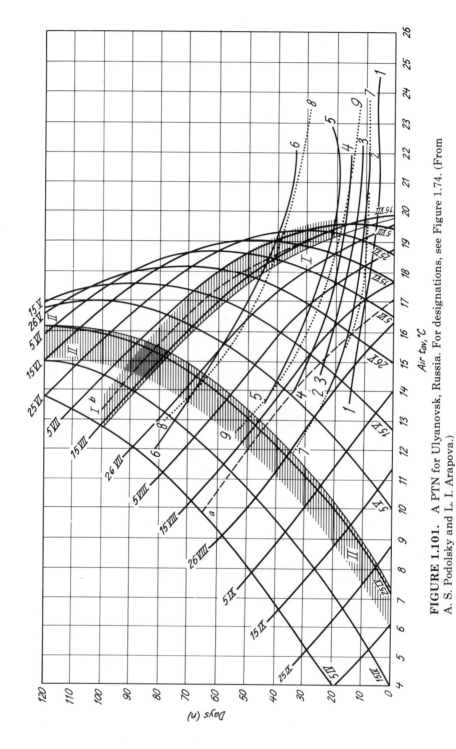

Air $t\omega$, °C

Days (n)

FIGURE 1.101. A PTN for Ulyanovsk, Russia. For designations, see Figure 1.74. (From A. S. Podolsky and L. I. Arapova.)

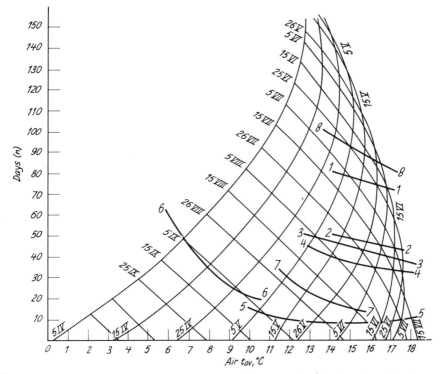

FIGURE 1.102. A PTN for Nerekhta in the Kostroma district, Russia, with phenological curves. Curve 1: mountain ash (*Sorbus*) from mass flowering to the beginning of ripening. Curve 2: red bilberry (*Vaccinium vitisidae*) from mass flowering to the beginning of ripening. Curve 3: bilberry (*Vacinium myrtillus*) from mass flowering to the beginning of ripening. Curve 4: raspberry (*Rubus*) from the beginning of flowering to the beginning of ripening. Curve 5: bog bilberry (*Vaccinium uliginosum*) from the beginning of flowering to mass flowering. Cranberry (*Oxycoccus*): curve 6: from the time when the average 24-hour air temperature steadily exceeds 0° C to opening buds; curve 7: from the opening of buds to flowering; curve 8: from flowering to ripening of fruits. (From A. F. Cherkasov and V. B. Gedikh, after A. S. Podolsky.)

1.50) in the form of developmental curve 10 from mass stalk shooting to mass heading; and (2) the equation for Yavan's heat resources, which is shown in Figure 1.50 as the line of average temperature marked March 26 (this date is near the horizontal axis, between 11 and 12° C). In terms of analytical geometry, the graphic solution of a system of two equations can be found at the intersection point of the two lines representing these equations, that is, in our case, the point where phenological curve 10 and the average-temperature line for March 26 intersect. In Figure 1.50 this point is marked by the circle R. The coordinates of this point, which can be read on the axes of the graph

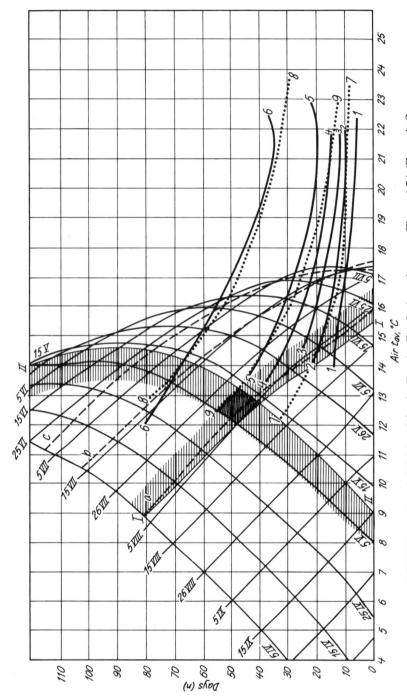

FIGURE 1.103. A PTN for Vologda, Russia. For designations, see Figure 1.74. (From A. S. Podolsky and L. I. Arapova.)

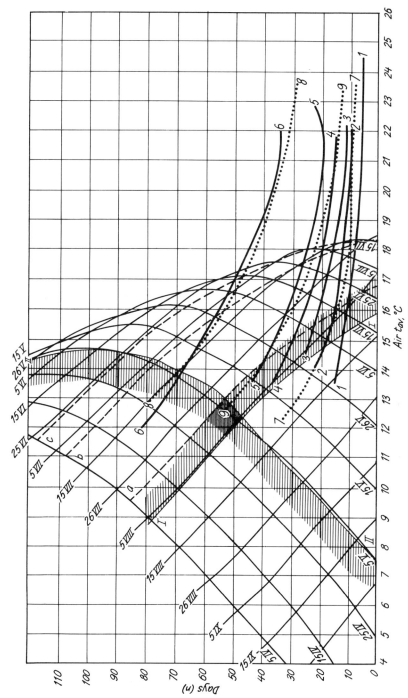

FIGURE 1.104. A PTN for Kirov, Russia. For designations, see Figure 1.74. (From A. S. Podolsky and L. I. Arapova.)

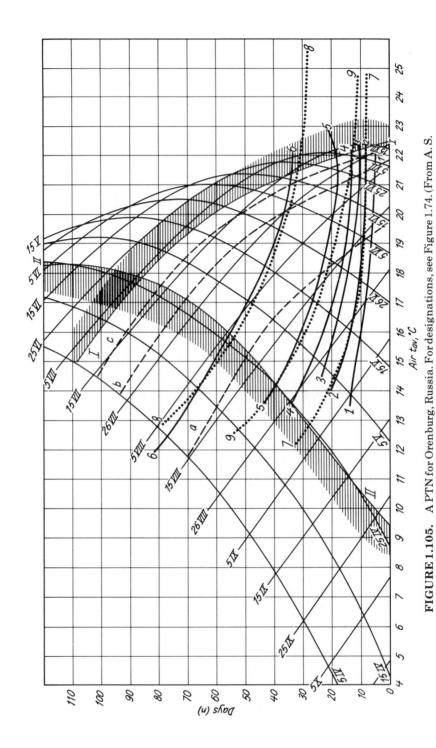

FIGURE 1.105. A PTN for Orenburg, Russia. For designations, see Figure 1.74. (From A. S. Podolsky and L. I. Arapova.)

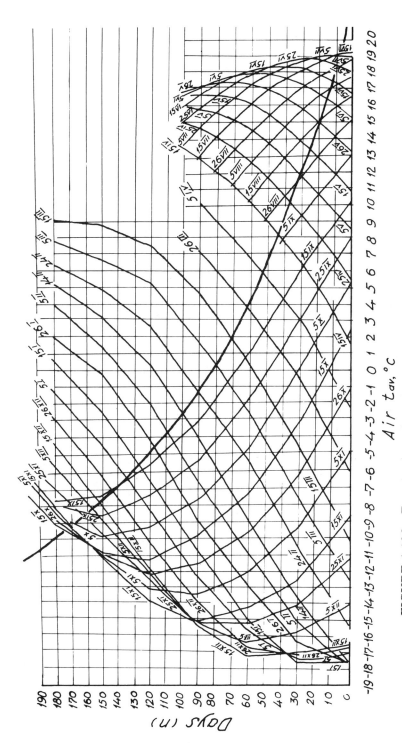

FIGURE 1.106. Temperature-ecological nomogram for Omsk (Western Siberia) with a phenological curve for the survival of foot-and-mouth disease virus A_{22} in environment (outside a living organism) in the shade. (The phenocurve was constructed taking into account the data for Western Siberia).

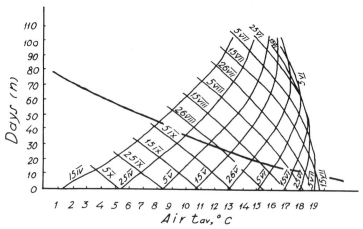

FIGURE 1.107. Temperature-ecological nomogram for Barnaul (Siberia) with a phenocurve for the survival of foot-and-mouth disease virus A_{22} on straw, in shadow.

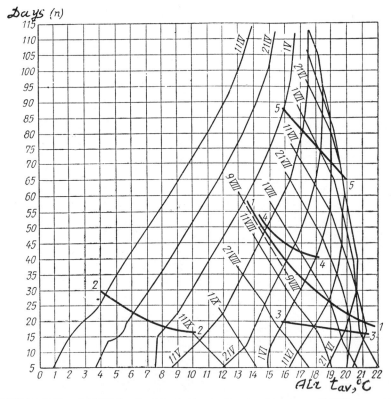

FIGURE 1.108. A PTN for Zavitinsk (Amur district, Far East, USSR; $h = 400$ m). Phenological curves: curve 1: *Anopheles maculipennis* from egg to the appearance of wings (after Beklemishev-Shelford, taken from Fig. 1.39). Spring wheat: curve 2: from sowing to sprouting; curve 3: from stalk shooting to heading; curve 4: from sprouting to heading; curve 5: from sprouting to wax ripeness. (From A. S. Podolsky and V. V. Golovin.)

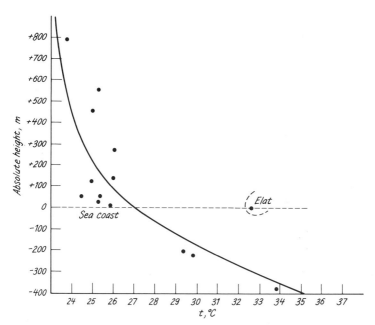

FIGURE 1.109. The relationship between the average multiannual air temperatures in the last 10-day period of July at the height above or below sea level (Israel).

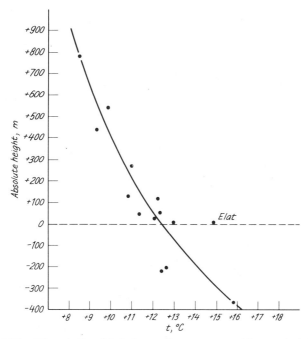

FIGURE 1.110. The relationship between the average multiannual air temperatures in the last 10-day period of January at the height above or below sea level (Israel).

111

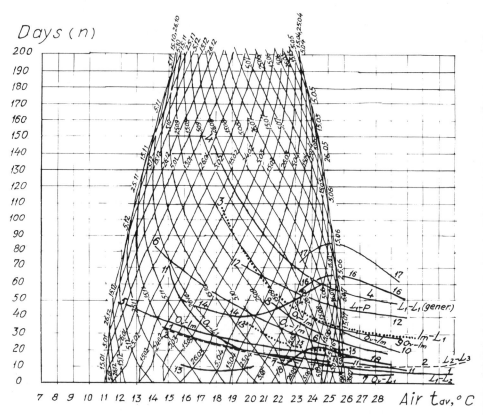

FIGURE 1.111. A PTN for Beer Sheva, Israel; $h = 270$ m. Phenological curves: Curves 1 to 4: **California Red Scale** (*Aonidiella aurantii*), laboratory and insectary. **Parasites of Ps. Comstocki:** curve 5: *Leptomastidea abnormis Girault* in grove; curve 6: *Anagurus sp.* **Predators of Ps. Comstocki:** curves 7 and 8: *Scymnus suturalis*, laboratory; curve 9: *S. quadrimaculatus*; curve 10: *S. includes.* Curves 11 and 12: *Thaumetopoea Jordana* (Staudinger), field and laboratory. **Cotton** variety Acala S.J.-2 (*Gossipium hirsutum L.*), Israel: curve 13: from planting to germination (50%); curve 14: from germination to budding (1st); curve 15: from budding to flowering (1st flower); curve 16: from 1st flower to ripening (when the number of open bolls accumulated per $m^2 = 10$); curve 17: from 1st flower to ripening (number of open bolls accumulated per $m^2 = 50$); curve 18: from ripening (= 10 bolls/m^2) to ripening (= 50 bolls/m^2).

(see the dashed lines leading from circle R), are $t_{av} = 15.6°$C and $n = 39$ days. Therefore, this method has given the solution of two equations with two unknowns—t_{av} and n.

In this case, we do not need the temperature, but the finding $n = 39$ days enables us to find the date that interests us—the multiannual average mass heading date—since 39 days is the length of the period

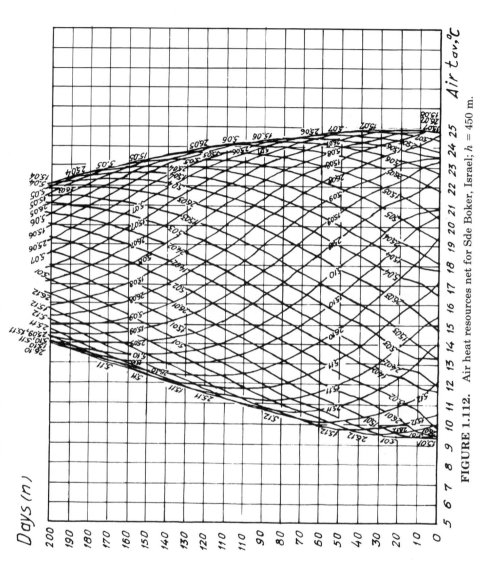

FIGURE 1.112. Air heat resources net for Sde Boker, Israel; $h = 450$ m.

113

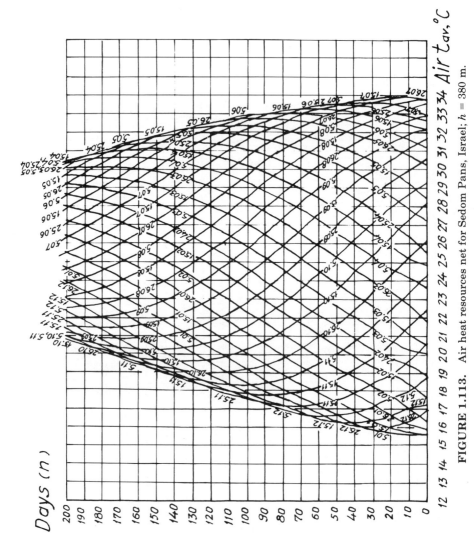

FIGURE 1.113. Air heat resources net for Sedom Pans, Israel; $h = 380$ m.

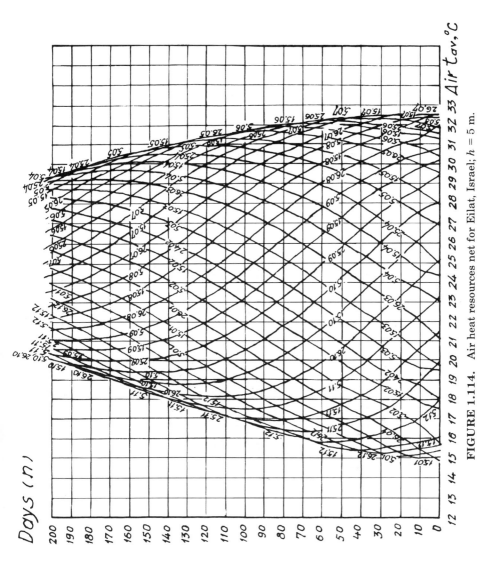

FIGURE 1.114. Air heat resources net for Eilat, Israel; $h = 5$ m.

115

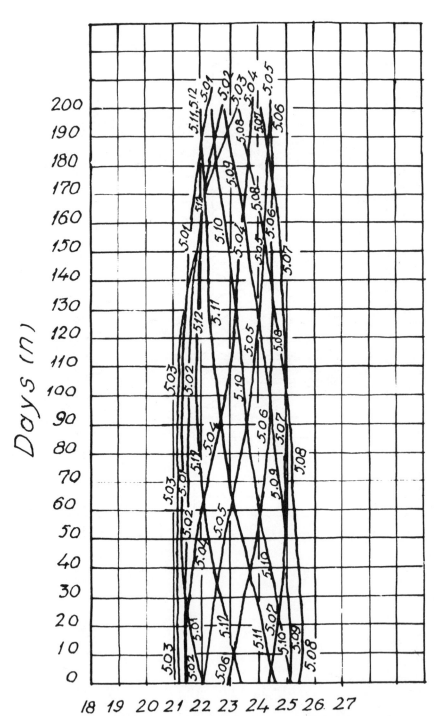

FIGURE 1.115. Net of heat resources on water surface for Eilat, Israel; $h = 5$ m.

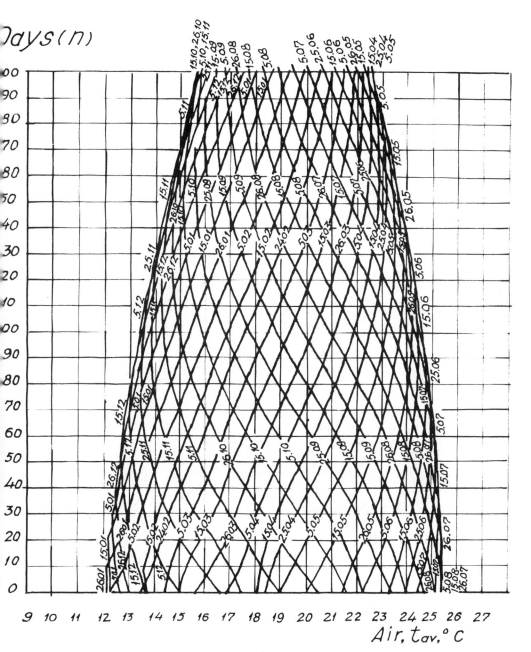

FIGURE 1.116. Air heat resources net for Beit-Dagan, Israel; $h = 30$ m.

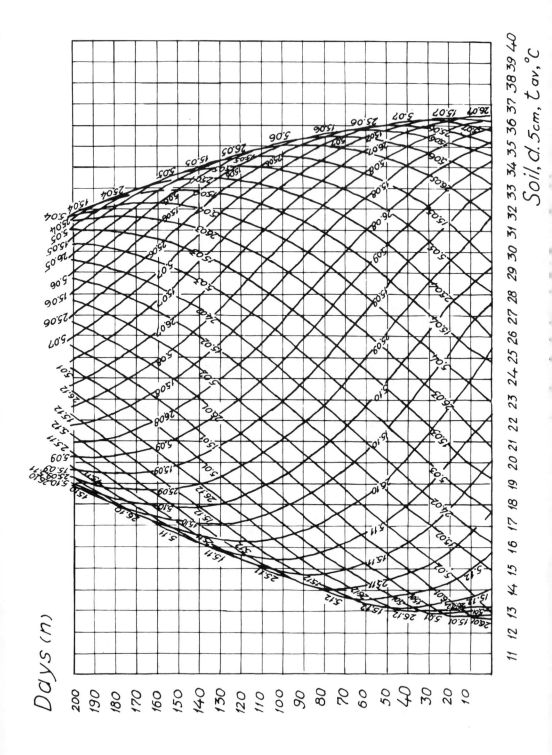

Days (n)

Soil, d. 5cm, t av, °C

118

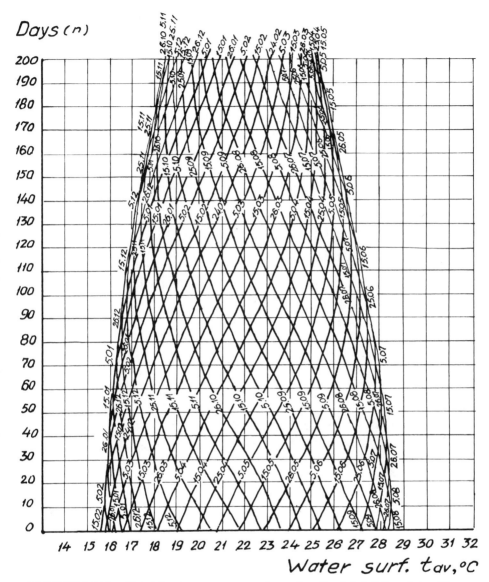

FIGURE 1.118. Net of heat resources on water surface for Ashdod, Israel; $h = 10$ m.

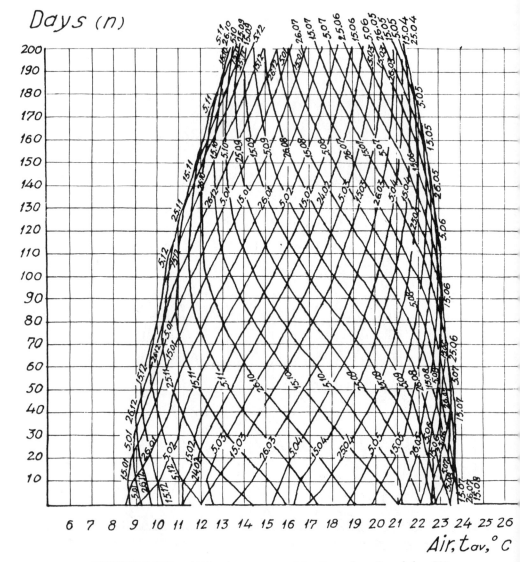

FIGURE 1.119. Air heat resources net for Jerusalem, Israel; h = 785 m.

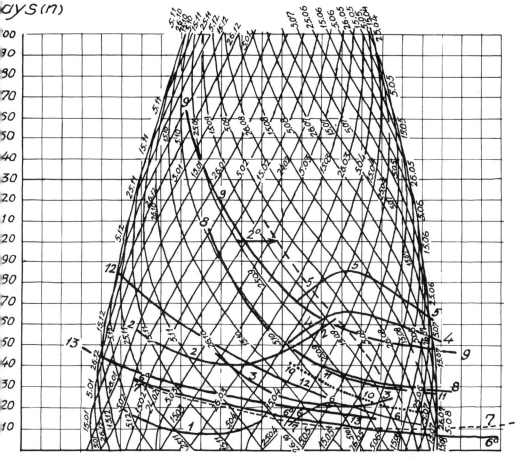

days (n)

Air t av,°C

FIGURE 1.120. A PTN for Tirat Zvi, Israel; $h = -220$ m. Phenological curves for **cotton** variety Acala S.J.-2 (*Gossipium hirsutum L.*), after data provided by Marani: curve 1: from planting to germination (50%); curve 2: from germination to budding (1st); curve 3: from budding to flowering (1st flower); curve 4: from 1st flower to ripening (when the number of open bolls accumulated per $m^2 = 10$); curve 5: from 1st flower to ripening (when the number of open bolls accumulated per $m^2 = 50$); curve 6: from ripening ($= 10$ bolls/m^2) to ripening ($= 50$ bolls/m^2). (Phenocurves 1 to 6 are transferred from Figs. 1.4 and 2.15). For **California red scale** (*Aonidiella aurantii*), female, laboratory and insectary: curve 6[a]: from first-age larva to second-age larva; curve 7: from second-age larva to third-age larva (curves 6[a] and 7, after data provided by Bliss, 1931, and Hunger, 1948); curve 8: from the appearance of a female to the birth of first-age larva from this female (Preoviposition period); curve 9: from first-age larva of one generation to first-age larva of the next generation (curves 8 and 9, after data provided by Bodenheimer, 1951, and Jones, 1935). (Phenocurves 6[a] to 9 are transferred from Fig. 1.45). **Predators and parasites of ps. Comstocki** in Palestine from egg to imago: curve 10: *Scymnus suturalis* (*Coccinellidae*); curve 11: *S. quadrimaculatus*; curve 12: *Anagurus sp.*; curve 13: *Leptomastidea abnormis Girault* (curves 10 to 13, after data from Rivnay and Perzelan, 1943). All phenocurves, as well as the net, were constructed by Podolsky.

121

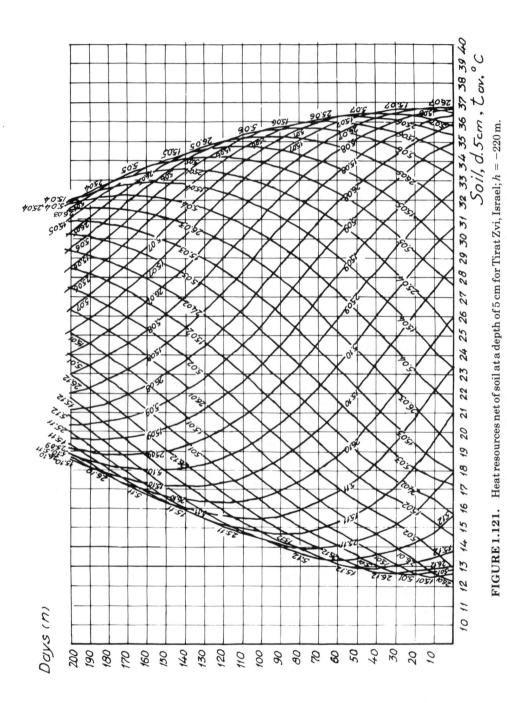

FIGURE 1.121. Heat resources net of soil at a depth of 5 cm for Tirat Zvi, Israel; $h = -220$ m.

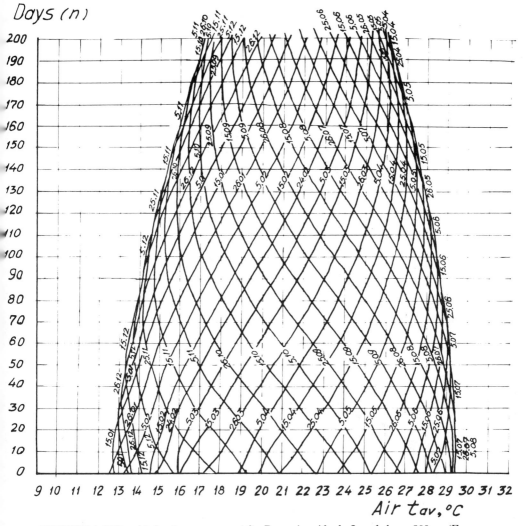

FIGURE 1.122. Air heat resources net for Deganiya Aleph, Israel; $h = -200$ m. (For soil. Fig. 1.123 can be used since the stations are near each other.)

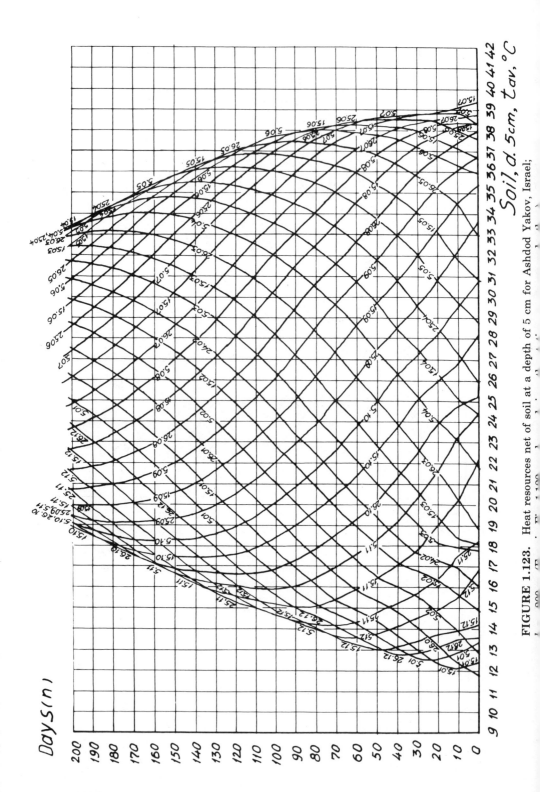

FIGURE 1.123. Heat resources net of soil at a depth of 5 cm for Ashdod Yakov, Israel;

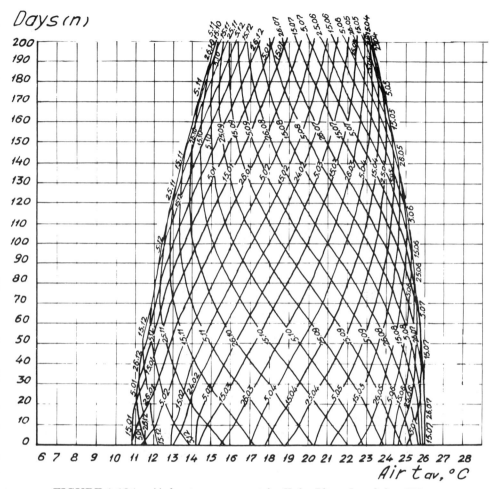

FIGURE 1.124. Air heat resources net for Kefar-Blum, Israel; $h = 130$ m.

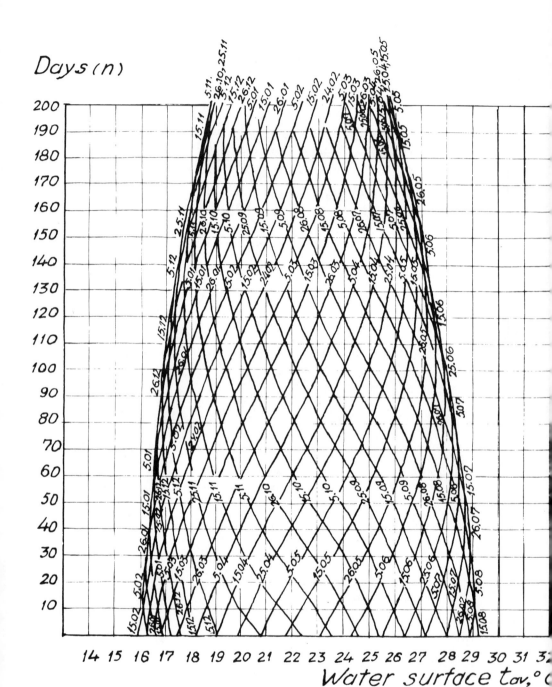

FIGURE 1.125. Net of heat resources on water surface for Haifa, Israel; $h = 10$ m.

126

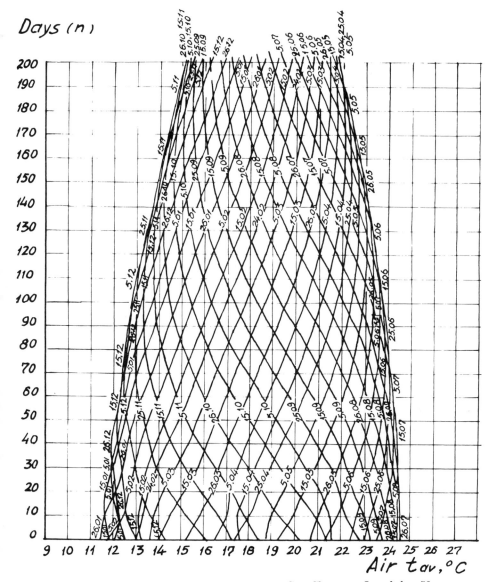

FIGURE 1.126. Air heat resources net for Gan Shomron, Israel; $h = 50$ m.

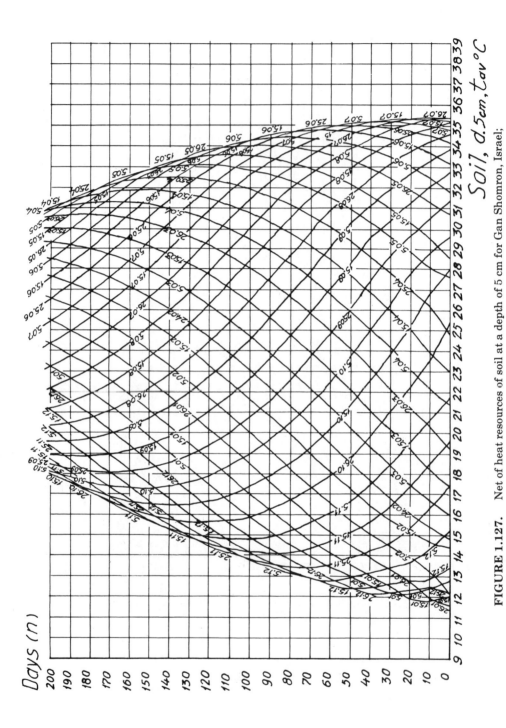

FIGURE 1.127. Net of heat resources of soil at a depth of 5 cm for Gan Shomron, Israel; $h = 50$ m.

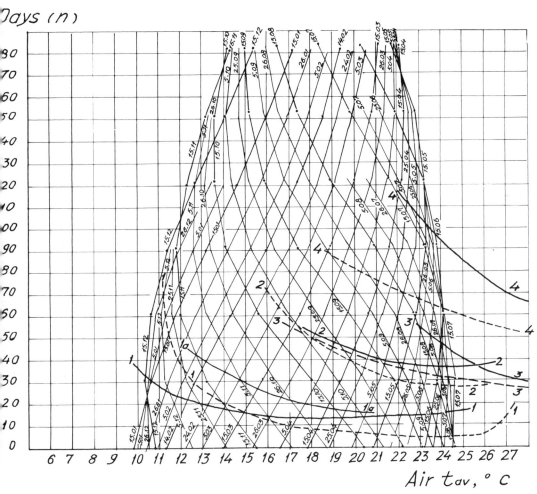

FIGURE 1.128. A PTN for Avdat, Israel; h = 400 m. Phenological curves for cotton: curve 1: from planting to mass germination (50%) with regular agricultural techniques; curve 1a; the same as curve 1, but with insufficient soil moisture and with soil crust; curve 2: from mass germination to mass budding; curve 3: from mass budding to mass flowering; curve 4: from mass flowering to mass opening of the first bolls. Solid lines are for variety 2I3 (2 Iolatan 3, USSR, descends from Egyptian cotton *Gossypium barbadense*). Dashed lines are for variety 108 F (108 Fergana, USSR, *G. hirsutum*).

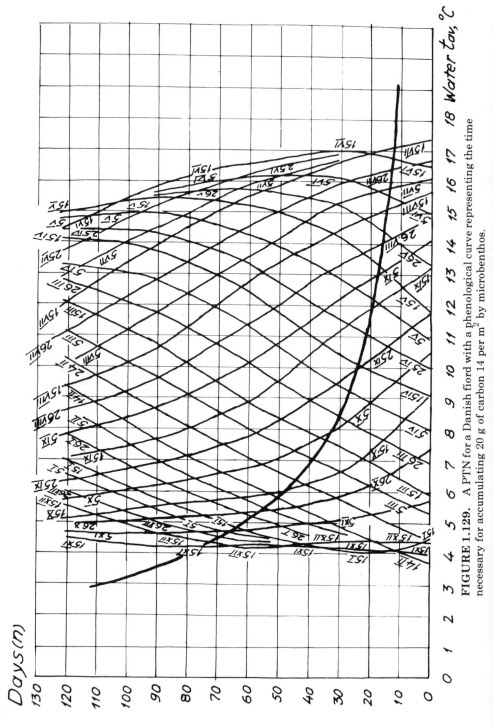

FIGURE 1.129. A PTN for a Danish fiord with a phenological curve representing the time necessary for accumulating 20 g of carbon 14 per m² by microbenthos.

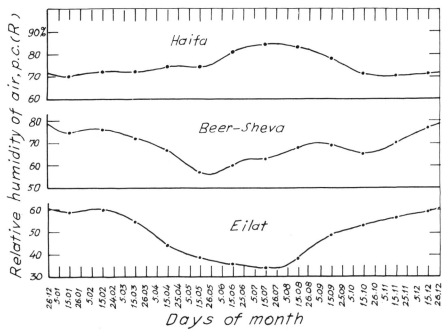

FIGURE 1.130. Annual trend of relative atmospheric humidity in Israel.

between stalk shooting (March 26) and heading (the whole of curve 10 and the necessary segment of the line marked March 26 refer to this period): March 26 + 39 days = May 4.

It is now easy to carry out a similar calculation, for example, Kurgan-Tyube (Fig. 1.55), when the starting date for mass stalk shooting is April 8. The only difference is that we must interpolate the line for April 8 (either with a pencil or mentally) between the lines for April 5 and 15 (obviously, the line for April 8 will be located slightly to the left of the midway point between April 5 and 15, that is, closer to April 5; see the fine dashed line *de* in Fig. 1.55). The intersection of this line for April 8 with curve 6 gives us n = 24 days, with which we can calculate the date of mass heading in Kurgan-Tyube as follows: April 8 + 24 days = May 2, and the average temperature during this period t_{av} = 17.8°C (abscissa of the point of intersection).

The agroclimatic data book for Tadzhikistan gives the following data about Kurgan-Tyube: when Surkhak 5688 wheat has mass stalk shooting on April 8, the multiannual average date for mass heading is May 4. Thus it can be seen that our semitheoretical calculation using the nomogram for Kurgan-Tyube gave results that fell within 2 days of this factual data.

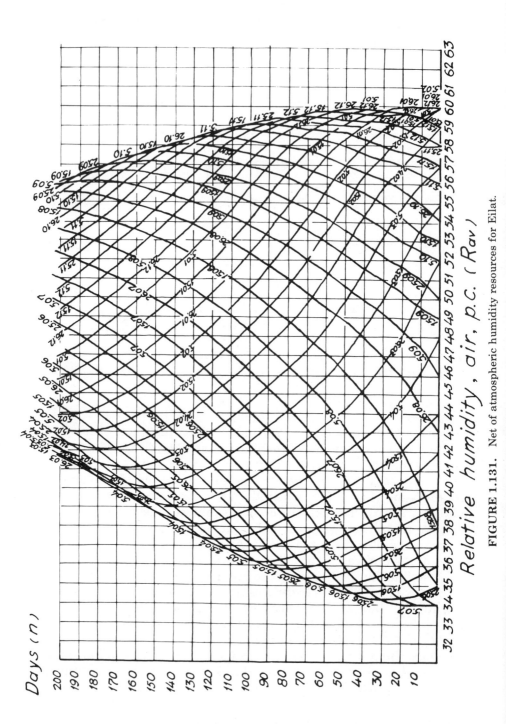

FIGURE 1.131. Net of atmospheric humidity resources for Eilat.

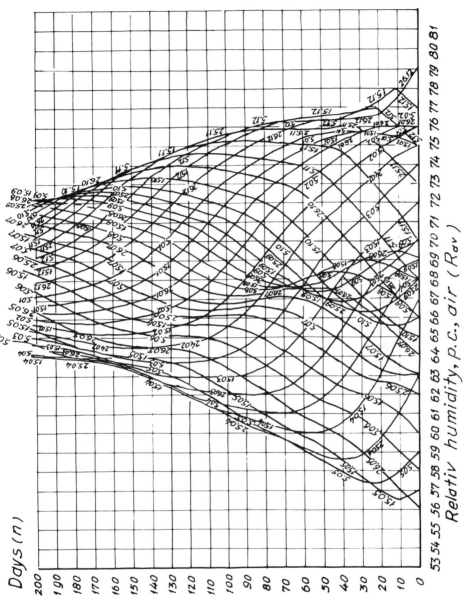

FIGURE 1.132. Net of atmospheric humidity resources for Beer Sheva.

133

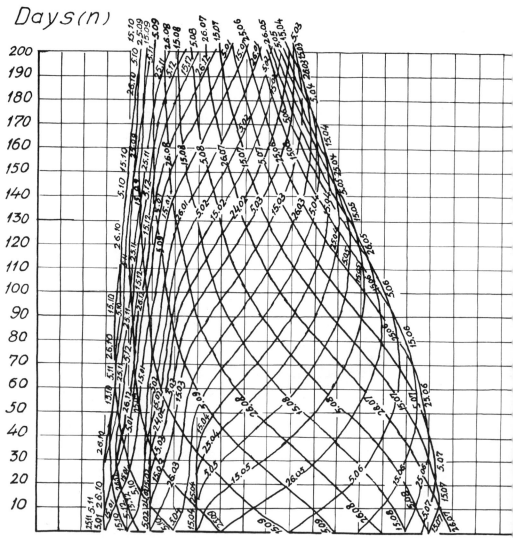

FIGURE 1.133. Net of atmospheric humidity resources for Haifa.

It can easily be calculated that if, using Yavan's nomogram, we begin with a stalk shooting date of April 8, then heading will occur on May 5.

We could proceed to solve a wide variety of ecological problems using these PTNs: operational forecasting of the dates of various developmental stages in crops, in pests, in medicinal plants and plant allergens, in crops sown after other crops in one and the same year and crops sown after hay; key dates in the life cycle of the carriers and the pathogens in various diseases affecting plants, animals, or human beings. We could also go on to consider the possibilities that this method provides in the spheres of microbiology and virology, selection, and the testing of crop varieties. Finally, we could consider the relationship between phenological forecasting and the study of biocenoses, since the world of plants and cold-blooded animals, to which we can apply this new method, is vast in extent. However, first we must spend some time discussing the general properties of the nomogram and its theoretical basis. This information will help us to make fuller use of the method.

CHAPTER TWO

Theory and Characteristics of Phenotemperature Nomograms (PTNs); a Brief Description of Forecasting the Yield, and Numbers of Parasites

2.1. MATHEMATICAL HEAT PHENOLOGY (MODEL) AND THE TEMPERATURE TOTALS METHOD

The direct mathematical meaning of the solutions described at the end of the previous chapter is clearly as follows. In the annual temperature trend system, a period of n days is found, such that the average temperature t_{av} over this period will be exactly sufficient for the completion of an organism's interdevelopmental stage during these n days. The theoretical difficulty of this problem results from the following circumstance: in order to obtain n from a phenological curve, we first have to know the condition t_{av} which is related to the annual temperature trend. However, in order to find the condition t_{av} from the annual trend, we first have to obtain the result n on the basis of the developmental curve, since, in the annual cycle, the average temperature over a certain period (t_{av}) changes continually in relation to the length of the period n over which the temperatures are taken. Thus we have a "vicious circle." The way out of it is given by a nomogram: the status of balance between t, n on the developmental curve, on the one hand, and t, n in the annual temperature trend system, on the other hand, is attained here automatically. In fact, the point where the line of average temperatures and the phenocurve intersect, and which has coordinates t,n, belongs both to the line and to the curve; therefore, the coordinates of such a point relate to both the line and the curve. This can be clearly understood by looking at Figures 1.18 and 2.1.

Let us assume that mass stalk shooting in wheat variety Surkhak 5688 occurs on April 8 according to the multiannual data in Yavan; we are therefore working with the same data as in the last example in the previous chapter. Using the date of April 8, we shall try to define the multiannual date of mass heading without using a nomogram.

This could be determined using phenocurve 2 (Fig. 1.18), provided that we know the average temperature t_{av} over the interdevelopmental stage that interests us here (on the horizontal axis in Fig. 1.18, only average temperatures over interdevelopmental stage are taken). Therefore, we assume that this average temperature is approximately equal to the temperature t' on the starting data of April 8. The latter (t') can be found easily from Figure 2.1, which represents the annual temperature trend; here $t' = 14.6°$ C (in Fig. 2.1, see the horizontal dotted line that is drawn from vertical line segment No. 1 on the annual temperature trend; this mark No. 1 corresponds to April 8—see the vertical dotted line). Then, using the temperature 14.6° C, we find on phenological curve 2 (Fig. 1.18) the approximate length of the period between two developmental stages, $n' = 49$ days. We mark the period of 49 days on Figure 2.1 by vertical line segment 2 designated by the date May 27, since April 8 + 49 days = May 27.

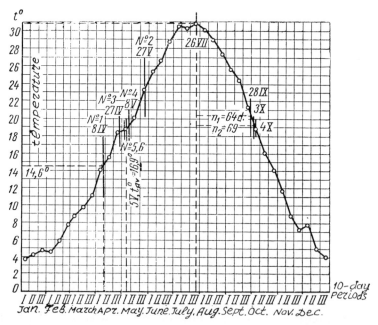

FIGURE 2.1. The annual trend of air temperature based on average data for 10-day periods, at Yavan meteorological station, USSR (average data over 11 years, 1951-1961).

But here, in the annual temperature trend, the average temperature over the period of 49 days from April 8 to May 27 is not $t' = 14.6°$C, as was assumed before, but $t''_{av} = 18.9°$C. It is easy to check this if, using the curve of the annual trend (Fig. 2.1), we total the temperatures from April 8 to May 27 every five days, for instance, and divide this total by the number of temperatures:

$$\frac{14.6°\,C + 15.2°\,C + 16.4°\,C + 17.8°\,C + 18.6°\,C + 18.8°\,C + 19.3°\,C + 19.9°\,C + 21.0°\,C + 22.4°\,C + 23.6°\,C}{11}$$

$$= 18.9°\,C$$

Once again we look at phenological curve 2 (Fig. 1.18) and find that, at the new temperature $t''_{av} = 18.9°$C, $n'' = 19$ days. We mark this period on Figure 2.1 by line segment No. 3 of April 27 (April 8 + 19 days = April 27). As can be seen, $n'' < n'$, since $18.9°$C $> 14.6°$C.

The average temperature over the period n'' from April 8 to April 27 once again turns out to be lower than $18.9°$C, namely, $t'''_{av} = 16.5°$C. After Fig. 2.1,

$$\frac{14.6°\,C + 15.2°\,C + 16.4°\,C + 17.8°\,C + 18.6°\,C}{5} = 16.5°\,C$$

In Figure 1.18 we see that, at $t'''_{av} = 16.5°$C, $n''' = 30$ days. We mark this period on Figure 2.1 by line segment No. 4 of May 8 (April 8 + 30 days = May 8). As can be seen, once again $n''' > n''$, but the space between line segments No. 1 and No. 2 is large, whereas between line segments No. 3 and No. 4 it is much smaller.

The average temperature over the period from April 8 to May 8, according to the annual trend, is $t''''_{av} = 17.2°$C. After Fig. 2.1,

$$\frac{14.6°\,C + 15.2°\,C + 16.4°\,C + 17.8°\,C + 18.6°\,C + 18.8°\,C + 19.3°\,C}{7}$$

$$= 17.2°\,C$$

At such a temperature, $n'''' = 26$ days (Fig. 1.18). Since April 8 + 26 days = May 4, vertical line segment No. 5 (Fig. 2.1) corresponds to May 4.

The average temperature over the period from April 8 to May 4, according to the annual trend, is $t'''''_{av} = 16.9°$C. After Fig. 2.1,

$$\frac{14.6°\,C + 15.2°\,C + 16.4°\,C + 17.8°\,C + 18.6°\,C + 18.8°\,C}{6} = 16.9°\,C$$

At such a temperature, $n''''' = 27$ days (Fig. 1.18), and therefore, vertical line segment No. 6 (Fig. 2.1) corresponds to May 5 (April 8 + 27 days = May 5). As can be seen, the space between line segments No. 5 and No. 6 is as little as one day.

However many additional calculations are made, the line segments will always swing between May 4 and 5.

Thus line segments 1 and 2, 3 and 4 and 5 and 6 (Fig. 2.1) swing around a particular boundary, to one side and the other, with diminishing amplitude, like the swing of a pendulum. Clearly, this boundary indicates the real period that interests us: $n = 26$ to 27 days between the developmental stages of wheat. The boundary also indicates the real average temperature over these 26 or 27 days, $t_{av} = 16.9°$ (see the calculation for the period from April 8 to May 4), around which t', t'', t''', and so on, swung to one side and the other.

In the case of some simple combinations of the shape of the annual temperature trend curve with the shape of an organism's developmental curve, the boundary that interests us may be located near the central point between line segments 1 and 2. Such a variant will correspond to solving problems relating to a swinging pendulum suspended from a fixed point. In the case of other combinations of the annual temperature trend curve with the developmental curve, the boundary may be located nearer to line segment No. 1 or No. 2. For example, in Figure 2.1 the boundary that interests us—May 4/May 5—is located in time nearer to line segment No. 2 by 5 to 6 days. Such a variant will correspond to solving problems relating to a swinging pendulum suspended from a moving point (such a pendulum can be called eccentric).

This is the mathematical model for the developmental rate of plants and poikilotherms in relation to the temperature trend.

Therefore, we have found that the period which interests us (n) equals 26 to 27 days, and the average temperature over this period (t_{av}) is slightly higher than 16.9° C.

It should be mentioned that in order to solve the given problem, we needed the average temperatures $t''_{av}, t'''_{av}, t''''_{av}, \ldots$, over various periods $n', n'', n''', \ldots$, as reckoned from a single starting date for wheat stalk shooting—April 8. It is clear that, were another starting date given, a new series of calculations similar to the previous ones would be needed, as well as a new series of average temperatures over the periods reckoned from the new starting date, and so on. It is therefore advisable, first of all, to construct a heat resources net on the basis of the annual temperature trend, so that this net can be used to obtain the average temperatures over any period as reckoned from any starting date. Next, the phenological curve for the organism's heat requirements must be superimposed onto the heat resources net, thus obtaining a PTN. Then the state of balance between t, n on the phenological developmental curve and t, n in the system of the annual temperature trend is easy to find, solving concurrently the requirements equation and the resources equation. This solution is reached at the point of intersection of the organism's developmental curve with the average temperature line marked by the date of the initial developmental stage. As a result, a solution that must be obtained using such a long and tedious process as the pendulum calculations is obtained automatically from the PTN of Yavan (Fig. 1.50): the point of intersection of curve 10 (wheat development from stalk shooting to heading) with the line of average temperatures from April 8 (date of stalk shooting) has coordinates $n = 27$ days, $t_{av} = 17.1°\text{C}$. As can be seen, this solution coincides almost completely with the previous result obtained using the pendulum method.

Thus the PTN is based on a new mathematical model of ecological processes in animate nature—in other words, on a new theory. The actualization of this model, that is, the direct (without temperature totals and thresholds) solution, on a nomogram, of the equations giving requirements and resources, becomes feasible owing to the special way of recording the heat resources of a region in the form of a net of average temperatures over certain periods, followed by the combination of the net with biological curves. If to all this we add the facts that recording complex empirical equations in an algebraical manner only, without using a special nomogram, is very complicated, and the exact analytical solving of these equations is almost impossible, it becomes clear that the words "pheno-temperature nomograms" have a very particular significance. Usually the word "nomogram" means a drawing that merely makes easier a calculation that could be done using another, accepted method, based on an idea common to the nomogram and to this other method; the PTN, however, is not a graphic representation of something known.

The mathematical model described above can be programmed on continuous-action analog computers with an oscillatory contour. But modeling machines are not necessary when the nomograms described above and circular plastic plates are available—it is sufficient to program only the calculation of a heat resources net on a discrete-action electronic digital computer (Chapter 1).

Phenological developmental curve 10 for wheat (the curve with which we have dealt up to now) is represented in Figure 1.50 by a branch related mainly to the below-optimal temperatures. Only the very right-hand side of curve 10 comes somewhat close to the above-optimal ascending branch. A number of other phenological curves in Figure 1.50 are the same. But curves 1, 6, 11, and 19 in Figure 1.50 and many curves in Figures 1.1 to 1.48 have ascending above-optimal branches for a number of organisms. Curve 5 for irrigated soya (Fig. 1.18), curve 2 for irrigated potato (Fig. 1.17), and some other curves have no below-optimal branches at all in the conditions of Middle Asia; they have only above-optimal branches. This is because the rise in temperature (high in this region as it is) results not in a shortening but in a prolongation of the period between two developmental stages of such crops as soya and potato. It is clear that the temperature totals method cannot be used in any of these cases, since its basis is a hyperbola, which requires that n decreases as t_{av} increases. The totals method is even less applicable when dealing with such complicated phenocurves as those in Fig. 2.15.

However, using the new method such problems can be solved without any difficulties. For example, the flowering of soya begins in Yavan (Fig. 1.50) on August 5. When will it ripen? As before we look for the line of average temperatures from August 5, and we find its intersection with phenological curve 16 representing soya's development from flowering to ripening. The intersection occurs at $n = 79$ days, therefore, ripening begins on August 5 + 79 days = October 23 (the nomogram for Dushanbe, where it is cooler than in Yavan, shows that ripening in Dushanbe begins 20 days earlier with the same flowering date).

It can be seen that the new method of phenological forecasting and bioclimatic estimation does not require, in contrast to the temperature totals method, that the temperature curve of the organism's development should be a hyperbola (which is the only curve shape that ensures the constancy of totals and thresholds of effective temperatures). For the purposes of the new method, this empirical curve may take any form found in nature, including a curve with two branches—below-optimal (when the temperature is lower than the optimal one) and above-optimal (when the temperature is higher than the optimal one)—a form which precludes the use of the temperature totals method. The temperature totals method solves only one particular case, whereas the new method gives the general solution.

In order to understand this, let us consider the history and essence of the temperature totals method. The French physicist and biologist Réaumur observed in 1735 that an agricultural crop ripens only after a certain total of positive air temperatures (reckoned from 0°) is accumulated; this total is more or less constant for the kind of crop. He observed a corresponding phenomenon in insects.

Later it was discovered that temperature totals, reckoned from the meteorological 0°, were not constant enough to serve as a basis for forecasting or for agroclimatological calculations.

Then, in 1844, another French scientist, Gasparen, introduced the notion of effective temperature, that is, a 24-hour average temperature that is above a certain lower temperature threshold, and from which this temperature threshold is deducted. The lower temperature threshold is the temperature below which a plant does not develop. Gasparen suggested +5° for this threshold (recently this same figure was suggested by A. A. Shigolev for all plants in the temperate zone of the terrestrial globe; the whole idea of effective temperature was repeated, but not discovered, by T. D. Lysenko and A. A. Shigolev).

The total of effective temperatures over a period between two developmental stages proved to be more constant for each biological object than the simple total of temperatures reckoned from 0°.

Designating the lower threshold of effective temperatures by B, and the total of effective temperatures for a given plant or cold-blooded organism by A, and *postulating* the constancy of B and A, we can write

$$(t_1 - B) + (t_2 - B) + (t_3 - B) + \cdots + (t_n - B) = A \qquad (9)$$

where $t_1, t_2, t_3, \ldots, t_n$ are 24-hour average temperatures over different days of a given interdevelopment stage of n days for a given organism.

The word "postulating" is used here intentionally, since there are no theoretical proofs for the constancy of A and B.

From Equation (9) we have

$$(t_1 + t_2 + t_3 + \cdots + t_n) - (B \cdot n) = A$$

or

$$\sum_1^n t - Bn = A$$

We divide both sides of the equation by n days:

$$\frac{1}{n}\sum_1^n t - \frac{B\not{n}}{\not{n}} = \frac{A}{n}$$

Since $1/n \sum_1^n t$ is the average temperature t_{av} over a period of n days, then $t_{av} - B = A/n$, hence $n(t_{av} - B) = A$ and

$$n = \frac{A}{t_{av} - B} \qquad (10)$$

An equation similar to (10) was obtained (in a somewhat different way) by the French scientist Babinet in 1851. This was the first analytical expression of the relationship of the developmental dates of plants and cold-blooded organisms to temperature. One could well say: "the first and last analytical expression. . . ," because the classical Equation (10) has not undergone any changes during the last 130 years; it has only been repeated, almost without change, by subsequent investigators.

As is known from analytical geometry, Equation (10) is an equilateral hyperbola equation that refers to asymptotes—the rough shape can be seen in Fig. 2.3. This equation indicates that the higher the average temperature over a certain period (t_{av}), the quicker the constant total A^0 is accumulated, and the shorter is the developmental period of an organism.

Since the hyperbola equation was deduced on the basis of the supposition that the totals of effective temperatures A and thresholds B are constant, then, vice versa, A and B are constant only when the relation between the duration of an organism's development in n days and the average temperature over this period (t_{av}) is expressed by a hyperbola.

In practice this hyperbola occurs only in a very limited range of temperatures t_{av}, and not for all plants and poikilotherms. Moreover, in a wide range of temperatures the development of a hyperbola is not possible at all, because Equation (10) demands an uninterrupted decrease of n as t_{av} increases infinitely. In fact, there is an optimal developmental temperature above which n does not decrease, but once again increases. This can be seen, for example, in Figure 1.1 (curve 5), Figure 1.2 (curve 1), Figure 1.4 (curve 2), Figure 1.8 (curve 1), Figure 1.20 (curve 2), Figure 1.27 (curve 2), Figure 1.29 (curve 7), Figure 1.43, and other figures.

In the optimum zone, the temperature totals and thresholds lose their biological sense. In fact, let us transform Equation (10), dividing one by both the right- and the left-hand sides of the equation:

$$\frac{1}{n} = \frac{t_{av} - B}{A}$$

or

$$\frac{1}{n} = \frac{1}{A} t_{av} - \frac{B}{A} \tag{11}$$

After this operation $1/n$ assumes the meaning of developmental rate: if the whole period between two developmental stages in n days is taken as one, then an organism advances in its development during one day by $1/n$ part of the whole interdevelopmental stage.

It is not difficult to see that Equation (11) is a straight line equation of the type $y = ax + b$, but with variable $1/n$ and t_{av}, with slope $a = 1/A$ and free term $b = B/A$.

The slope of a straight line is known to be the tangent of the line's angle of inclination to the x axis (abscissa), and the free term is a

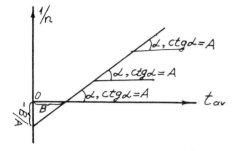

FIGURE 2.2. Developmental rate $1/n$ in relation to temperature t_{av}.

segment of the y axis (ordinate) which is cut off by the straight line. Since at the lower threshold ($t_{\mathrm{av}} = B$) the development rate approaches zero, the value of B is then read off on the x axis (Fig. 2.2). Then $\mathrm{tg}\alpha = 1/A$, hence $A = 1/\mathrm{tg}\alpha = \mathrm{Ctg}\alpha^*$ (tg = tangent and Ctg = cotangent). If $\alpha = constans$, then $\mathrm{Ctg}\alpha = constans$ also, and $A = const$. This will be true in the case of straight line—Figure 2.2. In nature, the situation is usually as presented in Figure 3.5, constructed by the author on the basis of field experiments. The shape of the curve confirms entirely the scheme given by B. B. Wiggles-Worth (1965).

As can be seen in Figure 3.5, in nature there is no inviolable constancy of temperature totals and thresholds; in the optimum zone, A approaches an infinite positive number of temperature degrees, while B approaches an infinite negative value. Hence we have an explanation for the well-known fact that phenological forecasts on the basis of temperature totals for the second and third generations of insects, especially in hot countries, are generally not successful; these generations occur within a period of optimal and above-optimal conditions.

In the review parts of the *Bibliography of Agricultural Meteorology* published in 1962 in the United States, it is written that the temperature totals method is less applicable in the United States than in northern latitudes because in many regions plant development occurs on the curvilinear sections of the curves $1/n = \varphi(t_{\mathrm{av}})$, and therefore investigators have to be careful when using the temperature totals method. P. J. H. Sharpe and Don W. DeMichele (1977) wrote about errors in forecasts at high temperatures.

The correctness of the theoretical supposition that temperature totals rise as they approach the optimum for development will be corroborated by the actual data given, for instance, in Table 4.4 for the piorcer. However, apart from these optimal zones, the total of effective

---

*Based on this equation we suggest, for admirers of the temperature totals method, an original method for calculating A: construct, on the basis of data from observations, a graph like Fig. 2.2; measure α on the graph; obtain $\mathrm{Ctg}\alpha$ from trigonometric tables; multiply this $\mathrm{Ctg}\alpha$ by the correlation of the scales t_{av} and $1/n$.

temperatures proves to be dependent on the level of the average temperature t_{av}, that is, on the level of the temperatures composing the total. Each point of the curve in Figure 3.5 has its own totals and thresholds, which are different from the totals and thresholds of other points. They are different not only in autumn and spring (when development takes place by virtue of the daytime temperature semi-wave), as thought by Yu. I. Chirkov (1969) and others. Thus if we consider + 5° to be the lower temperature threshold for wheat development, then at point R on curve 10 in Figure 1.50 (this empirical curve is taken from Fig. 1.18) we have, in accordance with Equation (10),

$$A = (t_{av} - B) \cdot n = (15.5° - 5°) \cdot 39 \text{ days} = 409.5°$$

where $A = 409.5°$ is the total effective temperatures, $t_{av} = 15.5°$ is abscissa of point R, and $n = 39$ days is ordinate of this point.

At another point on curve 10 with coordinates, say, $t_{av} = 14°$ and $n = 59$ days, we have

$$A = (14° - 5°) \cdot 59 \text{ days} = 531°$$

which is much higher than at the first point. On one and the same curve this total can be even less constant, if at different developmental temperatures the lower temperature thresholds B also prove different.

Therefore, those variable temperature totals that appear on empirical curves when t_{av} is multiplied by n days have nothing to do with the constant totals used by the temperature totals method.

The hyperbola-based theory is even less applicable to the development of plants and insects at above-optimal temperatures.

It should be noted that as long ago as the middle of the last century Bussengo showed in his experiments that the temperature total for potato fluctuates by more than 300° C, and that for maize by more than 400° C. It is clear that such vast fluctuations cannot be explained only by including in the total a number of temperatures lower than the threshold.

At the end of the last and at the beginning of the present centuries V. I. Palladin and E. G. Loske (1913) noted that the value of the temperature total is not constant for a given organism during a given interdevelopmental stage, but depends on concrete conditions and, obviously, *first of all on the levels of the average temperatures adding up to the total*. At the time, or somewhat later, the method was criticized for contradicting the law of optimum plant development (in nature, when the above-optimal temperatures rise, the interdevelopmental stage does not decrease, as the temperature totals method requires, but instead increases).

H. Geslin's experimental data, obtained in the 1930s, indicate the inconstancy of the effective temperature totals necessary for the germination of Vilmoren winter wheat (only effective temperatures are involved here, since the first signs of germination were observed by Geslin at $0°$ C).

P. I. Koloskov often pointed out (most strongly in 1947) the relation between the total of the effective temperatures necessary for an organism to complete its development and the structure of this total.

Even more definite is D. I. Shashko's statement about the dependence of A and B on t_{av}. In his opinion the effective temperatures method results in considerable errors and can be used only at certain temperature levels, but not for verifying phenological observations as was recommended by A. A. Shigolev (1955).

Many entomologists and parasitologists point out the inconstancy of the totals and thresholds of effective temperatures for the same organism during the same interdevelopmental stage; this will be described in Chapters 4 and 5.

In the last century, E. Wollny wrote less definitely about the relation of A and B to t_{av}, but more distinctly about the changeability of these constants under the influence of nontemperature factors. The temperature totals method was severely criticized in all its varieties by Agricultural Academy Member N. A. Maximov. The scientific importance of the temperature totals method was questioned by Dr. (Physics and Mathematics) V. I. Vitkevich (1960, 1965, and other publications): he was surprised that such a primitive method as the simple summing up of temperatures, suggested 250 years ago, is still in use today.

The temperature totals method is also severely criticized in later literature by E. A. Doroganevskaya (1953), V. P. Dmitrienko (1963), and J. D. Kudina (1971), and by one of the "patriarchs" of Soviet phenology, G. E. Shultz (1979).

For example, G. E. Shultz wrote that the doctrine of temperature totals proved to be untenable as a scientific theory, that temperature thresholds are not constant, and so on. I agree entirely with G. E. Shultz.

Therefore, the inconstancy of quantities A and B is the principal defect in the temperature totals method, because for this reason a more or less strict and exact solution to phenological problems is impossible.

Indeed, if it is not known up to what total A it is necessary to accumulate every 24-hour effective temperature in order to find the complete interdevelopmental stage n days, and if, moreover, even the effective temperature $(t - B)$ is not known because B is not known, then how can n days be obtained? Here we may mention that the temperature accumulation variant of the totals method was suggested by L. N. Babushkin (1949, 1951).

However, if we wish to use Equation (10) directly, without accumulative temperature totals, then it is necessary to know, at least, what function of t_{av} is A, and what function of t_{av} is B. This is clearly necessary, because now we have to write Equation (10) in the following form:

$$n = \frac{A(t_{av})}{t_{av} - B(t_{av})} \tag{12}$$

The author says "at least" because, when solving Equation (10) or (12) another difficulty arises: in order to find n days we have to know the average temperature t_{av} over this period n, but the average temperature over the period cannot be known as long as the length of the period itself is not known. This is a "vicious circle": to solve a problem it is necessary to known its condition, but the condition of the problem depends on its solution. In order to avoid this vicious circle, it is possible (as suggested by Shigolev and others) to use the average multiannual length and boundaries of the interdevelopmental stage of n days, but not the concrete ones for a year. It is clear that this deprives the solution of the equations of scientific strictness and exactness. It should be noted that, from this point of view, the variant of the totals method involving accumulating temperatures up to constant A is preferable to the variant with the equation, since it enables us to avoid the vicious circle.

What can be done with the variable totals? The use of "patches," that is, corrections to the totals, has been suggested (L. V. Sikura 1963; T. S. Druzhelubova and L. A. Makarova, 1968; Yu. I. Chirkov, 1969). However, these authors did not completely understand the cause of the variations in the totals: they considered that the totals are variable because of nighttime (in spring and in autumn) or daytime "ballast" temperatures. However, the inconstancy of the totals in fact occurs, first of all, because of the discrepancy between the function $n = f(t_{av})$, the form of a hyperbola.

It has even been suggested that it is possible to make corrections to totals which have been calculated incorrectly—that is, at temperatures higher than the optimal, the lower temperature threshold has been used instead of the higher one (Chirkov, 1969; Druzhelubova, 1972). (However, the higher threshold is necessary. See P. J. H. Sharpe and D. W. DeMichele (1977), as well as others.)

But what is the meaning of a "scientific" approach with the use of corrections? It would have been rather strange if modern physics had suggested empirical "corrections" to the results of Newtonian physics, instead of developing the quantum theory.

As can be seen, temperature totals and thresholds are very artificial. Once they had been proved wrong, attempts were made to use totals of squared temperatures, or combinations of effective temperature totals with active temperature totals (in particular, in entomology). Totals of active temperatures, in turn, are not constant even when the pheno-curve has a hyperbola form. It should be remembered that the active temperature is a temperature which is higher than the lower threshold, but inclusive of this threshold. The active temperature total can be obtained from the product $n \cdot t_{av}$, if t_{av} is higher than the threshold temperature.

Thus from Equation (10) we obtain

$$n \cdot t_{av} = A + n \cdot B$$

where $n \cdot B$ is a variable quantity, because n days changes in relation to the level of t_{av}; therefore, even at $A = constans$, $n \cdot t_{av} \neq constans$ and depends on t_{av}.*

Corrections to temperature totals, totals of squared temperatures, or various combinations of totals of effective temperatures with totals of active temperatures all constitute a purely empirical search that does not have any theory for its basis.

As was already mentioned, the new method does not require the phenological curve to be a hyperbola or the temperature totals to be constant: at the point R (Fig. 1.50), where the line of heat resources from March 26 intersects phenocurve 10 (see above, the first problem solved on a PTN), its own total and threshold are different from the total and threshold at another intersection (e.g., the intersection with the resources line dated March 5); however, we do not need to know this: the various totals and thresholds join together continuously and auto-matically in the solution, and they regulate the result.†

---

*A number of investigators, in particular A. F. Feodorov and E. S. Ulanova, point out that totals of active temperatures are stable to a lesser degree than totals of effective temperatures, for a given organism and for a given interdevelopmental stage. V. A. Smirnov shows the distinct dependence of totals of active temperatures on the level of temperatures composing these totals. However, none of these investigators mentions that this instability of totals is governed by a law and follows from the hyperbola equation; moreover, this instability could have been theoretically forecast at the time when the idea of the method of active temperatures first arose. However, the wide range of fluctuations in active temperature totals usually found in each plant was explained only by the influence of nontemperature factors, by peculiarities in the plant varieties, and, maybe, by the fact that the phenological curve is not a hyperbola. This circumstance accentuates once again the necessity of the above detailed description of the theory, practice, and criticism of the temperature totals method. It was definitely time for such a comprehensive discussion.

†On the basis of a phenotemperature nomogram, the "classical" (in agrometeorology) hyperbola equation $n = A/(t_{av} - B)$ could even be transformed into a nonhyperbola form by means of expressing the constants A (total of effective temperatures) and B (lower threshold of effective temperatures) as functions of t_{av} according to the conditions actually found in nature. However, there is no need for this equation or for transforming it, because the nomogram automatically takes into account the changeability of the quantities A and B.

Thus to express the matter using the terminology and notions of temperature totals, it can be said that the traditional method works with *constant* totals and thresholds, whereas the new method works with *variable*, "sliding" totals and thresholds that are automatically taken into account. It can also be said that the temperature totals method is one of many particular cases, whereas the new method deals with the general phenomenon.

However, it would be much more exact to say, directly: *the new method abolishes absolutely artificial notions about temperature totals and mathematical temperature thresholds.*

With this statement the majority of Soviet and Western investigators are in agreement (e.g., the American agrometerologist Jan-Ju-Wang, who gives a discussion of the new method in his university textbook, *Agricultural Meteorology* (1963, 1967, 1972).

Nevertheless, *biological* developmental thresholds exist, and these are taken into account by the PTN. For example, it is known that the lower temperature threshold for cotton development during the initial developmental stages is 10 to 12° C. This threshold can also be obtained from a nomogram using phenocurve 1, the most complete for cotton variety 108-F (Fig. 1.50). For example, the heat resources line from October 15 still intersects phenocurve 1, representing development from sowing to sprouting (when calculating, the lower intersection must be always taken); however, the line of October 26 does not intersect the biological curve. That means that cotton sowed at this late date (say, for special investigative purposes) will never produce sprouts, even if there are no early frosts—the temperature will not be high enough. The latest date on which cotton can be sowed, in order for mass sprouting to occur, is determined by the heat resources line of October 17, which is a tangent to curve 1. This line touches curve 1 at a point with ordinate $n = 26$ days. Therefore, the last sprouting will occur on November 12 (October 17 + 26 days)—(in Yavan early frosts begin at a later date). On the graph of the annual temperature trend in Yavan, or directly on the horizontal axis of the nomogram, where the average temperature line for November 12 begins, we see that the 24-hour average temperature for November 12 is near 10° C. Clearly, this is exactly the lower biological threshold for the sprouting period of cotton development, since at temperatures lower than 10° C (i.e., when cotton is sowed later than October 17) sprouting cannot take place. Our field experiments at low temperatures showed that the lower *biological* threshold for sprouting in cotton variety 180-F is indeed 9.8°.

Since sprouting is delayed by the soil crust,* curve 2 gives the lower developmental threshold as 11.6° C.

---

*After heavy showers, during the drought period, the soil is covered by a hardened layer which is 1-2 cm thick.

In the same way, we can obtain from the nomogram the generally known developmental threshold for the cotton bollworm (15°C) or for the malaria plasmodium (14.4°C). For example, in Figure 1.51, the last line that touches curve 5 is the line of average temperatures dated September 25 at $n = 30$ days. Therefore, the last sporogony cycle of *P. vivax* in Yavan finishes on October 25 at $B = 14.4°C$ (see October 25 and t on the horizontal axis). The same thresholds can be seen on the nomogram for other regions. These calculations will be clarified below.

All this is indicative of the fact that *the nomogram takes into account completely the concrete developmental temperature thresholds for each organism, although the nomogram is constructed on a scale beginning from 0°C* (and can equally be constructed on any other scale, including the Kelvin scale).

However, a number of investigators (P. I. Koloskov, D. I. Shashko, and others) consider 0°C to mark the beginning of the vital activity process, since the intensive course of biochemical reactions occurs only in liquid water solutions.

Since the new method does not contradict the developmental "optimum law," and since it does not depend on the constancy or inconstancy of temperature totals and thresholds (as well as because of other advantages), the experimental and mathematical temperature phenology method proves to be more accurate than the temperature totals method (Gaplevskaya, 1960; Ostapenko, 1962; Mustafayev, 1967; Cherkosov, 1974; Podolsky, 1967, 1974, and others—for more details see, in particular Sections 3.1.a and 4.1.c).

Returning to the question of the inconstancy of temperature totals, it may be noted that an attempt was made to overcome this shortcoming in a short article by V. K. Abramov (1972), who is acquainted with the author's method. He expressed the phenological curve not by the coordinates $n \to t_{av}$, but by the coordinates $n \cdot t \to t_{av}$, that is, on the vertical axis he put not n, but the temperature totals that, because of this, turn out to be dependent on the t_{av} constituents of these totals. Then, instead of lines of average temperatures, Abramov used lines of accumulating totals on the resources net. It is not difficult to see that this is again the author's method, but expressed by more complicated (and more congenial for those accustomed to the terminology of the temperature totals method) language.

There is more sense in the attempt to avoid the inconstancy of temperature totals by using the known method of accumulating development rate values $1/n$, as was done in works by Livingston (1913-21), Bodine (1925), B. P. Uvarov (1927-1929), A. A. Skvortzov (1930), Oganov and Rayevsky (1940), G. Z. Venzkevitz (1960), and V. N. Juravlyov (1966).

Nevertheless, the main virtue of the new method is not that it is more precise than the temperature totals method, but that it enables us to solve a system of two or more equations, as well as to solve problems about developmental stage/temperture combinations in the organism's development, cases which the temperature totals method or the developmental rate totals method are in principle powerless to solve.

For example, on what date is it necessary, in most years, to sow cotton variety 108-F at the Vakhsh zone experimental station (Kurgan-Tyube), so that the interdevelopmental stage between mass sprouting and mass budding will proceed over this period at an average temperature of, for instance, 17.5° C? Is such a developmental stage/temperature combination possible at all in the climatic conditions of Kurgan-Tyube?

Such problems with developmental stage/temperature combinations are important for experiments, and specifically for crop breeding. For instance, it is known that the breeding of fast-ripening and cold-resistant varieties of cotton is connected with raising cotton at lower temperatures during the budding period.

In Figure 1.55 we see that the vertical line drawn from t_{av} = 17.5° C up to phenological curve 8 (representing cotton development at the period sprouting/budding) intersects this curve beyond the autumn-frosts zone; therefore, the mentioned developmental stage/temperature combination is possible, that is, budding comes before the autumn frosts. Through the intersection pass two average temperature lines, dated March 23 and September 3. This means that it would be possible to obtain the necessary developmental stage/temperature combination twice a year, with mass sprouting on March 23 and September 3. A retrospective phenological calculation with the aid of development curve 7 shows that such early mass sprouting as March 23 is impossible in normal years, whereas sprouting on September 3 could be obtained if sowing were carried out on August 28 (August 28 + 6 days = September 3).

Thus in most years, the average temperature over the interdevelopmental stage of sprouting/budding in Kurgan-Tyube can be 17.5° C or close to this only if sowing occurred on August 28. Such calculations were corroborated by special experiments in Dushanbe: having done a preliminary theoretical calculation, we obtained repeatedly, in autumn field experiments with cotton, the process from sprouting to budding at t_{av} during this process of 17° to 18° C. Therefore, the budding itself proceeded at even lower temperatures, although the possibility of this has been disclaimed theoretically by some investigators (e.g., by T. D. Lysenko).

Such problems cannot be solved using the temperature totals method. If the annual trend is curvilinear, as it usually is, then an approximate solution to the problem described above using the temperature totals method could be achieved only by means of repeated attempts (i.e., we take a test date for sowing and accumulate the temperature total necessary up to sprouting. In this way we define the sprouting date; then we accumulate the temperature total necessary from sprouting to budding, which defines the budding date; we calculate the average temperature over the period between the obtained dates of sprouting and budding and verify the results. If it does not prove to be the date we need, then we make all the calculations over again, trying another sowing date, etc.). However, since totals and thresholds of effective temperatures do not remain constant, but instead depend on the level of the temperatures constituting these totals, then even such a solution, which is mathematically not strict, is not achievable by the temperature totals method.

It should be noted that *if one method solves a given category of problems and another method does not, this once again emphasizes the difference in principle between the methods*.

However, the most important point is that from the starting point of temperature totals (even if they were constant), and *especially from the starting point of totals of developmental rates*, many problems that contain several interdependent variables remain unsolvable. For example, it is known that the effective infectibility of mosquitoes (from people who are ill with malaria)* occurs not only when *Plasmodium* (malaria pathogen) has the necessary temperature conditions for completing sporogony (i.e., propagation in the body of a carrier mosquito), but also when the malarial mosquito, having been infected soon after its wings have formed, is enabled by the temperature conditions to live until the end of the sporogony. Therefore, it is necessary to solve a system of four rather complicated empirical equations with four unknowns: the equation giving the dependence of the length of the sporogony period on temperature; the equation giving the dependent of the lifetime of female mosquitoes on temperature; spring and autumn equations giving the relationship of the regions temperature resources to time. That is, it is necessary to find the date of wing formation and the corresponding date of infection of the mosquito, such that the end of sporogony will coincide exactly with the end of the mosquito's life; all of this must be calculated for the temperature conditions of Iolotan, for example. With the aid of Figure 1.62 this complicated problem can be solved very simply. The intersection of biological curves 2 and 3 shows that, at the point of intersection, the sporogony and the lifetime (which started simultaneously) have the same duration. All that remains is to find which lines of average temperatures pass through this same intersection. These lines are dated April 27 and September 3. Therefore, female mosquitoes, which form wings on April 27 and September 3, live till the end of sporogony

---

*Effective infectibility of mosquitoes refers to an infection that mosquitoes can transfer to healthy people.

in the outdoor air temperature conditions of Iolotan. Therefore, the mosquito's effective infectibility season in Iolotan begins, on the average, soon after April 27 and finishes not long before September 3.

Determining the dates of the mosquito's effective infectibility season is one of the main problems in modern malariology.

This problem belongs, in essence, to the field of biogeocenoses (ecosystems). A number of other similar problems have been solved using the author's method, in the fields of medical and veterinary parasitology and geography, by Z. I. Martinova (1968), A. V. Kondrashin (1970), I. F. Pustovoi (1970), N. T. Kunitzkaya (1976), Yakunin (1979), and others.

The method described was also used successfully in Kazakhstan on an area of 5 million hectares (12.5 million acres) by V. K. Adzhbenov (1971) for combined forecasting of the heading date of spring wheat and the flight date of *Apamea anceps Schiff-Hadena sordida Bkh.* As is known, this pest lays eggs in the heads of corn, thus causing the death of the crop over vast areas, when heading coincides with the flight of the butterflies.

The same problem occurs with the mass death of oats in Western Siberia caused by the pupation virus that develops in the leafhopper *Calligypona striatella Fall marginata F.*

The Colorado potato beetle/potato combination was investigated with the help of the author's method in Byelorussia by L. I. Arapova (1972). This will be described in more detail in later chapters of this book.

As can be seen, such combined phenological forecasts are of great importance for studying and managing various biogeocenoses.

Thus the next defect of the temperature totals method and the developmental rates method lies in the fact that they are not able to solve a number of problems that are important to science and to practical living.

Finally, it should be mentioned that calcualtions on PTNs are much less laborious than calculations with the temperature totals method or with the developmental rates method. This is especially true if phenological forecasting calendars are constructed on the basis of nomograms: in the first column of such a calendar, the actual date of the initial developmental stage is found; and in the succeeding columns, the forecast for the succeeding developmental stages, with various temperature anomalies, can be read (Podolsky, 1957, 1967, 1974—phenoforcasting calendars for cotton; Podolsky and Yermakov, 1973—for cereal crops and pests; Podolsky, 1974, 1975—for the Colorado potato beetle in the central regions of the USSR; Podolsky and Sadomov, 1976, 1977—for the fall webworm, a quarantine object; Podolsky and Shatunova, 1977—for medicinal herbs; Porodenko, 1972—for pathogens of plant diseases; Ryabov, 1978—for wild mel-

liferous plants, medicinal herbs, and plant allergens; Arapova, 1978—
for the Colorado potato beetle in Byelorussia; Kharchenko, Bunyakin,
and Kononutchenko, 1978, 1979—for vegetable pests; and others).

The final criticism is as follows: $t_{av} = 20°$, for instance, chracterizes
an objective reality, that is, a certain level of kinetic energy in
molecules, or a certain level of biochemical and biophysical processes
in a living cell (according to van't-Hoff); what, in turn, does the
temperature total 3500° characterize (as a temperature total, not a
calorie total) in plant growing? Or a total of squared temperatures,
27,000° in entomology??

The above criticism does not mean that the temperature totals
method or developmental rates method have proved to be absolutely
fruitless. From Reaumur's time to the present day a great number of
useful works have been carried out using these methods. In particular,
G. T. Selyaninov and other scientists carried out a wide range of works,
dividing into agroclimatical regions both the Soviet Union and the
entire world. A. A. Shigolev (1949, 1954) used the totals method in
phenological forecasting for cereal crops and wild plants (USSR). F. F.
Davitaya (1948) used the totals method for determining the possible
distribution of grapes, and for a general calculation of climatic
resources for the agriculture of the USSR (the latter in collaboration
with S. A. Sapojnikova). L. N. Babushkin (1949, 1951) used the totals
method for describing the development of cotton and other crops in
Middle Asian USSR.

A great deal of work has also been done outside the USSR concerning
the temperature totals method and the developmental rates totals
method, as well as methods derived from them. It is sufficient to
mention Gasparen (1844), Babinet (1851), Livingston (1913), Appleman
and Eaton (1921), Winkier and Williams (1939), Bomalaski (1948),
Lindsey and Newman (1956), Wang (1958), Holmes and Robertson
(1959), and others.

However, times progress and methods improve. The most interesting, in this respect, is the work of P. J. H. Sharpe and D. W. DeMichele
(1977). They managed to bring to a common model the day/degrees
conception (i.e., conception of temperature totals) and Arrhenius
hypothesis expressed in H. Eyring's equation (1939). As a result, a
stochastic thermodynamic model of the development of poikilothermic
organisms is obtained. This model enables us to calculate the development not only at the medium temperature range (when the
relation between the developmental rate and the temperature is
approximately linear), but also at temperatures close to the lower and
higher developmental thresholds, as well as at temperatures close to
the optimum. In this model the shortcomings of the temperature totals
method are overcome to a considerable degree. The work has been
written on a high biochemical, biophysical, and thermodynamical

level. The applicability of the work for microorganisms, pests, and higher plants has been shown. But the authors' model is not entirely free from the temperature totals and does not take into account the nontemperature factors. It is also desirable for the model to be simplified for practical application. In these respects, the method proposed by the author of this book might prove to be useful.

R. E. Stinner et al. (1975) also proposed an interesting algorithm for the simulation of variability in temperature-dependent development.

Thus the PTN gives a graphic solution of equations. But would an analytical solution be possible on the basis of nomograms?

Of course, if the graphic equation giving the average temperature line and the phenological curve of the organism's development could be expressed, with the necessary accuracy, by an algebraic or analytical function, then it would be possible with a certain approximation to obtain an analytical solution to this equation system. Since, however, it is rather difficult not only to solve (analytically), but also to express each of the abovementioned graphs by any single function, then it is apparently impossible to achieve such a solution without a PTN.

Nevertheless, the problem can be simplified: let us express the equation of the organism's heat requirements by a hyperbola equation, and the equation of a region's heat resources from a given date by the equation of a straight line, two-intercept—for instance, for the second half of summer (Fig. 2.3). Then it would be possible to demonstrate that the combined algebraical solution of these two equations with two unknowns—t_{av} and n—results in a single squared (quadratic) equation with one unknown, as follows:

$$t_{av}^2 - (a + B)\, t_{av} + \frac{aA + abB}{b} = 0 \qquad (13)$$

where A and B are the total and the lower threshold of effective temperature; they are constants here because (above) a hyperbola has been assumed. The quantities a and b are constants, in accordance with Figure 2.3. Having solved this squared equation with regard to t_{av}, and having put this obtained result into one of the two initial equations (the hyperbola or the straight line), we can obtain the corresponding n. Thus both the unknowns are obtained—t_{av} and n.

However, a squared equation has two roots. Therefore, we obtain two

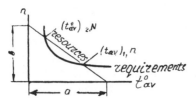

FIGURE 2.3. On the algebraic solution of equations for the heat requirements of an organism and heat resources of a region.

solutions: $(t_{av})_1$, n and $(t_{av})_2$, N. As often happens to the roots of a squared equation, only one of the two roots has a real sense. In our case the biological sense is always peculiar to the solution with the smaller n, that is, to the lower point of intersection of the requirements curve with the resources line. This happens since the organism will not delay its development until the expiration of N days, if all the necessary conditions for the completion of the development process are available during the first n days, which are a part of N.

For example, on the PTN for Yavan (Fig. 1.50), it can be seen that the line of average temperatures from March 26 or from April 8 and so on intersects phenological curve 10 (representing wheat development from stalk shooting to heading) at one point. But the line of average temperatures from September 25 intersects this curve at two points—at a lower point with ordinate $n = 18$ days and at a higher point with ordinate $n = 70$ days. It is clear that even if autumn frosts did not interfere, heading would still occur in the wheat at 18 days after stalk shooting. In this example the developmental dates are taken for wheat that can be sown both in autumn and in spring. These dates are unusual for production, but may perhaps be found in crop-breeding experiments. However, on other nomograms and with other organisms we can meet more practical situations with two intersections. In all such cases (with two intersections) the solution of a phenological problem is always to be found at the lower intersection.

If the line of average temperatures does not intersect the development curve at all, this means that the next developmental stage will never occur. For example, in Figure 1.50 the lines of average temperatures dated September 15, September 20, September 25 still intersect developmental curve 10 (at two points which gradually draw closer together); the line dated September 29 (which must be interpolated) just touches curve 10 at one point; the line dated October 5 does not touch it at all. Therefore, the last stalk-shooting date that ensures the heading is September 29.

Since we also dealt with phenological problems concerning the second half of summer, it is interesting to understand how the mathematical mechanism of the nomogram works when solving such problems. This mechanism differs from the "mathematical pendulum" described for the first half of summer.

Indeed, let us assume that mass flowering in cotton variety 5904-I occurs in Yavan on 26 July.* When does mass boll unfolding (ripening) occur? On the curve giving the annual temperature trend in Yavan (Fig. 2.1) we mark the date July 26 with a vertical line segment. As before, we take the temperature $t_1 = 30.9°$ C for this initial date (see the dashed horizontal line in Fig. 2.1) as an average (highly approximate)

---

*Late flowering is also possible when cotton is resown.

temperature over the whole interdevelopmental stage that interests us here. Then, on the basis of phenological curve 2, which represents the development of cotton variety 5904-I from flowering to ripening (in Fig. 1.1), we determine that at $t_1 = 30.9°$ C the period between these two stages is $n_1 = 64$ days. We mark the period of 64 days in Figure 2.1 with a vertical line segment at September 28 (July 26 + 64 days = September 28). Here we find the actual average temperature over the period from July 26 to September 28: $t_{av} = 26.6°$ C (for this purpose we take from Fig. 2.1 the temperatures every 5th day, and divide their total by the number of component temperatures:

$$\frac{30.9°\,C + 30.5°\,C + 30.2°\,C + 29.5°\,C + 29.0°\,C + 28.1°\,C + 27.5°\,C}{14}$$

$$\frac{+\ 26.6°\,C + 25.7°\,C + 25.1°\,C + 24.5°\,C + 22.9°\,C + 21.3°\,C + 20.6°\,C}{14}$$

$$= \frac{372.4°\,C}{14} = 26.6°\,C)$$

We return to Figure 1.1 and, on the basis of curve 2, we find that at $t_{av2} = 26.6°$C, $n_2 = 69$ days. We mark this period also on Figure 2.1 with a vertical line segment at October 3 (July 26 + 69 days = October 3). A new average temperature appears, from July 26 to October 3: $t_{av3} = 26.1°$ C [in accordance with Fig. 2.1 and the previous calculation $(372.4°\,C + 19.3°\,C)/15 = 26.1°\,C$].

From Figure 1.1 at $t_{av3} = 26.1°$ C we obtain $n_3 = 69.5$ days. Now the difference between n_2 and n_3 has become so small that the vertical line segment marking the next period is located very close (Fig. 2.1) to the line segment dated October 3. As can be seen, the distances between the vertical line segments diminish regularly; they approach a certain boundary which obviously will be the exact time (that concerns us) when the next developmental stage begins. In our problem this is the ripening stage of cotton, hence it occurs on October 3-4, 69.5 days after mass flowering (July 26), when the average temperature over the interdevelopmental stage is $t_{av3} = 26.1°$ C.

This laborious way of solving the problem by means of a series of every-approaching results can be replaced by an instantaneous solution on the PTN for Yavan (Fig. 1.50): the point where phenological developmental curve 5 (which represents the development of cotton variety 5904-I between flowering and ripening) intersects the line of average temperatures for July 26 has coordinates that are almost exactly equal to the previous solution: $n = 69$ days, $t_{av} = 26.2°$ C.

2.2 HOW THE PHENOTEMPERATURE NOMOGRAM TAKES INTO ACCOUNT THE INFLUENCE OF NONTEMPERATURE FACTORS (AGRICULTURAL TECHNIQUES, HUMIDITY, LIGHT FACTOR) ON THE DEVELOPMENT OF PLANTS AND POIKILOTHERMS

Although temperature is the decisive factor for the developmental dates of plants and poikilotherms, a certain influence is exerted by other factors, too.

Let us first estimate the importance of individual environmental factors with regard to the dates of plant development.

A phenological line is given in Figure 1.8 which represents the relationship between barley development during the period from heading to full ripeness and the temperature. This line is calculated using a linear regression equation, and on the basis of our experiments with crops sowed at different times of the year in Kursk. The correlation coefficient r in these experiments proved to be -0.8. Therefore, the coefficient of determination is $D = r^2 = 0.64$. This means that 64% of all variations in the interdevelopmental stage are caused by temperature variations, and only 36% fall to the share of all other factors (humidity, light, agricultural techniques, etc.).

An astonishingly close agreement with these data can be seen in N. E. Buligin's work (1979). He dealt with small-leaved linden in the Leningrad suburbs, and he obtained, with the use of N. Dreiper and G. Smith's method, the figure of 65.04% for temperature-dependent variability in flowering phenodates. Only 3% of the variability was due to the influence of relative air humidity.

For oats, from the appearance of the panicles to wax ripeness (Fig. 1.9), we obtained data that favor the heat factor even more: $r = -0.84$, $D = 71\%$.

Nevertheless, nontemperature factors do exert an influence on the developmental rate, and it is necessary to know how to take these factors into account. For example, during the period from sowing to sprouting, in agricultural crops, the agricultural techniques exert a very considerable influence*. Indeed, in Figure 1.2 we see two experimental phenological curves (1 and 2) representing the development of cotton variety 108-F during the period from sowing to mass sprouting. Curve 1 was obtained by the author using techniques that obey good

---

*It should be noted that, in the subsequent developmental stages, the agricultural techniques exert more influence on the growth and the accumulation of the yield than on the dates of plant development. However, before or after sprouting it is advisable to regulate the developmental rate with such techniques as fertilization of fields, planning of crop densities, selection of favorable direction for the rows, irrigation with water warmed in the sun, and so on.

agricultural practice, whereas curve 2 was obtained in the presence of soil crust and with insufficient water in the soil. A comparison of these curves demonstrates that when the temperature is 14° C, the soil crust and insufficient soil moisture can delay cotton sprouting by 22 days, whereas with a temperature of 20° C, when rapid sprouting destroys the soil crust, sprouting is delayed by only 2 days.

On the PTN, this agrotechnical factor is taken into account very simply. Let us assume that we wish to determine when mass sprouting of cotton variety 108-F occurs in Yavan if the cotton is sown on April 5 in a field kept in good condition, as well as in a field with soil crust and insufficient soil moisture. In the first case, we trace a line of average temperatures from April 5 on the nomogram for Yavan up to the intersection with phenocurve 1; in the second case, we trace the line up to the intersection with phenocurve 2 (Fig. 1.50). We obtain in the first case April 5 + 16 days = April 21; in the second case, April 5 + 22 days = April 27, where $n = 16$ days and $n = 22$ days are ordinates of the first and the second intersections. As can be seen, in the second case mass sprouting occurs 6 days later than in the first.

The length of the period from sowing to sprouting is also affected by the soil type (potato, phenocurves 7-8, Fig. 1.73). This nontemperature factor is taken into account in the same way described above.

Sometimes in cereals, in the period after sprouting but before tillering (bushing), the developmental rate is slightly dependent on soil moisture (Fig. 1.40); this can also be easily taken into account when forecasting.

Accounting for the moisture of soil and subsoil when forecasting the growth and development of plants is particularly necessary, as is clear, for arid zones. In this respect, important initial data were published by J. F. Angus et al. (1980).

The development of poikilotherms, such as ticks, is affected by another nontemperature factor—atmospheric humidity. For example, in Figure 1.50 we see two biological curves that represent, for the tick *Hyalomma detritum*, the length of the period of development from egg to larva, in relation to temperatures: 20—with relative humidity 20-38%; and 21—with humidity 50% (the curves were constructed by the author on the basis of actual data from I. G. Galuzo; see Fig. 1.36).

In Figure 1.63 we see two phenocurves for fleas—carriers of plague bacteria—with 75-80% and 90-91% relative humidity in the soil air. These nontemperature factors are taken into account on the nomogram in a manner analogous to the foregoing. In the subsequent forecasting, the author will take into account nontemperature factors where necessary.

Therefore, in order to take into account the influence of non-temperature factors on the duration of development in a given organism, it is necessary to construct, on the basis of actual observations (field, thermohydrostatic, etc.), phenological temperature curves of development (i.e., relating the length of interdevelopmental stages to temperatures) for different levels of the nontemperature component that concerns us. Then, in the calculations, we use that developmental curve which corresponds to the state of the nontemperature component.

Since the changes caused by nontemperature elements in the duration of an organism's development are in many cases comparatively small, it is possible to simplify this process of taking into account the nontemperature factors. For example, it is possible to "glue together" (as mathematicians sometimes say) phenological curves 20 and 21 for the tick's development (Fig. 1.50) into a single curve: the upper section (with the low temperatures) of such a complex curve will be represented by the upper segment of curve 21 for 50% humidity; the lower section (high temperatures) will be represented by the lower segment of curve 20 for 20-38% humidity; this is because the low temperatures in winter, spring, and autumn are usually accompanied by higher relative atmospheric humidity, whereas the high temperatures in summer are accompanied by lower relative atmospheric humidity. The curves for ticks in Figures 1.55 to 1.57 are "glued" together in exactly the same way. If the average values of humidity exceed 20-50%, then developmental curves that differ more sharply will have to be "glued" together. It should be noted that at higher altitudes, as well as near sea coast, the maximum relative humidity can alter from winter to summer, and that this will require the reverse gluing of the curves (e.g., the high-temperature section of the complex curve should be represented by the lower segment of curve 21, not curve 20).

In a similar way, but automatically, the influence of seasonal variations in the light factors on the duration of development in plants and in cold-blooded organisms is taken into account. For example, phenocurve 1, representing the development of cotton variety 108-F from sprouting to budding (Fig. 1.3), is constructed from the results of our experiments with different sowing dates, from March to August, in one and the same place—the agrometeorological station of Dushanbe. It is clear, therefore, that curve 1 is also "complex," and that it reflects not only the influence of the temperature on cotton development, but also the influence of different day lengths on this development in different seasons.

Indeed, the left-hand, low-temperature section of the curve is based on data from spring experiments, with relatively short days; on the other hand, the right-hand, high-temperature section is based on summer requirements, with longer days. If the day length affects the development rate, then this influence should be reflected on the shape of curve 1 and in its position on the coordinate net or on the nomogram net. On the nomogram net the curve will "work," with its low-temperature, short-day section, in spring and autumn, when the curve is crossed by spring and autumn lines of average temperatures; on the other hand, the high-temperature, long-day section of the curve will "work" in summertime, when the curve is crossed by summer lines of average temperatures (see, e.g., Fig. 1.50, where the same phenological curve is transferred as curve 3).

Therefore, in order to take into account seasonal variations of the effect through a phenocurve of a plant's development, it is necessary to have the results of field experiments with sowing at different times in one and the same place. When, on the other hand, a curve covers a wide range of temperatures, because it is constructed on the basis of approximately simultaneous sowings (but in years with different temperatures, and in different places), then it will be more difficult to take seasonal changes in the light factors into account.

Phenological curves representing the development of cold-blooded organisms, and based on field experiments (Figs. 1.1, 1.34 and others), also take into account automatically seasonal changes in the photo-effect, humidity, and so on, if the length of the curve is produced by data for different generations that develop in different seasons and at different temperatures. What often occurs is that the difference in temperature conditions between the development of one generation and another can exceed the difference in temperature conditions between the development of the same generation in different places and in different years.

The latitudinal change in day length and its effect on the development of plants and cold-blooded organisms can be taken into account, if for the same organism there are several phenological developmental curves in relation to temperature, namely for different latitudes.

In the case of a number of agricultural crops, such curves can be constructed easily with the use of isopleths as drawn by V. A. Smirnow (1960) on the basis of phenological field observations in different latitudes (Figs. 1.98 and 2.4).

It is possible to take into account changes in the light factors by using phenological curves for plants based on observations in a phytotron, in a climatic chamber or in greenhouses with artificially regulated daylight length (or with regulated light and temperature).

Moreover, data for constructing similar curves for the development of cold-blooded organisms can also be obtained from field observations in different latitudes, or from observations of poikilotherms' development in chambers with artificially regulated temperature and daylight length.

Having available "spectra" of such phenological curves, where each curve corresponds to the invariable average day length for an inter-developmental stage in a given latitude, we choose the phenocurve that corresponds in day length to the geographical latitude on the PTN.

This simplified method gave excellent accuracy, for example, in phenoforecasts for buckwheat, and it is entirely sufficient in practice. Theoretically, however, the simplified method is not strict enough, since the day length during an interdevelopmental stage is variable.

A simultaneous and theoretically strict method for taking into account latitudinal and seasonal changes in the light factors was used by Podolsky and Yermakov (1973) as follows.

First, however, it must be noted that the development of millet (Fig. 2.4) slows down as the day length increases: the longer the day, the higher the curve lies. Barley, on the other hand, accelerates its development (Fig. 1.98): the longer the day, the lower the curve lies. This reaction by the crops to the day length corresponds to their nature: millet is a short-day plant and barley is a long-day plant.

From the *Agroclimatic Reference Book* it is known that at the meteorological station of Karachev in the Bryansk district of the USSR, the multiannual average sprouting date for barley of the Vinner variety is May 20. We now wish to define the multiannual heading date. In order to do so, as usual, we find on the PTN for Karachev (Fig. 1.98) the line of average temperatures from the initial date of May 20. This line is marked in Figure 1.98 by the dotted line. We obtain seven points where the dotted line intersects each of the phenocurves of the spectrum. In order to choose one of these points, which will correspond to the heading date, we have to know the average day length over the period from sprouting to heading; this day length could be calculated using astronomic tables, if the heading date were known. For the present it is unknown, and this creates a vicious circle.

However, a way out can be found with relative ease. For this purpose, on each line of multiannual average temperatures we make notches at

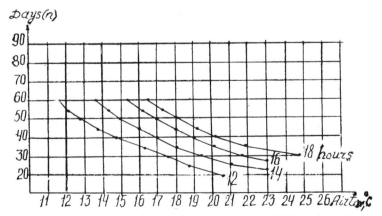

FIGURE 2.4. Phenological curves for the development of millet variety Saratovskoye 748 from sprouting to the appearance of panicles, with different day lengths (from O. S. Yermakov, A. S. Podolsky, and V. A. Smirnov).

intervals of 10 days, and at these notches, we mark the average day length from the initial date. The first two figures signify hours, and the second two signify minutes; the figures are written across the line to which they relate, thus avoiding confusion.

For example, on the heat resources net for Karachev (Fig. 1.98), it can be seen that the average day length over the 10-day period from May 5 to May 15 is 15 hours and 32 minutes; over the period of 21 days it is 15 hours and 50 minutes; over the period of 31 days it is 16 hours and 03 minutes; and so on. The average day length over the 10-day period from September 5 to September 15 is 13 hours and 04 minutes; over 20 days it is 12 hours and 44 minutes; over 30 days it is 12 hours and 23 minutes; and so on.

Having plotted onto the net the values of the average day length over a certain period, we can, by interpolation, calculate this day length for any period and from any initial date. Table 2.1 presents the average day length calculated in this way.

In Figure 1.98 we see, for example, that the line of multiannual average temperatures dated May 20 intersects the phenological curve corresponding to a day length of 20 hours, at the point with ordinate 43 days.

It is possible, in this case, that the heading stage occurs on the 43rd day? It would occur if, at the Karachev station, the average day length over the 43-day period from May 20 were 20 hours. But, since the actual day length over this period is 16 hours and 41 minutes (Table 2.1), the heading stage will not occur. For the same reason, the heading stage will not occur at the points of intersection of the multiannual average line from May 20 with the phenological curves for day lengths of 19 and 18 hours.

On approximately the 48th day the line of average temperatures intersects the phenological curve for a 17-hour day length. In fact, the average day length over the 48-day period from May 20 is 16 hours and 43 minutes (Table 2.1). As we see, this factual day length is close to 17 hours.

On approximately the 51st day the line of average temperatures intersects the phenological curve for a 16-hour day length. The real average day length over the 51-day period from May 20 is 16 hours and 43 minutes (or a little less). As can be seen, this day length exceeds the "value" of the 16-hour phenological curve. Therefore, the solution lies between the penultimate intersection at $n = 48$ days and the last intersection at $n = 51$ days, that is, approximately at $n = 49$ days.

Thus at the meteorological station of Karachev in the Bryansk district, given the multiannual average sprouting date for barley of the Vinner variety as May 20, we have the corresponding multiannual average heading date: May 20 + 49 days = July 8. The factual heading date for this type of barley at the Karachev station, according to the

TABLE 2.1. Average day length for the period *n* days from May 20 at the station of Karachev in the Bryansk district.

Period *n* Days	Average Day Length for the Period (hours, minutes)	Period *n* Days	Average Day Length for the Period (hours, minutes)	Period *n* Days	Average Day Length for the Period (hours, minutes)	Period *n* Days	Average Day Length for the Period (hours, minutes)
10	16.18	24	16.31	38	16.40	52	16.42
11	16.19	25	16.32	39	16.40	53	16.42
12	16.20	26	16.33	40	16.41	54	16.42
13	16.21	27	16.34	41	16.41	55	16.42
14	16.22	28	16.34	42	16.41	56	16.41
15	16.23	29	16.35	43	16.41	57	16.41
16	16.24	30	16.36	44	16.42	58	16.41
17	16.25	31	16.36	45	16.42	59	16.40
18	16.26	32	16.37	46	16.42	60	16.40
19	16.27	33	16.38	47	16.42	61	16.40
20	16.28	34	16.38	48	16.43	62	16.39
21	16.29	35	16.38	49	16.43	63	16.39
22	16.30	36	16.39	50	16.43	64	16.38
23	16.30	37	16.39	51	16.43	65	16.38
						66	16.37
						67	16.37
						68	16.36
						69	16.36
						70	16.35
						71	16.34
						72	16.34

Agroclimatic Reference Book, is July 6—close to the date that we have calculated.

The spectra of phenological curves given here are a graphic representation of the equation of statistical relationships between three variables: the length of the interdevelopmental stage, its average temperature, and the average day length. The meteorological curves of the net express the relationship between the length of the period and its average temperature (this relationship is also statistical, because the average multiannual temperatures of the heat resources net are obtained from the temperatures of different years), as well as the relationship between the length of the period and the average day length (this relationship is functional) in the annual cycle. The agroclimatic forecast described above is a graphic solution of this system of equations. Since the majority of the above mentioned equations are statistical, it can also be said that the principle described is an original graphic way of solving a multiple regression equation.

It should be noted that in order to compare the "value" of a phenological curve with the actual average day length, it is not necessary to draw up a table such as Table 2.1, since the only purpose of this table is to avoid the investigator's having to interpolate the average day length, and to facilitate understanding the problem.

The method described above for taking into account the average day length directly on the heat resources net is analogous to the one used by A. S. Podolsky (1967) for taking into account critical values of seasonal factors (see below).

Calculations of the multiannual average dates of barley and oat heading (appearance of panicles), carried out by O. S. Yermakov (1973) for different locations in the USSR, resulted in good agreement with the observed data (Table 2.2).

The principle proposed above for taking into account the day length can be used, of course, not only for bioclimatical estimations, but for all operational forecasts.

The principle for taking into account the light factor can be used for other nontemperature factors: for soil moisture during the period between sowing and tillering, or, for example, for the relative atmospheric humidity, when forecasting the tick's developmental dates.

It was generally supposed that the temperature is the key factor causing the developmental dates of plants and cold-blooded organisms. Nevertheless, in a small number of cases it may be found that the key factor is not the temperature, but any other element with a more or less systematic seasonal trend, such as relative atmospheric humidity. In this case, the phenological developmental curves (moisture requirements of an organism) will represent the dependence of the length of the interdevelopmental stage in *n* days, not on the temperature, but on the relative atmospheric humidity; at the same time, the humidity

TABLE 2.2 Comparison of multiannual average dates of heading (appearance of panicles) of spring barley and oats (forecast takes into account the day length) with actual multiannual average dates as found in the *Agroclimatic Reference Book*, USSR.

Station	Latitude	Crop, Variety	Actual Average Multiannual Date of Sprouting[a]	Actual Multiannual Average Date of heading (appearance of panicles)[a]	Forecast Multiannual Average Date of heading (appearance of panicles)	Divergence (in days) of the Actual Date of heading (appearance of panicles) from the Forecast Date
1	2	3	4	5	6	7
Karachev, Bryansk region	53°08′	Barley Viner	20 V	6 VII	8 VII	−2
Karachev, Bryansk region	53°08′	Oats Lohovsky	15 V	3 VII	6 VII	−3
Trubchevsk, Bryansk district	52°35′	Barley Viner	11 V	30 VI	2 VII	−2
Shchors, Chernigov district, Ukraine	51°48′	Oats Lohovsky	29 VI	23 VI	25 VI	−2
Baranovichy, Brest district	53°08′	Barley Viner	12 V	29 VI	2 VII	−3
Purekh, Gorky district	56°39′	Oats Lohovsky	20 V	6 VII	9 VII	−3
Nikopol, Dnepropetrovsk district, Ukraine	47°35′	Oats Lohovsky	24 IV	15 VI	13 VI	+2

[a]According to the *Agroclimatic Reference Book*.

resources of the region will be represented by a family of lines (a net) giving average relative humidity, but not temperatures. Such nets have already been given (in Figs. 1.131-1.133). But since the temperature, even in this case, must also influence the dates of an organism's development, then, as before, it is necessary to have several humidity curves for different temperature gradations for each organism.

Obviously, such a method can also be used when the developmental rates of an organism are greatly dependent on the seasonal changes in the photo period; by analogy with a phenotemperature or a phenohumidity nomogram, it is possible to construct a phenophoto nomogram. On the horizontal axis of such a nomogram will be not temperature or humidity, but the averae (over a certain period) number of daylight hours. For example, the line of the "average light effect" over a certain period from April 15 will show the average number of daylight hours over the 10 days starting from April 15; then the average number of daylight hours over the 20 days starting from the same April 15; and so on. Correspondingly, the phenological developmental curve of an organism should be represented in the form of the relationship of the length of an interdevelopment stage (in days) to the average number of daylight hours over the interdevelopment stage.

However, the fact is that hitherto we have not come across any organism for which the relationship of n to any nontemperature factor is closer than the relationship of n to the temperature factor.

From the above, it is clear how to proceed in the simplest case when the length of an organism's interdevelopment stage n is a function of more than two factors. For example, how should we proceed if the development of ticks depends on temperatures (first of all) and on atmospheric humidity, and is also affected by light factors (which, in turn, depend on geographical latitude)? We then need not two, but four curves, to represent the relationship of n to t_{av}, that is, four phenological temperature curves that give the tick's development, if we are using only two gradations of each nontemperature factor:

$$
\begin{array}{lll}
\textbf{1.} & n = f_1(t_{av}) & \text{at } r_1\% = C_1 \\
\textbf{2.} & n = f_2(t_{av}) & \text{at } r_2\% = C_2
\end{array} \Bigg\} \quad \varphi_1 = C'
$$
$$
\begin{array}{lll}
\textbf{3.} & n = f_3(t_{av}) & \text{at } r_1\% = C_1 \\
\textbf{4.} & n = f_4(t_{av}) & \text{at } r_2\% = C_2
\end{array} \Bigg\} \quad \varphi_2 = C''
$$

(14)

where f_1, f_2, f_3, and f_4 are functions of the interdevelopment stage (n days), depending on the average temperature over this period (t_{av}); $r_1\% = C_1$ and $r_2\% = C_2$ are two gradations of relative atmospheric humidity (in each gradation the humidity is constant: $C = const$); $\varphi_1 = C'$ and $\varphi_2 = C''$ are two constant geographical latitudes.

Onto a given heat resources net of an area, we put either the first and second or the third and fourth curves, according to the latitude of the location.

2.3. TAKING INTO ACCOUNT, ON A PHENOTEMPERATURE NOMOGRAM, THE CRITICAL VALUES (FOR THE LIFE OF ORGANISMS) OF SEASONAL FACTORS

The heat resources nets on all the nomograms are constructed on the basis of average temperatures (over different periods). However, an average temperature that is favorable to an organism's development may include individual values that would prove disastrous for the plant or cold-blooded organism. It is therefore necessary to know, when working with a nomogram, how to take into account these harmful temperature values, as well as the critical states of other factors.

As was noted before, in Figure 1.50, to the left of the line with the hatching, there is a temporal zone of autumn frosts. Its essence and construction were described above in Chapter 1.

Let us assume that we wish to determine whether the beginning of lucerne flowering will occur in Yavan if the successive hay harvest was on September 15 (as before, the problems are solved on a multiannual basis). The solution is that the beginning of flowering will indeed occur, because the line of average temperatures marked September 15 (Fig. 1.50) intersects phenocurve 14 for lucerne's development (from the successive hay harvest to flowering) at $n = 44$ days. This means that the beginning of flowering occurs before the autumn frost: September 15 + 44 days = October 29 (the first autumn frosts in Yavan are on November 17). Indeed, in Figure 1.50 we see that this intersection lies outside the shaded frosts zone.

Now let us solve the same problem, but when the successive hay harvest was made, for example, on September 23. According to the average temperatures, it seems possible for lucerne to flower in this case, too: the line of average temperatures from September 23 intersects the developmental curve 14 at $n = 70$ days. But then the flowering would occur on September 23 + 70 days = December 2, that is, much later than the beginning of autumn frosts. In all probability, the flowering organs and leaves on the lucerne would be damaged by frost, before flowering occurred. That is, the line of average temperatures from September 23 intersects developmental curve 14 in the autumn frosts zone.

Hence it is clear that in all further such cases there is no need to make a calculation: *when the summer or autumn line of average temperatures, marked by the date of the initial developmental stage for the calculation, intersects the phenological curve of an organism's develop-*

ment (during the period from the initial developmental stage to the next stage) in the autumn frosts zone, then the next developmental stage does not occur (or it occurs on the eve of the organism's death) if the organism is sensitive to frosts.

For example, the last (in the year) eggs of the tick *Hyalomma detritum* that are still able to form larvae will be laid, in the conditions of Yavan, no later than September 6, since the line of average temperatures from September 6 is the last that touches phenological curve 21 for *H. detritum*'s development from egg to larva (Fig. 1.50).* However, these larvae will not survive: as is known, at the slightest cooling they lose their ability to cling to their host, and the point where the curve touches the temperature line occurs in the autumn frosts zone.

It is also important to know how to take spring frosts into account (a method for constructing the spring frosts zone was described in Chapter 1). For example, the early flowers of fruiters can perish from spring frosts. Therefore, if the line of average temperatures intersects the corresponding phenocurve for fruiters in the spring frosts zone (Fig. 1.50), a good yield cannot be expected.

In the same way, it is possible to take into account the critical values of high temperatures. For example, it is known that the majority of *H. detritum* ticks hatching from the nymph (that has fallen from cattle) perish when the temperature in the grass (where the tick attempts to avoid high temperatures) reaches +45°, which corresponds to a temperature of approximately +60°C on a sunlit soil surface. Such temperatures occur in Yavan around June 20. This will be the date when the summer purification of pastures from pubescent ticks begins. Then (by analogy with the reasoning for constructing the autumn frosts zone) the line of average temperatures from June 15 will consist of two parts: above-critical and below-critical. A five-day part, from June 15 to June 20, that is, up to the time when critically high temperatures occur, and the rest of the line, that is the above-critical part (see the mark x on the line from June 15 in Fig. 1.51). The line of June 5 will consist of both the 15-day below-critical part (from June 5 to June 20) and the above-critical part (see the mark y on the line from June 5) and so on (z, w). Now it can be seen that phenological curve 10, representing *H. detritum*'s development during the period from the falling away of the nymph to the imago, is intersected only by those below-critical parts of the average temperature line which are dated not later than May 15-20 (Fig. 1.51). This means that nymphs which fall from pasturing cattle after May 15-20 will not produce pubescent ticks, because the latter will perish from the high tempertures. On the other

---

*We take curve 21, but not 20, because in autumn there is higher relative atmospheric humidity.

hand, nymphs that fall away, for instance, on May 5 will produce imagos (May 5 + 38 days = June 12) before the critical temperatures occur. Here n = 38 days is the ordinate of the point where the below-critical part of the average temperature line from May 5 intersects developmental curve 10.

If we now imagine a line drawn through points $x, y, z,$ and $w,$ and we imagine a shaded band on the right along this line, then we would have a zone not of early frosts, but of high temperatures (this high temperature zone has not been shaded, in order to avoid overloading Fig. 1.51).

Taking into account the critical states of nontemperature factors does not differ much from the above process, provided these critical states have a seasonal trend. For example, it is known that malarial mosquitos from the egg to the imago develop in water. If the reservoirs dry up in the summer, then the flight of the mosquitos is impossible at this time. Let us assume that in Yavan the reservoirs dry up by June 20. Which egg layings will give malarial mosquitos *Anopheles superpictus* with wings?

The below-critical (before June 20) parts of the average temperature line from June 5 (15-day part), from May 26 (25-day part), from May 15, from May 5, and so on intersect curve 4 for the mosquito's development from egg to imago in Figure 1.51; the below-critical, five-day part of the line from June 15 does not intersect this curve. Moreover, this curve is not intersected by the lines dated June 25, July 5, and so on, which have no below-critical parts at all. Therefore, egg layings before June 5-10 will lead to the imago state; egg layings after June 5-10 will not give imagoes, because the reservoirs dry up during the stage of metamorphosis. When the reservoirs fill with water again, egg layings by females that survived or that came from other regions will again give imagoes.

It is known that for a number of organisms the critical day length is also important. For example, as was shown in the experiments by L. I. Arapova in Pruzhany in Byelorussia (1974), the percentage of individual organisms of Colorado potato beetle that go into the diapause approximates 100 percent, if the daylight hours, including twilight, are reduced to approximately 17 hours 15 minutes. In astronomical calendars, it is easy to see that at the latitude of Pruzhany (52.5° northern latitude) such a day length occurs around July 21. Thus the average temperature line with an initial date of, for instance, May 5 (Fig. 1.73) will consist of the below-critical part, which is 77 days (July 21 minus May 5), and the above-critical part; the line from May 15 will consist of the below-critical part, which is 67 days (July 21 minus May 15), and the above-critical part; and so on. Through the points marking these bounds we draw a line, and we shade the area on the left along this line, thus obtaining zone I, which gives critical day length.

If in Pruzhany egg laying by the Colorado potato beetle is recorded, for instance, on May 26, then the young beetles will emerge on May $26 + 50$ days = July 15, where $n = 50$ days is the ordinate of the intersection of the May 26 line with phenocurve 6 (Fig. 1.73). Since this intersection is outside the critical zone of the photoeffect, the young beetles have time to become pubescent and to lay eggs. Therefore, the next generation is possible. However, if the egg laying occured later, for example, on June 15, then the young beetles would not become pubescent and lay eggs, because the intersection of the meteorological line of June 15 with the biological line 6 occurred in the shaded zone I.

It is clear that for other geographical latitudes the critical day length of 17 hours 15 minutes (see above) occurs earlier or later than in Pruzhany.

The actual date can easily be determined by using astronomical calendars or well-known graphs (such as cited by A. S. Danilevsky). For example, in Kaliningrad (formerly Königsberg), with latitude 54.8°, this critical day length occurs approximately on July 10. Thus in Figure 1.84, the line of average temperatures with critical date May 5 consists of the below-critical part of 66 days (July 10 minus May 5) and the above-critical part; the line from May 15 has a below-critical part of 56 days (July 10 minus May 15); and so on. Thus we obtain zone I, which is somewhat different from that plotted for Pruzhany.

Generally speaking, the decrease of daylight hours is only a "signal" for cold-blooded organisms to prepare for the unfavorable conditions of autumn and winter by laying no further eggs. The fundamental cause is most probably the heat factor. This will be discussed in more detail in Section 6.6.b about the diapause, there we will show how to calculate the dates of the diapause without using the light factor. But here we should mention that according to L. I. Arapova (1974), the percentage of beetles going into the diapause is related not only to the day length but also to temperature (Fig. 2.19). When the temperature decreases, even if the day length does not alter, the diapause occurs at earlier dates, and a greater percentage of beetles go into the diapause. This circumstance was accounted for when constructing zone I in the majority of PTNs (i.e., the graph shown in Figure 2.19 was taken into consideration).

In order to protect crops from the Colorado potato beetle, it is important to know the dates of its development, not so much on wild *solanaceae* as on the potato, to which the beetle moves after the appearance of the potato sprouts and on which it lays its eggs. The sprouting dates can be easily forecast on the basis of phenocurves 7 and 8 (Fig. 1.73), and using the dates of early, intermediate, and late sowings. The sprouting date is taken as the date of egg laying, and from this date the subsequent developmental stages of the beetle are calculated for early, intermediate, and late potato sowings, respectively.

Besides this, it is important to know with which developmental stages of the potato the developmental stages of the pest coincide. For this purpose, we can choose in the *Agroclimatic Reference Book*, the average multiannual dates of potato development, for example, of mass sowing (on intermediate dates). In Pruzhany the mass sowing gives sprouting on June 3, budding on June 28, flowering on July 12, and fading of tops on August 5. By analogy with the construction of a zone of critical day length, we can construct zones relating the periods after sprouting, after budding, after flowering, and after fading of tops. This is done in Figure 1.73 in the form of dashed lines *a, b, c,* and *d*. Thus when the Colorado potato beetle lays eggs on June 5, the larvae of the first, second, and third ages will appear on the sprouts, because the line of average temperature from June 5 intersects phenocurves 1, 2, and 3 at points that lie after dashed line *a*. The appearance of larvae of the fourth age will coincide with the beginning of budding, because the meteorological line from June 5 intersects phenocurve 4 exactly on dashed line *b*. The larvae will begin to go into the soil for pupation during the period when the potato is budding (intersection with phenocurve 5 after dashed line *b*), and young beetles of the summer generation will appear on the flowering potato, because the line of June 5 intersects the phenocurve 6 after the dashed line *c*.

This is an example of one kind of combined phenocalculation. In this case it enables us to investigate such a "biogeocenosis," or combination of the Colorado potato beetle and the potato, and even to manage this biogeocenosis. The management involves not only taking measures directed against the Colorado potato beetle, but also varying the dates of potato sowing, especially if we take into consideration the beetle's requirement for a certain day length for reproduction.

Problems of this kind appear repeatedly in real life, as well as in the pages of this book.

We would like to point out that in the USSR many investigators and entire research institutions are making contributions to the application and development of Podolsky's method. Among them the Byelorussian Research Institute for Plant Protection is investing the greatest efforts in this direction (particularly involved are L. I. Arapova, N. N. Kharchenko, and Academy Member N. A. Dorozhkin). Unfortunately, the USSR's Hydrometeorological Service impedes interest in the new method in both central and provincial organizations.

2.4. TAKING INTO ACCOUNT THE MICROCLIMATE OF THE ORGANISM'S HABITAT

Many of the heat resources nets given in this book are related to the air temperatures recorded by meteorological stations at the generally accepted height of 2 m above the soil surface. Of course, if we have information about temperatures in the crop stems medium, in the ploughed layer of the soil, in the water where mosquito larvae live, or in houses where adult mosquito live, we can construct analogous nets for these media. This has sometimes been done. However, in this case, the temperature curves for the organism's development must be constructed using not the temperatures recorded at meteorological stations, but the temperatures in these media, because requirements and resources equations are to be solved together on the nomogram.

It is not always necessary to use the temperature data for the biotope. There are two cases when a special effort to take into account the microclimate of the biotope is not necessary:

1.   When both temperture developmental curves and the heat resources net are constructed using air temperatures, which were recorded at a meteorological station in a psychrometric box (Stevenson screen), and there is a correlative connection between the temperatures in the box and the temperatures (not equal to them) in a biotope of the same region. Such a connection almost always exists (Fig. 2.5). The microclimate of the biotope may then be ignored because when we use the temperatures recorded at a height of 2 m, the temperature requirements of an organism are overstated or understated to the same extent as the temperature resources of the biotope are overstated or understated.

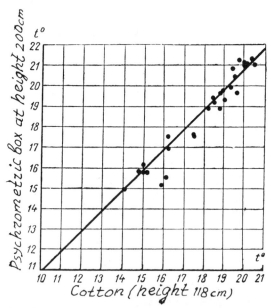

FIGURE 2.5. The relationship between the air temperatures at a standard height of 200 cm and in the foliage of cotton plants, according to the author's observations at the agricultural meteorological station in Dushanbe, 1953. The temperatures in the cotton are lower because of evaporation and transpiration.

It is thus clear that here the microclimate is indirectly taken into account. A specific example of this is the dates until which the oncospheres of beef tapeworm survive in different biotopes (Section 6.3).

2. When the resources net is constructed on the basis of the atmospheric temperatures at a meteorological station, and the phenological curve of an organism's development—on the basis of the temperatures in a biotope, but the biotope, in regards to its 24-hour average temperatures, is "isothermically" (according to the classification by V. N. Beklemishev, 1944) identical to the atmosphere.

It should be pointed out that we have to relate to those phenological curves of an organism's development, which are constructed using biotope data, these curves obtained using data from controlled-temperature chambers. This is feasible since the temperature of the chamber is approximately equal to the temperature of the organism's body, and the latter is approximately equal to the temperature of the biotope where ticks or insects live naturally. For example, in Figures 1.36, 1.38, and 1.45 the curves for ticks, the *Plasmodium*, virus, and citrus pests are constructed on the basis of observations in a controlled temperature and humidity chamber.

On the other hand, it is necessary to take into account the microclimate of an organism's habitat on the heat resources net, if the temperatures of the biotope differ considerably from the atmospheric temperatures, even in the case of the 24-hour average values, and the phenological curve of the organism's development is constructed using the temperatures in the biotope or in a temperature-controlled chamber. For instance, it has already been mentioned that the imago of the malarial mosquito lives, basically, in human dwellings. The temperatures in houses (even when not heated) can sometimes differ considerably from the temperatures of the outdoor air. However, the length of the maturation period for malarial parasites in the mosquito's body (sporogony) in relation to temperature (Fig. 1.38) is given by well-known data prepared by V. P. Nikolayev and other authors, for a controlled-temperature chamber (these can be considered equivalent to data for houses, as regards the equality of temperatures in the mosquito's body and in the environment). The same can be said about the curve giving the length of the malarial mosquito's gonotrophic cycles in relation to temperature (as a rule the temperature used is that in a controlled-temperature chamber). It is clear that in such cases it is better not to use the heat resources net based on the outdoor air temperatures. However, this does not mean that it is necessary to construct a special net for the temperatures inside the houses. Considering that the seasonal trend of temperatures in unheated houses is almost identical to the temperature trend of the air, varying only several degrees higher or lower,* it is enough, mentally or with a pencil, to move the required part of the heat resources net for outdoor air to the right or to the left along the axis of abscissas by that same number of degrees.

Thus in order to take into account the temperature specificity of a biotope, when working with a nomogram where the heat resources net is based on the air temperatures at meteorological stations and the phenocurve of the organism's development is based on the temperatures in the biotope (controlled-temperature chamber), we have to move along the axis of abscissas those average temperature lines required for the calculation, to the number of degrees equivalent to the anomaly in the 24-hour average temperatures of the biotope (in comparison to the meterological station). We can also do the reverse: the net remains fixed, but the biological curve is moved to the extent of the temperature anomaly. This is exactly what has been done with curve 2 in Figure 1.40, and on a number of nets.

If there are observations of the temperature in a biotope over a specific and relatively short period, and a given phenological problem is to be solved in relation to this period only, then we can construct a small fragment of the net using actual temperatures in the biotope, and we apply this fragment together with the "biotope" or "controlled-temperature chamber" developmental curve (in the same way as the temperature totals obtained from these curves are sometimes compared with the temperatures observed over a specific period in the biotope).

2.5. TAKING INTO ACCOUNT THE 24-HOUR TEMPERATURE OSCILLATIONS

It is known that the 24-hour temperature oscillations have a specific influence on the developmental rate of plants and cold-blooded organisms, basically in spring and in autumn, when the nighttime and the 24-hour average temperatures fall below the biological threshold, and development continues to a limited extent under the influence of the daytime temperatures. However, the curve giving the relationship of the plant's interdevelopmental stage to the average temperatures over this period may be constructed using data from experiments with sowing at different times of the year (unfortunately, such experiments are not always carried out). Then this curve, in its low-temperature part, which concerns the spring and autumn development of the plant, must take into account the integrated result of the acceleration and retardation of development under the influence of high daytime

---

*For example, in Russia it is accepted that the 24-hour average temperature of indoor air is 1-2° higher than that of outdoor air.

temperatures and low nighttime temperatures, respectively. The same can also be said about cold-blooded organisms with several generations per season: the low-temperature part of the developmental curves based on field observations of these organisms relates to the spring and autumn generations (Figs. 1.1, 1.36, and others).

For an illustration of taking into account the influence of 24-hour temperature oscillations on development, see Figures 1.34 and 1.35. Here the lower temperature threshold, for example, for the geohelminth *Haemonchus contortus*, seems to be understated in comparison to the limits mentioned in the literature, and this fact reflects the constant temperature in a controlled-temperature chamber. This discrepancy can be explained by the fact that in Figure 1.34 (as well as in Fig. 1.35) the average temperatures over a certain period in the field are given, as obtained from 24-hour temperatures; although *H. contortus* does not develop at a constant temperature of, for example, +4°C, the spring (or autumn) average temperature of +4°C is associated with comparatively high daytime temperatures on the soil surface in Tadzhikistan, to such an extent that the development from egg to third-age larva is completed within 24-26 days (in Fig. 1.34, the points at the extreme left). It is clear that when we put the phenological curve of *H. contortus*'s development onto a PTN, the latter will take into account the influence of 24-hour temperature oscillations on the dates of *Haemonchus*'s development. This influence will be taken into account in terms of the usual values, even though the values differ somewhat for each concrete year and place. However, any developmental curve based on field observations reflects to a certain extent the influence of 24-hour temperature oscillations on the organism in the northern and southern sections of a given region.

2.6. FORECASTING THE YIELD OF PLANTS AND THE NUMBERS OF PARASITES

Attempts have been made in the past to estimate the yield of agricultural crops on the basis of temperature totals and precipitation or, more usually, temperatures and humidity; such attempts are still made today. But in all probability the problem of estimating and forecasting the yield for various plants can be solved more accurately on the basis of the biomathematical method described above. First let us focus on solving the simpler problem: to find the correspondence between the yield and the length of time it takes to accumulate. First we will discuss how to calculate the prefrost yield of cotton as a percentage of the whole yield (this will aid later discussions in this book). After this, other questions will be discussed.

The longer the period between the date of mass unfolding of the first

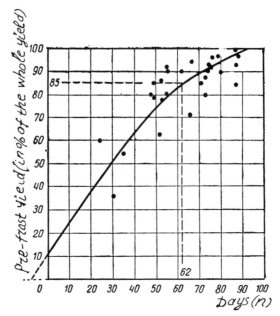

FIGURE 2.6. Dynamics of the accumulation of prefrost yield (as a percentage of the whole yield) in fine-fiber cotton variety 5904-I (*Gossypium barbadense*) in relation to the length of the period (*n*) between mass unfolding of the first seed bolls and cessation of vegetation (according to the actual data from Tadzhikistan's NTV over the period 1955-1961).

cotton bolls and the first autumn frosts (in the regions where there are autumn frosts), which usually arrest the vegetation of the cotton, the greater the prefrost yield of a given variety. Of course, this is true if other conditions are equal.* Therefore, it is advisable to construct, using actual observations, the graphs shown in Figures 2.6 to 2.14 for various cotton varieties (Tadzhikistan).

It is quite simple to use graphs such as these together with a PTN for long-term forecasting of the prefrost yield of raw cotton. For example, the mass flowering of cotton variety 5904-I (Fig. 2.6) at the Kurgan-Tyube NTV center took place in 1961 on June 22. What prefrost cotton yield, as a percentage of the whole yield, is to be expected, if it is known that the average multiannual date of the first autumn frost in the air of Kurgan-Tyube is October 29?†

Using the actual flowering date of June 22, and with the help of the PTN for Kurgan-Tyube (Fig. 1.55), we obtain a forecast for the date of mass unfolding of the first seed bolls: June 22 + 67 days = August 28

*However, other conditions influence the yield in metric centner per hectare more than the prefrost yield in per cent.

†The First autumn frost comes on the soil surface earlier than in the air.

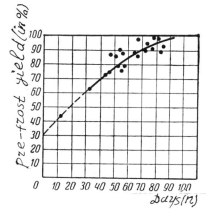

FIGURE 2.7. Dynamics of the accumulation of prefrost yield (as a percentage of the whole yield) in fine-fiber cotton variety 504-V (*G. barbadense*) in relation to the length of the period (*n*) between mass unfolding of the first bolls and cessation of vegetation (data for the period 1955-1961).

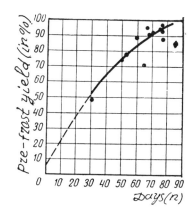

FIGURE 2.8. Dynamics of the accumulation of prefrost yield (as a percentage of the whole yield) in fine-fiber cotton variety 5595-V (*G. barbadense*) in relation to the length of the period (*n*) between mass unfolding of the first bolls and cessation of vegetation (data for the period 1958-1961).

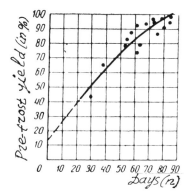

FIGURE 2.9. Dynamics of the accumulation of prefrost yield (as a percentage of the whole yield) in fine-fiber cotton variety 9123-I (*G. barbadense*) in relation to the length of the period (*n*) between mass unfolding of the first bolls and cessation of vegetation (data for the period 1958-1961).

179

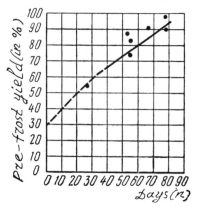

FIGURE 2.10. Dynamics of the accumulation of prefrost yield (as a percentage of the whole yield) in fine-fiber cotton variety 2I3 (*G. barbadense*) in relation to the length of the period (*n*) between mass unfolding of the first bolls and cessation of vegetation (data for the period 1955-1958).

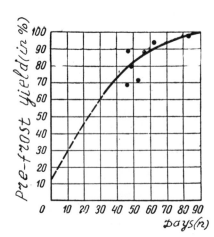

FIGURE 2.11. Dynamics of the accumulation of prefrost yield (as a percentage of the whole yield) in fine-fiber cotton variety 5769-V (*G. barbadense*) in relation to the length of the period (*n*) between mass unfolding of the first bolls and cessation of vegetation (data for the period 1960-1961).

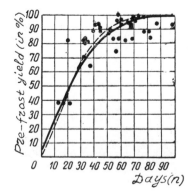

FIGURE 2.12. Dynamics of the accumulation of prefrost yield (as a percentage of the whole yield) in regular cotton variety 108-F (*G. hirsutum*) in relation to the length of the period (*n*) between mass unfolding of the first bolls and cessation of vegetation (data for Tadzhikistan for the period 1955-1961, from A. S. Podolsky). The dashed curve without points is for Azerbaidzhan (from A. R. Mustafayev).

180

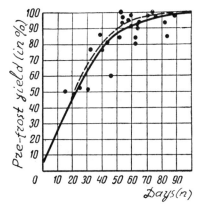

FIGURE 2.13. Dynamics of the accumulation of prefrost yield (as a percentage of the whole yield) in regular cotton variety S-47 27 (*G. hirsutum*) in relation to the length of the period (*n*) between unfolding of the first bolls and cessation of vegetation (data for Tadzhikistan for the period 1955-1961, from A. S. Podolsky). The dashed curve without points is for Azerbaidzhan (from A. R. Mustafayev).

(where $n = 67$ days is the ordinate of intersection of the average temperature line for June 22, with phenological curve 9 for the development of cotton variety 5904-I from flowering to boll unfolding).

If in the year for which the forecast is made the first autumn frosts occur (and damage the cotton) at a date close to the multiannual date (October 29), then the prefrost yield will be accumulated after the mass ripening, during a period of 62 days (from August 28 to October 29). We find on the axis of abscissas in Figure 2.6 the mark for 62 days, and from this mark we trace a vertical line up to the intersection with the curve; from the intersection we trace a horizontal line to the axis of ordinates, where we read 85% (see dashed lines). This will be a long-term forecast of the yield. In actual fact, at the NTV center in Kurgan-Tyube in 1961 a 89.5% prefrost yield of cotton variety 5904-I was harvested, with flowering occurring on June 22 (the forecast for ripening—August 28—coincides exactly with the real date of ripening).

This forecast was made a very long time before the date required—

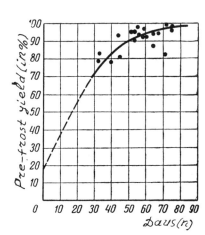

FIGURE 2.14. Dynamics of the accumulation of prefrost yield (as a percentage of the whole yield) in regular cotton variety 149-F (*G. hirsutum*) in relation to the length of the period (*n*) between mass unfolding of the first bolls and cessation of vegetation (data for the period 1956-1961).

more than four months (67 days from the actual date of flowering to the forecast date of cotton ripening, plus 62 days from ripening to autumn frosts; in total 129 days). We can reduce the length of this period by using not the forecast date of cotton ripening, but the actual date; in this way the yield forecast can be defined more exactly. In addition, a reliable long-term weather forecast will spare us the necessity of using the multiannual date for the first autumn frosts, thus giving us a more accurate forecast.

Without any special calculation of the root-mean-square error σ of yield forecasts, we see in Figures 2.6 to 2.14 that the rough value of this error (a result of not taking into account conditions unrelated to the length of the period during which the yield accumulates) is $\pm$ 5-6% (since about 0.7 of all the points on each graph fall within a belt $\pm$ 5-6% wide which follows the path of the regression curve and is equally wide on both sides of this curve). The probable error $\sigma_w = 0.6745\sigma = \pm$ 3-4% (50% of all actual errors are greater than the probable error; the other 50% are smaller). Greater accuracy in the yield forecast can be achieved by taking into account on the PTN the heat and humidity conditions of the autumn during the harvest period.

As to the relationship between the percentage yield and the length of the period over which it is accumulated, Figures 2.6 to 2.14 can be used for forecasting cotton yield in different years and in different regions. Changing a region or a year will result only in changes in the period of n days from ripening to frost, which we find on the axis of abscissas, however, the curves for Tadzhikistan and Azerbaidzhan, for example, remain almost invariable (Figs. 2.12 and 2.13).

Is it possible, on the axis of ordinates in Figures 2.6 to 2.14, to put the prefrost yield not in percent, but in metric centners per hectare? It is obviously possible, since the cotton yield (in centners per hectare) is influenced not only by the agricultural techniques and the heat and humidity conditions of autumn, but also by the length of time during which the yield accumulated after ripening. However, for one and the same cotton variety we should have several graphs of the relationship between the yield (in centners per hectare) and the length of the period from mass unfolding of the first bolls to the halt in vegetation: the different graphs will relate to different agricultural techniques (different crop densities, different fertilization regimes, different levels of care, etc.). Graphs of the latter kind could probably be constructed for the purpose of yield forecasting not only for cotton, but also for a number of other agricultural crops that accumulate their yield little by little (e.g., potato, beet).

In this connection, the work on yield programming done by A. A. Nichiporovich and I. S. Shatilov may be useful. In the process of such programming, when the photo synthetically active radiation (PAR) and moisture and fertilizers are taken into account, it would be useful for a number of crops to take into account the length of the period of yield accumulation by means of the method described above.

There is also another method for forecasting cotton yield. In Fig. 2.15 the yield is taken into account not in percent, but as the number of open bolls per square meter. The figure shows in how many days, beginning from the flowering date, 10 open bolls will be accumulated, and then 50 bolls (from the flowering date). It also shows when 50 open bolls will be accumulated, beginning from the date when 10 open bolls were accumulated, on area of 1 m².

Figure 2.15 is based on data provided by Professor A. Marani (1981). Marani and his collaborators counted periodically the accumulation of the number of open cotton bolls per square meter in different places in Israel. By plotting these data on a graph in relation to the dates of counting (for each place), it is possible to work out the dates when 10 bolls and 50 bolls are accumulated. In the case of Figure 2.15, the number of days necessary for this accumulation was calculated, beginning from the first flower, or from the date when 10 bolls were accumulated. Then the average temperatures for these periods was determined. Finally the results were plotted. As can be seen, the yield is transformed to a time factor. This idea will often recur below.

In Figure 2.15 the left-hand branches of the curves diverge from the

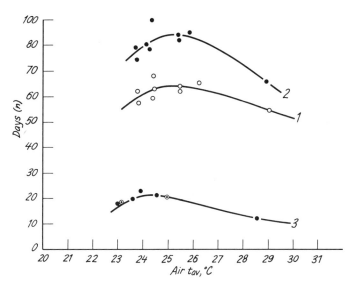

FIGURE 2.15. Phenological curves for cotton variety Acala S.J.-2 (*Gossypium hirsutum* L.). Curve 1: from the first flower to ripening (when the number of open bolls accumulated per m² = 10); curve 2: from the first flower to ripening (when the number of open bolls accumulated per m² = 50); curve 3: from ripening (= 10 bolls/m²) to ripening (= 50 bolls/m²). Phenological curves were constructed by Podolsky using the data provided by A. Marani.

usual pattern: here the temperature increase (though over a small interval) is accompanied not by a decrease in the length of the interdevelopmental stages, but by an increase. Apparently this is a result of the intensive irrigation used in Israel at the end of flowering and the beginning of ripening in cotton: the intensive evaporation of moisture cools off the fields (Fig. 2.5), although the air temperature at the meteorological station increases. Only very high air temperatures cause the field to dry quickly, thus leading to an increase in the temperatures of the field, and thus to the usual decrease of the interdevelopment stage (see the right-hand branches of the curves).

If we now transfer the phenocurves from Figure 2.15 onto a heat resources net of one region or another, we can make a corresponding calculation of the yield. For example, in 1978 near Tirat-Zvi (Israel) the first flowers on cotton variety Acala S.J.-2 appeared on June 9. We wish to determine when 10 open bolls per square meter will be accumulated.

The solution follows: in Figure 1.120 we find the line of average temperatures over the period from June 9 (the line will be near the middle between lines June 5 and 15). The ordinate n of the intersection of this line with phenocurve 4 is 56 days. Then the yield contained in 10 open bolls will be accumulated by August 4 (June 9 + 56 days). The actual date was, in fact, August 3, and so the error in the forecast is 1 day.

Now let us determine, using date of the first flowers, the date when 50 open bolls per square meter will be accumulated. For this purpose we find the intersection of the meteorological line June 9 with phenocurve 5. We obtain June 9 + 66 days = August 14. The forecast turned out to have no error. If, for the initial date of the forecast, we take the actual date when 10 bolls per square meter are accumulated, that is August 3, then the calculation using phenocurve 6 gives an error of −1 day.

The average errors (whether positive or negative) equal /2/ days when using the same data as that used for constructing the pheno-curves, and are somewhat greater when other data are used. When the forecast is made 70 days before the forecast date, excellent accuracy is obtained (see Table 3.3).

It should be noted that the temperature totals method is incapable of coping with such complicated phenocurves as those appearing in Fig. 2.15.

We should mention that graphs of yield accumulation help in determining the most advantageous dates for sowing.

. . .

In the article by F. E. Round (1964), the author was particularly interested in the experimental graph shown in Figure 2.16.

Although some investigators think that only the intensity of solar radiation can explain the rate of carbon accumulation in plant biomass, the author was not surprised by this graph: temperature is

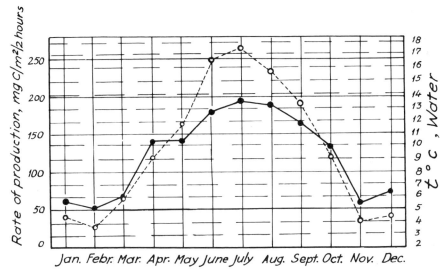

FIGURE 2.16.   Rate of potential gross production of microbenthos in a Danish fiord.
Average data from all experiments. The solid line represents the rate of production; the
dashed line represents the temperature. (From Round, 1964, after Grøntved, 1960.)

quite an exact reflection of solar radiation. Since only a few observations of solar radiation have been made throughout the world, successful attempts are now being made to calculate the radiation balance on the basis of the duration of sunshine and of temperature (T. G. Berland, USSR).

If we assume that the "2 hours" indicated on the ordinate of Figure 2.16 are "average" for the daylight part of the 24-hour period in regard to the intensity of carbon assimilation, then we can calculate the productivity in milligrams of carbon per square meter (mg C/m^2) for the daylight part of the 24-hour period. The length of the daylight part of the 24-hour period in different months at the latitude of Denmark can easily be obtained from an astronomical calendar or from a relevant graph.

How many days are necessary at different times of the year for the accumulation of 20 g C/m^2? Why 20 g? This is not important. We could just as easily take 30 or 40 g. However, 20 g C/m^2 will ensure a sufficient "yield" of microbenthos mass per hectare at the latitude of Denmark in order to give an acceptable harvest for experimental or commercial purposes.

Having determined the periods necessary for accumulating the "yield" of C^{14}, the average water temperatures over a month were extracted from the temperature graph given in Fig. 2.16. This is represented in Table 2.3.

Data from columns 5 and 6 of Table 2.3 were transferred to Figure

TABLE 2.3.   Periods necessary for accumulating 20 g C^{14}/m^2 by microbenthos, and water temperatures in a Danish fiord.

Conditions	Productivity (mg C/m^2 per 2 hours)	Length of Daylight Portion of 24-hour Period (hours)	Productivity (mg C/m^2 per 24-hour period)	Number of Days Necessary for Accumulating 20 g^a	Average Water Temperature over a Month (°C)
1	2	3	4	5	6
In Jan.	64	8.0	256.0	78.0	4.5
In Feb.	53	9.8	259.7	77.0	3.7
In Mar.	67	12.0	402.0	50.0	5.9
In Apr.	140	14.2	994.0	20.0	9.1
In May	141	16.0	1128.0	18.0	11.8
In June	180	17.0	1530.0	13.0	16.5
In July	192	16.6	1593.6	12.6	17.4
In Aug.	187	14.9	1393.2	14.4	15.7
In Sept.	160	12.4	992.0	20.0	13.2
In Oct.	134	10.4	696.8	29.0	9.1
In Nov.	60	8.6	258.0	77.5	4.0
In Dec.	74	7.4	273.8	73.0	4.4

^{a}C/m^2 if the conditions of the corresponding month were to continue unchanged.

2.17. The curve from Figure 2.17 was transferred onto the net of water heat resources in a Danish fiord (Fig. 1.129). This net was constructed using the temperature curve from Figure 2.16.

Now, with the help of Figure 1.129, we can carry out various calculations. For example, when will 20 g C/m² be accumulated, if a Danish fiord is cleaned of algae (microbenthos) on March 5? We find the line of average temperatures for March 5. The intersection of this line with the phenocurve gives $n = 39$ days. Therefore, the date that interests us is March 5 + 39 days = April 13.

Let us compare this result with Table 2.3. It is shown in the table that under the conditions existing *in March* (column 1), that is, with a *constant* temperature of 5.9° (column 6) and a *constant* day length of 12 hours (column 3), 20 g C/m² will be accumulated in 67 days (column 2). The nomogram gives the figure of only 39 days. This happens because the nomogram takes into account the increase in temperatures and in day length after March 5.

It would be rather difficult or impossible to obtain this result directly from Figure 2.16. From Figure 2.16 we can obtain periods necessary for accumulating only very small amounts of carbon. For example, we could say that if in the middle of March approximately 70 mg C/m² can be accumulated every 2 hours, then in the middle of July approximately 40 minutes will be necessary for accumulating the same amount.

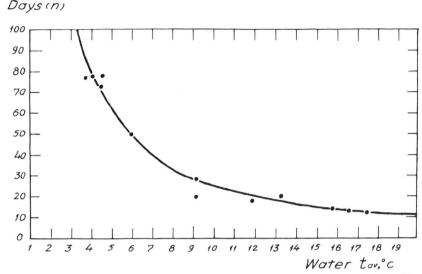

FIGURE 2.17. Phenological curve for the time necessary to accumulate 20 g C^{14}/m^2 by microbenthos in a Danish fiord, in relation to water temperature.

Let us continue with solving problems on the nomogram (Fig. 1.129).

If the reservoir is cleaned of water plants on July 20, then 20 g C/m^2 will be accumulated in $n = 13$ days, that is, by July 20 + 13 days = August 2, where n is the ordinate of the intersection of the average temperature line from the initial date of July 20 with the phenocurve.

How many times a year can 20 g C/m^2 be harvested in a Danish fiord? If we assume March 5 to be the date of the first harvest, then the second harvest will be, as has been already calculated, on April 13; the third harvest will be April 13 + 25 = May 8; and so on. It is possible to have 15 harvests, yielding overall 15 harvests × 20 g = 300 g C/m^2.

It is possible to take into account the temperature anomalies or the long-term weather forecast for a given year by transferring the line of average temperatures to the right or to the left to the same number of degrees as the temperature anomaly (this will be discussed in more detail in the relevant sections of this book).

It is also possible to superimpose our phenocurve on the water heat resources net for the Netherlands, or, for example, for Germany and maybe for Sweden, since there are analogous species of algae (although we should not place too much faith in the transferability of the phenocurve, because each latitude has its own day length).

If reservoirs are fertilized artificially, thus making it possible to construct two or three phenocurves for different kinds and quantities of fertilizer, then all of these phenological calculations can be made for different fertilization regimes.

Last, it is possible to construct phenoforecasting calendars (which were already mentioned and which will be discussed further).

As can be seen, the idea of dealing with yield in terms of the time factor is a very fruitful one. This idea is also applicable to inanimate nature (Chapter 10).

. . .

A lot of works are devoted to forecasting yields of cereal crops, but they usually relate to normal sowing dates. Not enough has been written about harvesting losses when sowing is carried out later than normal or at excessively late dates. To investigate this, we must see the distinct systematic pattern of the graphs giving the relationship between harvest and sowing dates. This is possible on the basis of data from field experiments with unusual sowing dates in Kursk, USSR. Our first interest here is one of the basic components of yield—the weight of 1000 grains of wheat, barley, and oats. This alone does not give the yield, since here we have not yet considered such factors as losses caused by pests (these losses are especially great after late sowing dates), losses caused by low crop densities due to lack of moisture, and so on. However, the decrease in the weight of 1000 grains after late sowing dates can be explained by the reduction in the period of intensive assimilation activity in the crops; this is one of the most important causes of the yield decrease after late sowing.

The results of our experiments (Podolsky, Yermakov, Dolgikh) are shown in Fig. 2.18. As can be seen, the systematic decrease in yield begins from the sowing date of May 3. In regard to spring wheat, the decrease in the weight of 1000 grains caused by each 10-day postponement of the sowing date, that is, $\Delta y/\Delta \tau$, averages 15%. The optimum sowing dates are April 22 to May 2, when the graph for spring wheat (as well as for spring barley) does not show any decrease in yield. Indeed, in the *Agroclimatic Reference Book* for the Kursk region, the multiannual average sowing date for spring wheat is April 22 to 27. Therefore, agronomy and practical folk experience both succeeded in finding the optimum dates.

Knowing the optimum sowing date and by how much sowing is late in a given field, we can forecast the approximate percentage of decrease in yield using the value $\Delta y/\Delta \tau = 15\%/10$ days. This is a useful contribution to methods of yield forecasting.

As to spring barley and oats, it can be seen in Figure 2.18 that their yield losses are less than for wheat, they are less demanding of light and heat.

. . .

The new method for phenological forecasting and bioclimatic estimations provides a number of possibilities for forecasting the numbers of poikilotherms. Thus in the section where malaria is discussed, we will see how the number of malarial mosquitos increases

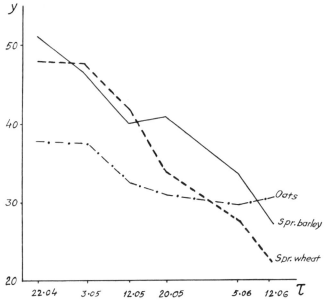

FIGURE 2.18. Sowing dates (τ) and the weight of 1000 grains (y) of spring wheat, spring barley, and oats in field experiments with sowings at different times (Kursk, Russia, 1971).

because of the coincidence of different generations. The developmental dates of individual generations, and their coincidence, can be forecast on PTNs.

Forecasting the number of possible generations for various poikilotherms under given climatic conditions (cotton boll worm, spider mite, fall webworm, plague-infected fleas, etc.) is also connected with the numbers of individuals. This will be discussed in the corresponding sections of the book.

Finally, in Section 2.3 where we discussed how to take into account the critical values of seasonal factors, it was shown that for the Colorado potato beetle, zone I is critical on all PTNs. It should be remembered that if the intersection of any line of average temperatures with the phenocurve from egg to imago occurs in zone I, then 100% of the young female beetles will be unable to lay eggs, and thus will not produce another generation. This happens because the day becomes shorter than is necessary for the beetle to become pubescent and lay eggs.

By analogy with zone I, and to the left of it, we could calculate a prezone. Intersection in this prezone would show that, for example, 80% of the young beetles will go into the diapause, but 20% are able to produce another generation. Of course, in this second case the Colorado potato beetle presents a greater threat.

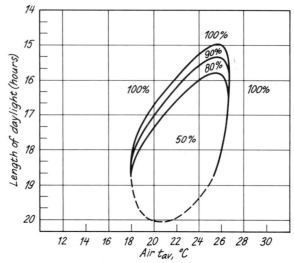

FIGURE 2.19.   Percentage of Colorado potato beetles going into diapause, in relation to day length and air temperature.

To plot such a prezone, we have to know what critical day length corresponds to 80% of females going into the diapause. It is clear that this day length will be greater for 100% of females gonig into the diapause.

The relationship between the percentage of beetles going into the diapause and the day length and temperature can be seen in Figure 2.19, which is constructed based on the work of L. I. Arapova.

Figure 2.19 is to be used for calculating and plotting on PTNs the subzones of critical day length which will lead to different numbers of individual Colorado potato beetles.

CHAPTER THREE
Phytophenological Forecasts and Bioclimatic Estimations

3.1. PHENOFORECASTING USING VARIOUS METHODS

3.1a. Phenological Forecasts with Prolonged Foresight for Cotton.

The first phenological forecast for cotton was given at the beginning of the section about yield forecasting. Using the actual (but not multi-annual) date of flowering for cotton variety 5904-I in Kurgan-Tyube on June 22, 1961, we calculated the ripening date to be August 28: June 22 + 67 days = August 28, where $n = 67$ days is the ordinate of intersection of the average temperature line from June 22 with phenocurve 9 on Figure 1.55.

The calculation was called "forecast," although the PTN is constructed on the basis of average multiannual data, and therefore, the temperature conditions during the interdevelopment stage do not relate the year of 1961 specifically.

Nevertheless, the calculation given above is specific to the given year and for the given farm, since its initial date integrates all influences of place and time. Thus the mass flowering date of June 22, which was given above for the purpose of finding the ripening date, was obtained as a result of the previous (before June 22) specific temperature trend in 1961, and as a result of the influence of other meteorological factors, as well as of the agricultural techniques used at the particular farm.

However, since such calculations are made on the basis of a multiannual net of average temperatures, they can also be called "phenoclimatic." To distinguish from the forecasts that will be discussed below, let us call the above calculation (made from the date of the initial developmental stage to the next stage) "forecast with prolonged foresight."

"Phenoclimatic" forecasts with prolonged foresight are the simplest kind of phenoforecasts, and they can be calculated very quickly.

In spite of this, and in spite of the climatic data used, the multi-annual experience of various authors shows that the exactness of phenoclimatic forecasts with prolonged foresight based on PTN is entirely acceptable for science (and for practical purposes) and considerably exceeds the exactness of analogous forecasts made by using the temperature totals method. This is also confirmed by the results of trials in Dushanbe (which are extensive and have undergone a careful mathematical appraisal).

On the basis of the PTN for the Dushanbe region, 188 phenoforecasts were calculated for the dates of sprouting, budding, flowering, and ripening in cotton variety 108-F *G. hirsutum* (the forecasts relate to 1947-1953, to the Hissar Valley, which has the most changeable weather conditions in Tadzhikistan, and to different sowing dates in each year). Then, all the calculations were compared with the actual dates observed on the experimental plots of the Dushanbe agricultural meteorological station or on neighboring fields.

The errors in the phenoforecasts, obtained as a result of these mass comparisons, were distributed according to their values, without relation to the stages of development, and the recurrence (frequency) of errors of each value was calculated. The results are represented in the form of a frequency spectrum of errors (Fig. 3.1).

Figure 3.1 is to be understood in the following way: among the 188 errors of different values, the error +3 days occurs 10 times (see point α), zero error occurs 40 times, and so on; the total of all frequencies is 188. As can be seen, the majority of actual errors in the phenoforecasts with prolonged foresight do not exceed ± 3-4 days, and almost a quarter of all forecasts (40 out of 188) are completely free from error. The root-mean-square error of phenoclimatic forecasts with prolonged foresight is $\sigma = \pm 3.7$ days.*

*Here, and everywhere following, standard (root-mean-square) deviations σ from average arithmetical errors of phenoforecasts are given. Since the average arithmetical errors approximated to zero, σ becomes the root-mean-square error of forecasts. Nevertheless, it is more expedient to use average errors, obtained as a result of first totaling all the errors without *distinction* of *positive* and *negative*, and then dividing this total by the number of errors. Such a calculation will be especially appropriate if the average arithmetical error of phenoforecasts (taking into account plus and minus) is not equal to zero. The average error (without taking into account plus and minus) is usually less than σ. This is why *the real level of average errors in phenoforecasts is even less (and often considerably less) than the values of σ which are indicated everywhere in this book.*

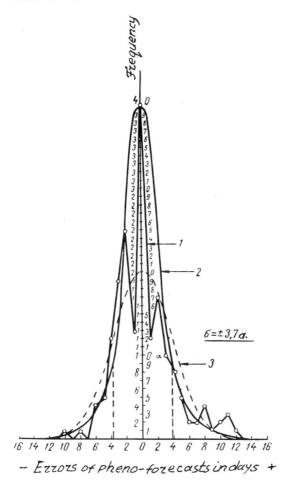

FIGURE 3.1.   Distribution of error frequencies (recurrence) in phenological forecasts with prolonged foresight (phenoclimatic) for cotton variety 108-F, on the basis of the new method. Curve 1: actual results of testings; curve 2: a smoothed-out curve of actual results; curve 3: theoretical curve of the normal distribution of frequencies, according to Gauss at $N = 188$ and $\sigma = \pm 3.7$ days.

According to the normal theoretical distribution of probabilities, the number of results that deviate from the arithmetical average by no more than $\pm \sigma$ is 68.26%.

If we find, using our actual data, the number of forecasts with errors not greater than ± 3.7 days, then it turns out that they make up a greater percentage of all cases—namely, 79.7%. Thus the actual curve showing the distribution of the frequency of errors is favorably different (in regard to the exactness of forecasts) from the curve of

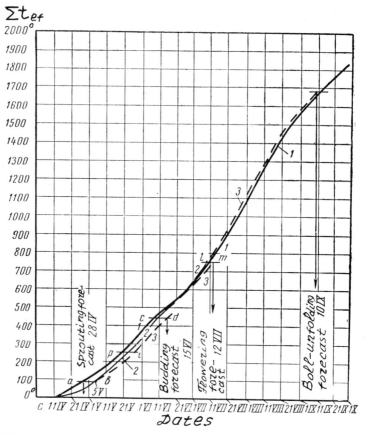

FIGURE 3.2. Forecasting of developmental dates for cotton variety 108-F using Babushkin's method; accumulation of the totals of effective temperatures in the Dushanbe district from April 11. Curve 1: using multiannual data; curve 2: using data for 1950; curve 3: extrapolated accumulation.

normal distribution. We can verify this by calculating the curve of normal distribution, using Gauss's equation (Fig. 3.1)*.

Thus the root-mean-square error in phenoforecasts with prolonged foresight made by using the new method is ± 3-4 days (with a "guarantee" that in 80% of the forecasts, the error will be equal to this or smaller), and the probable error $\sigma_w = \pm$ 1-2 days (with a 50%

*If we except large errors that occur as a result of poor agricultural techniques, and sometimes as a result of great temperature anomalies in individual years, then σ would be less, and the curve of the normal distribution of probabilities would more closely approximate the actual curve. Therefore, the first cause of the difference is the heterogeneity of erros, composing σ. The second cause is that errors of ± 0.5 days are approximated to 0 (and ± 1.5 days to 2), as is customary in meteorology and mathematics.

"guarantee"); the average foresight period of the forecast is 38 days (this is the length of the average interdevelopmental stage).

It should be noted that "forecasts" based on a simple augmentation of average multiannual intervals give errors up to ± 20 days.

Now we must compare all these results with those obtained by effective temperature totals method, on the basis of the same data (an example of a calculation based on the totals method is given in Fig. 3.2).

If we compare Figure 3.1 with Figures 3.3 and 3.4, it can be seen that by using the new method, out of a total 188 forecasts, there are 40 forecasts that have absolutely no error; using L. N. Babushkin's method, there are 31 with no error; using A. A. Shigolev's method [i.e., based on Equation (10)] there are 25 with no error. If we then look at the "tails" of the frequency spectra, in Figure 3.1 they are limited by errors of ± 10-12 days, in Figure 3.3—+25 and −14 days, in Figure 3.4—+18 and −37 days; it is characteristic that these latter, greater errors in the

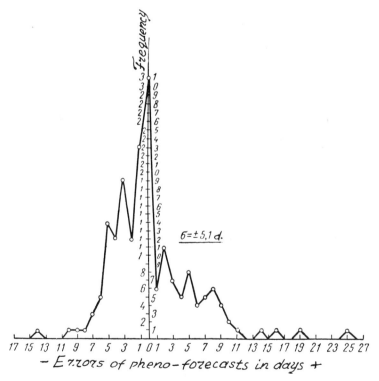

FIGURE 3.3. Distribution of error frequencies (recurrence) in phenological forecasts for cotton variety 108-F, calculated on the basis of effective temperature totals, (in Babushkin's version).

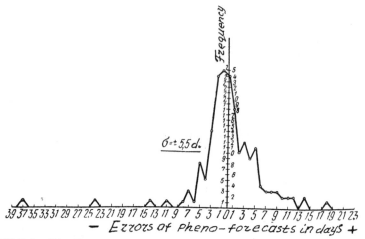

FIGURE 3.4. Distribution of error frequencies (recurrence) in phenological forecasts calculated on the basis of effective temperature totals (in Shigolev's version).

forecasts correspond to the small errors in forecasts made by using the new method.

The root-mean-square errors σ turned out to be equal to ± 5.1 days for Babushkin's method, and ± 5.5 days (or more) for Shigolev's method, in contrast to ± 3.7 days for the proposed method. It should be noted that if we take into consideration a number of additional cases which were not included in the above analysis, the root-mean-square error of forecasts using Babushkin's method increases to ± 5.9 days, and using Shigolev's method it increases to ± 7.1 days.

Thus the new method of phenological forecasts is more exact than the method of effective temperature totals. It should be pointed out that about 200 of the forecasts compared relate to normal sowing dates; with late and early sowing dates, the difference in the exactness of the methods is much greater.

The sources and causes of the greater exactness of the new method in comparison with the totals method are clear. Regarding the Babushkin method, the proposed method is more exact because it:

1. Automatically takes into account the inconstancy of totals and thresholds of effective temperatures within the bounds of one and the same interdevelopmental stage.

2. Does not come into conflict with the "optimum law" of an organism's development.

3. Automatically takes into account the effect of seasonal changes in light by means of phenological developmental curves based on experiments with sowing at different times.

4. Is based on the actual dates of the initial developmental stage (this cause is of lesser importance, see Section 3.1g).

5. Does not require the interpolation of the forecast developmental stage (the midpoint between two sowing dates is not necessarily correlated with the midpoint between the two corresponding dates for flowering or ripening, as is assumed in Babushkin's method).

As to Equation (10), by Shigolev and Babinet, the reduced accuracy in comparison with the proposed method occurs because of the causes indicated in terms 1, 2, and 3, as well as because the calculations are based not on the concrete, but on the average multiannual length and average multiannual limits of the interdevelopmental stage.

An appraisal of phenoforecasts made using Shigolev's method was also made by A. V. Protzerov and E. S. Ulanova (1961), using a large amount of data from different places in the USSR. This test showed that for a number of plants, developmental stages, and regions, the results of forecasting are unsatisfactory even with very favorable permissibility. It was thus considered excellent if errors not exceeding 20% of the foresight period of the forecast occurred in 80% of all forecasts. In that case, for our forecasts of the dates of cotton boll opening, on the basis of flowering dates, the permissibility of errors is as great as ± 14 days, because the foresight period in such forecasts is about 70 days. In fact the root-mean-square error of forecasts for cotton, using the new method, is ± 3.7 days, with an 80% guarantee, as was seen above.

The results of tests on the new method made after 1953 are given in Table 3.1. These forecasts correspond to the development of cotton variety 108-F in the fields of farms in the area of Dushanbe, Tadzhikistan. The trials were conducted by A. F. Podolyan-Brudnaya (Hydrometeorological Observatory, Tadzhikistan). All the forecasts given in Table 3.1 can be easily repeated on the nomogram for Dushanbe (Fig. 1.56). It is even simpler to compare the data of Table 3.1 with phenological calendars (Table 3.7-3.10).

In Table 3.1 the success of the phenoforecasts is appraised on a five-point scale, according to the instructions of the Hydrometeorological Service in the USSR. One type of mark or another is given in accordance with the correlation between the extent of the error and the foresight period of forecast (in the present case the latter is equal to the length of the interdevelopmental stage).

As can be seen, in the majority of cases an excellent level is achieved in the forecasts of budding, flowering, and ripening (including crops that were sowed unusually late; see, e.g, sowing on May 15, 1951). Forecasts of sprouting dates are less successful (the average mark is somewhat higher than fair). However, it is interesting that in Table 3.1

TABLE 3.1. Appraisal of the success of phenoclimatic forecasts with prolonged foresight for cotton variety I08-F using Podolsky's method (Lenin region in Tadzhikistan, according to A. F. Podolyan-Brudnaya, Hydrometeorological Service).

	Developmental Stage				
	Sowing	Sprouting	Budding	Flowering	Ripening
		1951			
Forecast	—	24 IV	5 VI	13 VII	21 IX
Actual date	4 IV	26 IV	10 VI	16 VII	16 IX
Deviation (in days)	—	+2	+5	+3	−5
Appraisal (in points)	—	5	4	5	5
		1951			
Forecast	—	27 IV	8 VI	12 VII	21 IX
Actual date	11 IV	28 IV	9 VI	15 VII	20 IX
Deviation (in days)	—	+1	+1	+3	−1
Appraisal (in points)	—	5	5	5	5
		1951			
Forecast	—	23 V	22 VI	21 VII	2 X
Actual date	15 V	21 V	21 VI	26 VII	5 X
Deviation (in days)	—	−2	−1	+5	+3
Appraisal (in points)	—	1	5	4	5
		1952			
Forecast	—	23 IV	1 VI	11 VII	7 IX
Actual date	1 IV	18 IV	8 VI	8 VII	3 IX
Deviation (in days)	—	−5	+7	−3	−4
Appraisal (in points)	—	1	4	5	5
		1953			
Forecast	—	27 IV	6 VI	10 VII	7 IX
Actual date	11 IV	29 IV	6 VI	8 VII	10 IX
Deviation (in days)	—	+2	0	−2	+3
Appraisal (in points)	—	4	5	5	5
		1954			
Forecast	—	24 IV	5 VI	8 VII	7 IX
Actual date	3 IV	26 IV	3 VI	7 VII	30 VIII
Deviation (in days)	—	+2	−2	−1	−8

TABLE 3.1. *(Continued)*

	Developmental Stage				
	Sowing	Sprouting	Budding	Flowering	Ripening
1954					
Appraisal (in points)	—	5	5	5	4
1955					
Forecast	—	28 IV	10 VI	16 VII	23 IX
Actual date	11 IV	6 V	14 VI	16 VII	20 IX
Deviation (in days)	—	+8	+4	0	−3
Appraisal (in points)	—	1	5	5	5
1956					
Forecast	—	27 IV	2 VI	30 VI	27 VIII
Actual date	12 IV	22 IV	26 V	30 VI	26 VIII
Deviation (in days)	—	−5	−7	0	−1
Appraisal (in points)	—	1	3	5	5
1957					
Forecast	—	7 V	12 VI	14 VII	11 IX
Actual date	26 IV	10 V	12 VI	10 VII	16 IX
Deviations (in days)	—	+3	0	−4	+5
Appraisal (in points)	—	2	5	4	5
1958					
Forecast	—	29 IV	7 VI	12 VII	11 IX
Actual date	14 IV	30 IV	10 VI	10 VII	8 IX
Deviation (in days)	—	+1	+3	−2	−3
Appraisal (in points)	—	5	5	5	5
1959					
Forecast	—	5 V	10 VI	12 VII	12 IX
Actual date	26 IV	4 V	12 VI	10 VII	12 IX
Deviation (in days)	—	−1	+2	−2	0
Appraisal (in points)	—	4	5	5	5

Average points given to a phenoforecast, excluding forecasts for sprouting: 4.8 (out of a possible 5). Average points given to a phenoforecast, including forecasts for sprouting: 4.4.

TABLE 3.2. Actual dates of occurrence of developmental stages for various crops in Iolotan (Turkmenistan) in 1960, forecast dates obtained using two methods, and an appraisal of the accuracy of the forecast (according to G. M. Ostapenko, 1962).

Date when Forecast Was Made (date of the stage preceding the forecast stage)	Forecast Stage	Date of the Forecast Stage		Actual Date of Occurrence of Stage	Foresight Period of Forecast (in days)	Appraisal of Accuracy (in points according to scale)	
		Using Totals of Effective Temperatures	Using Podolsky's Method			For Effective Temperature Totals Method	For Podolsky's Method
1	2	3	4	5	6	7	8
12 IV	Sprouting of cotton variety 213 (*G. barbadense*)	27 IV	24 IV	24 IV	12	2	5
24 IV	Budding of cotton	4 VI	1 VI	10 VI	47	4	3

Event						
Flowering of cotton	10 VI	2 VII	10 VII	10 VII	30	1	5
Boll unfolding of cotton	10 VII	6 IX	13 IX	8 IX	60	5	5
Flowering of early-ripening apricot	31 I	13 III	7 III	18 II	18	1	1
Unfolding of the first leaves of mulberry tree	22 II	20 III	3 IV	8 IV	46	1	5
Unfolding of the fifth leaves of mulberry tree	22 II	30 III	10 IV	—	—	—	—
				Total points		14	24

positive errors predominate in the forecasts of sproutings. This means that in the fields, sprouting occurred later than was forecast. The causes of the late sproutings are probably soil crust and insufficient soil moisture in the fields during the postsowing period. If Podolyan-Brudnaya had known the agrotechnical condition of a specific field, she would have used for forecasting, in the presence of soil crust, not phenological curve 1, but curve 2 (Fig. 1.2), which was constructed especially for such fields. Then the forecasts of the occurrence of sprouting would have been as accurate as for other developmental stages.

In Table 3.2 we give the results of the application of the experimental-mathematical method to the analysis of crop development in Turkmenistan (Ostapenko, 1962). There the method was tested and compared with the temperature totals method, not only for cotton, but also for other crops. The forecasts were phenoclimatic with prolonged foresight, made by using the PTN specifically constructed for Iolotan.

As can be seen from Table 3.2, the best results were given by Podolsky's method.

Podolsky's new method was also tested by A. R. Mustafayev, in the Kura-Arax lowland in Azerbaidzhan, for cotton variety S-4727 in the period between sprouting and flowering, using data from 75 forecasts. With an average foresight period of 58 days, $\sigma = \pm 4$ days. Phenoforecasts for apricots and mulberry trees were equally accurate.

Let us look at the results of testing the new method of forecasting for cotton in Israel. For this purpose the phenocurves from Figures 1.4 and 2.15 were transferred onto tracing paper, using the same scale as the net (better still, they can be transferred from Fig. 1.120, where the scales are already the same), and the tracing paper was superimposed successively onto the heat resources nets for different regions in Israel. Calculations were made in the same way as described above. The results are given in Table 3.3 (for practice, readers can make all these calculations themselves, checking their results against Table 3.3).

As can be seen from Table 3.3, phenoforecasts with prolonged foresight (phenoclimatic forecasts) using the new method are quite successful in Israel, too. The tests were made using related data (i.e., data used for constructing phenocurves). Unrelated data have not been accumulated yet. However, the first forecasting calculations for unrelated data show that the errors remain small, although increasing slightly.

3.1b. Phenoforecasts with Prolonged Foresight for Cereal Crops.

In Kazakhstan—which borders on Middle Asia—V. K. Adzhbenov (1972) carried out 430 phenoforecasts for the development of spring wheat variety Saratovskaya-29, after Podolsky's method, and using data from 1957-1968 (Figs. 1.5 and 1.66). The root-mean-square error $\sigma = \pm 3.9$ days. The guarantee of such an accuracy in the forecasts is 80%. Out of 430 forecasts, 94 were error free.

V. K. Adzhbenov (1979) continued testing the new method in 1974-1976 for wheat varieties Saratovskaya-29, Bezenchukskaya-98, and Kharkovskaya-46. Some results are shown in Table 3.4.

Unfortunately, Adzhbenov did not compare the forecasts based on the new method with forecasts based on the temperature totals method, because for the varieties mentioned above the heat constants (totals) are unknown, and the previously used varieties have now been superseded.

Phenoforecasts for spring wheat enabled Adzhbenov to solve a number of urgent problems, such as defining the degree of coincidence of wheat heading with the flight of *Apamea anceps Schiff-Hadena sordida Bkh*, which lay its eggs in heads of wheat; defining permissible dates for possible chemical treatment of crops, so as to avoid any risk of toxic substances appearing in bread; calculating the latest sowing dates that will still give a yield before the frosts; making everyday estimates of spring wheat development. We shall meet these problems in our subsequent discussion.

High-accuracy forecasts with prolonged foresight (without taking into account the actual temperatures in a specific year) in Middle Asia and Kazakhstan are favored by the settled summer weather. The errors in such forecasts for the European section of the USSR (ESU) are somewhat larger, however, errors in phenoforecasts based on the new method are smaller than those based on the temperature totals method.

O. S. Yermakov compared these two methods in Table 3.5.

Forecasts using Shigolev's method were carried out (Table 3.5) on the basis of totals of effective temperatures given by Shigolev: in the period from heading to wax ripeness for spring wheat variety Lutescens-62, $A = 490°$, and for spring barley variety Viner, $A = 410°$; in the period from the appearance of panicles to wax ripeness for oat variety Pobeda, $A = 378°$. The lower thresholds of effective temperatures are $+5°$ (according to Shigolev-Gasparen).

As can be seen in the total lines of Table 3.5, the new method is more successful than the temperature totals method (more successful even for wheat, if we take into consideration the many errors of 12-18 days made when using the temperature totals method).

It should be noted that O. S. Yermakov calculated a record number of different phenoforecasts, based on the new method, for many crop varieties and regions of ESU (about 1100 forecasts).

Successful testings of the new method were made by departments of the Hydrometeorological Service in the Ukraine, in the Central Black-Earth Zone (CBEZ), and in the Trans-Baikal region (Siberia, Far East)

TABLE 3.3. Phenoforecasts based on Podolsky's method, and the actual (*Gossypium hirsutum L.*) (Israel, 1978).

District	Air Heat Resources Net based on Data from the Nearest Representative Meteorological Station	Actual Date of Planting	Forecast of Germination (50%) based on Actual Date of Planting	Actual Date of Germination (50%)	Error in the Forecast of Germination	Forecast of Budding (1st bud) based on Actual Date of Germination	Actual Date of Budding (1st bud)	Error in the Forecast of Budding	Forecast of Flowering (1st flower) based on Actual Date of Budding	Actual Date of Flowering (1st flower)
1	2	3	4	5	6	7	8	9	10	11
1. Maagan Michael	Gan Shomeron	2 IV	+8 = 10 IV	10 IV	0	+42 = 22 V	22 V	0	+27 = 18 VI	19 VI
2. Maagan Michael	Gan Shomeron	3 IV	+8 = 11 IV	10 IV	−1	+42 = 22 V	26 V	+4	+26 = 21 VI	19 VI
3. Magal	Gan Shomeron	1 IV	+9 = 10 IV	10 IV	0	+42 = 22 V	21 V	−1	+27 = 17 VI	14 VI
4. Magal	Gan Shomeron	4 IV	+8 = 12 IV	12 IV	0	+42 = 24 V	24 V	0	+27 = 20 VI	20 VI
5. Gan Shomeron	Gan Shomeron	31 III	+9 = 9 IV	9 IV	0	+43 = 22 V	22 V	0	+27 = 18 VI	20 VI
6. Negba	Negba	31 III	+9 = 9 IV	8 IV	−1	+41 = 19 V [a]	No data	—	+26 = 14 VI	21 VI
7. Negba	Negba	5 IV	+8 = 13 IV	12 IV	−1	+40 = 22 V [a]	No data	—	+25 = 16 VI	19 VI
8. Ein-Harod	Tirat Zvi	28 III	+7 = 4 IV*	No data	—	+43 = 17 V	18 V	+1	+21 = 8 VI	No data
9. Ein-Harod	Tirat Zvi	2 IV	+7 = 9 IV*	No data	—	+48 = 27 V	19 V	−8	+20 = 8 VI	11 VI
10. Ramat David	Kefar-Blum	3 IV	+8 = 11 IV	9 IV	−2	+40 = 19 V	No data	—	—	No data
11. Ramat David	Kefar-Blum	4 IV	+8 = 12 IV	10 IV	−2	+40 = 20 V	3 VI	+14	+20 = 23 VI	25 VI
12. Ramat David	Kefar-Blum	18 IV	+7 = 25 IV	25 IV	0	+43 = 7 VI*	No data	—	+20 = 27 VI	26 VI
13. Beit-Shean	Tirat Zvi	22 III	+8 = 30 III	No data	—	—	No data		—	9 VI
14. Hulata	Kefar-Blum	6 IV	+8 = 14 IV	13 IV	−1	+40 = 23 V	20 V	−3	+23 = 12 VI	20 VI
15. DAN	Kefar-Blum	13 IV	+7 = 20 IV	19 IV	−1	+42 = 31 V	14 VI	+14	+20 = 4 VII	29 VI
16. Amir	Kefar-Blum	5 IV	+7 = 12 IV	11 IV	−1	+40 = 21 V	2 VI	+12	+20 = 22 VI	19 VI
17. Gonen	Kefar-Blum	13 IV	+7 = 20 IV	23 IV	+3	+43 = 5 VI	7 VI	+2	+20 = 27 VI	3 VII
18. Gadot	Kefar-Blum	7 IV	+7 = 14 IV	14 IV	0	+40 = 24 V	24 V	0	+22 = 15 VI	16 VI
19. Lehavot Habashan	Kefar-Blum	7 IV	+7 = 14 IV	15 IV	+1	+40 = 25 V	26 V	+1	+20 = 15 VI	15 VI

Average errors, without distinction of positive and negative	/0.9/	/4.3/
Average foresight period of pheno-forecasts (days)	7.8	41.9
Percentage of errors in forecasts in relation to foresight period (%%)	11.5	10.2
Percentage of excellent and good forecasts (%%)	94	79

[a] Calculated dates that are taken as actual (for lack of actual data) for purposes of further calculation and comparison.

for spring wheat, spring barley, and oats, and in the Tadzhikistan department of the Hydrometeorological Service for cotton, all during recent years. These were mass testings, using multiannual data. In every region the Technical Councils decided "to recommend Podolsky's method for application." The department of the CBEZ made a negative decision only in the case of oats. However, for the same oats in the Ukraine a positive decision was reached (probably in the Ukraine the quality of the observations was better).

Forecasts of the developmental dates of crops are necessary for planning agricultural work, especially for the allocation of machines and people at harvest time. For example, according to data from V. Korbut and V. Kovdan (1976), the harvesting losses of wheat are 2%

dates of development (obtained by A. Marani) of cotton variety Acala S.J.-2

Error in the Forecast of Flowering	Forecast of ripening (10 open bolls/m²) based on actual date of 1st Flower	Actual Date of Ripening (10 bolls/m²)	Error in the Forecast of Ripening (10 bolls/m²)	Forecast of Ripening (50 bolls/m²) based on Actual Date of 1st Flower	Actual Date of Ripening (50 bolls/m²)	Error in the Forecast of Ripening (50 bolls/m²)	Forecast of Ripening (50 bolls/m²) based on Actual Date of Ripening (10 bolls/m²)	Actual Date of Ripening 50 (bolls/m²)	Error in the Forecast of Ripening (50 bolls/m²)
12	13	14	15	16	17	18	19	20	21
+1	+60 = 18 VIII	16 VIII	−2	+80 = 7 IX	2 IX	−5	+21 = 6 IX	2 IX	−4
−2	+60 = 18 VIII	21 VIII	−3	+80 = 7 IX	7 IX	0	+20 = 10 IX	7 IX	−3
−3	+60 = 13 VIII	22 VIII	+9	+79 = 1 IX	13 IX	+12	+20 = 11 IX	13 IX	+2
0	+61 = 20 VIII	19 VIII	−1	+80 = 8 IX	7 IX	−1	+20 = 8 IX	7 IX	−1
+2	+61 = 20 VIII	23 VIII	+3	+80 = 8 IX	9 IX	+1	+20 = 12 IX	9 IX	−3
+7	+64 = 24 VIII	23 VIII	−1	+85 = 14 IX	12 IX	−2	+21 = 13 IX	12 IX	−1
+3	+65 = 23 VIII	23 VIII	0	+85 = 12 IX	12 IX	0	+20 = 12 IX	12 IX	0
—		No data			No data			No data	
+3		No data			No data			No data	
—		No data			No data			No data	
+2		No data			No data			No data	
−1	+66 = 31 VIII	31 VIII	0	+86 = 20 IX	20 IX	0	+20 = 20 IX	20 IX	0
—	+56 = 4 VIII	3 VIII	−1	+66 = 14 VIII	14 VIII	0	+10 = 13 VIII	14 VIII	+1
+8		No data			No data			No data	
−5		No data			No data			No data	
−3		No data			No data			No data	
+6		No data			No data			No data	
+1		No data			No data			No data	
0		No data			No data			No data	
/2.9/		/2.2/			/2.3/				/1.7/
23.1		61.4			80.1				19.1
12.6		3.6			2.9				8.9
81		88.9			88.9				88.9

when harvesting is 3 days late and 20% when harvesting is 12 days late. Besides this, forecasting the development and ripening dates of crops helps to estimate future yield: the phenoexpression of cereal crops in arid years does not indicate the probability of large yields, whereas early ripening in cotton gives the most profitable yield.

Damage to crops by pests is often connected with specific crop developmental stages, and it is possible, therefore, to calculate beforehand when the crop should be sown in order to avoid pests. In the same way, the sowing date can be selected to avoid other harmful influences, particularly drought.

Comparing forecasts for the multiannual develomental dates of agricultural crops with the actual development in a given field in a

TABLE 3.4. Phenoforecasts for spring wheat compared with the actual data for the Turgai region in Kazakhstan.

Years	Mass Heading		Wax Ripeness	
	Forecast	Actual Date	Forecast	Actual Date
1974	7 VII	6 VII	5 VIII	4 VIII
1975	7 VII	7 VII	6 VIII	6 VIII
1976	5 VII	10 VII	17 VIII	15 VIII

given year, and ascertaining the causes of slower or faster development in a given year help the agronomist to orient himself in his field work, and so on.

In conclusion, in Table 3.6 we give the results of a 10-year testing of the mathematical phenoforecasting method. These results were obtained and published in the Far East by V. V. Golovin (1959). Golovin, like all the authors mentioned above, based his testing on calculations of phenoclimatic forecasts with prolonged foresights (even superfluously prolonged).

Golovin summarizes as follows: "As can be seen from the Table, over the last 10 years the deviation of calculated date of spring wheat's wax ripeness from the actual date does not exceed 1-3 days when the average foresight period is 78 days. Only in 1954 was the deviation 4 days. In 1954 in the third 10-day period of May, the average air temperature was 2° lower than the multi-annual, and in the first and second 10-day periods of June, 0.7° lower than the multi-annual temperature."

How can we explain the high accuracy of phenoforecasts with prolonged foresight obtained by taking into account the actual (for a specific year and a specific farm) date of an initial developmental stage, but without taking into account concrete (for a given year) temperatures during the interval between the initial developmental stage and the forecast developmental stage (because the net is constructed on the basis of multiannual temperatures)? Apparently, this can be explained by "biological inertia" in crop development. As already mentioned, the actual date of an initial developmental stage is the result of preceding conditions and of the concrete conditions for a specific year. These conditions "dictate" the development rate of the plants. Plants cannot change their developmental rate suddenly, even when the weather changes sharply (or they risk death), just as a loaded cart, moving in a certain direction at a certain speed, cannot turn suddenly if it receives a bump from one side (if the bump is very forceful, there is a risk that the cart will turn over).

TABLE 3.5. Comparison of phenoforecasts with prolonged foresight based on the new method and on the temperature totals method (Shigolev's variant) for the European section of the USSR

Station, District	Crop, Variety	Year	Actual Date of the Initial Developmental stage (heading)	Actual Date of Wax Ripening	Forecast for Wax Ripening based on Temperature Totals Method	Error Using the Totals Method (days)	Forecast for Wax Ripening based on the New Method	Error Using the New Method (days)
1	2	3	4	5	6	7	8	9
Krotovka, Kuybishevsky district	Spring wheat, Lutescens-62	1948	18 VI	20 VII	19 VII	+1	18 VII	+2
"	"	1950	20 VI	25 VII	21 VII	+4	20 VII	+5
"	"	1952	30 VI	31 VII	31 VII	0	30 VII	+1
Sernovodsk, Kuybishevsky district	Spring wheat, Lutescens-62	1948	26 VI	22 VII	28 VII	−6	27 VII	−5
"	"	1950	26 VI	6 VIII	28 VII	+9	27 VII	+10
"	"	1951	24 VI	20 VII	26 VII	−6	25 VII	−5
"	"	1952	30 VI	31 VII	30 VII	+1	1 VIII	−1
Krasnoe poseleniye, Kuybishevsky district	Spring wheat, Lutescens-62	1948	22 VI	24 VII	25 VII	−1	24 VII	0
"	"	1949	20 VI	20 VII	7 VIII	−18	22 VII	−2
"	"	1950	22 VI	26 VII	28 VII	−2	24 VII	+2
"	"	1951	26 VI	30 VII	29 VII	+1	28 VII	+2
"	"	1952	24 VI	31 VII	27 VII	+4	26 VII	+5
Kuybishev,	Spring wheat,	1948	20 VI	16 VII	22 VII	−6	21 VII	−5

TABLE 3.5. (*Continued*)

Station, District	Crop, Variety	Year	Actual Date of the Initial Developmental stage (heading)	Actual Date of Wax Ripening	Forecast for Wax Ripening based on Temperature Totals Method	Error Using the Totals Method (days)	Forecast for Wax Ripening based on the New Method	Error Using the New Method (days)
1	2	3	4	5	6	7	8	9
Kuybishevsky district	Lutescens-62							
"	"	1950	28 VI	12 VII	30 VII	−18	2 VII	+10
"	"	1951	26 VI	31 VII	28 VII	+3	27 VII	+4
"	"	1952	26 VI	24 VII	28 VII	−4	27 VII	−3
Inza, Ulyanovsky district	Spring wheat, Lutescens-62	1948	18 VI	24 VII	23 VII	+1	21 VII	+3
"	"	1952	30 VI	31 VII	3 VIII	−3	3 VIII	−3
Melekess, Ulyanovsky district	Spring wheat, Lutescens-62	1949	25 VI	28 VII	28 VII	0	26 VII	+2
"	"	1950	5 VII	15 VIII	6 VIII	+9	5 VIII	+10
"	"	1951	25 VI	25 VII	28 VII	−3	26 VII	−1
"	"	1952	20 VI	25 VII	23 VII	+2	22 VII	+1
Rochchino, Leningradsky district	Spring wheat, Lutescens-62	1952	20 VII	14 IX	5 IX	+9	4 IX	+10
"	"	1953	30 VI	14 VIII	16 VIII	−2	9 VIII	+5
"	"	1954	4 VII	16 VIII	20 VIII	−4	15 VIII	+1
Krasnoe poseleniye, Kuybishevsky district	Oats, Pobeda	1949	30 VI	31 VII	30 VII	+1	29 VII	+2
"	"	1950	22 VI	22 VII	22 VII	0	21 VII	+1
"	"	1951	4 VII	2 VIII	3 VIII	−1	1 VIII	+1

Location	Crop	Year						
Krotovka, Kuybishevsky district	Oats, Pobeda	1949	30 VI	31 VII	28 VII	+3	30 VII	+1
Gorkiy, Gorkiy district	Oats, Pobeda	1948	29 VI	31 VII	2 VIII	−2	2 VIII	−2
"	"	1949	28 VI	30 VII	1 VIII	−2	1 VIII	−2
"	"	1950	3 VII	12 VIII	6 VIII	+6	6 VIII	+6
"	"	1951	2 VII	31 VII	5 VIII	−5	5 VIII	−5
Vologda-Molochnoe, Vologodsky district	Spring barley, Viner	1952	10 VII	2 VIII	15 VIII	−13	10 VIII	−8
Starodub, Bryansky district	Spring barley, Viner	1964	20 VI	10 VII	22 VII	−12	19 VII	−9
"	"	1963	30 VI	20 VII	1 VIII	−12	27 VII	−7
"	"	1962	24 VI	20 VII	26 VII	−6	23 VII	−3
"	"	1961	20 VI	20 VII	22 VII	−2	20 VII	0
"	"	1960	16 VI	10 VII	19 VII	−9	15 VII	−5
"	"	1959	22 VI	18 VII	24 VII	−6	22 VII	−4
"	"	1958	26 VI	26 VII	28 VII	−2	25 VII	+1
"	"	1957	18 VI	18 VII	21 VII	−3	18 VII	0
"	"	1955	30 VI	26 VII	1 VIII	−6	29 VII	−3
"	"	1950	16 VI	18 VII	9 VII	+9	17 VII	+1
"	"	1948	10 VI	2 VII	13 VII	−11	10 VII	−8
"	"	1969	6 VII	2 VIII	5 VIII	−3	4 VIII	−2
Karachev, Bryansky district	Spring barley, Viner	1959	4 VII	28 VII	5 VIII	−8	31 VII	−3
"	"	1956	30 VI	3 VIII	31 VII	+2	30 VII	+4
"	"	1955	14 VII	10 VIII	16 VIII	−6	11 VIII	−1
"	"	1954	6 VII	30 VII	7 VIII	−8	3 VIII	−4
"	"	1953	8 VII	6 VIII	9 VIII	−3	5 VIII	+1
"	"	1952	6 VII	28 VII	7 VIII	−10	3 VIII	−6

TABLE 3.5. *(Continued)*

Station, District	Crop, Variety	Year	Actual Date of the Initial Developmental stage (heading)	Actual Date of Wax Ripening	Forecast for Wax Ripening based on Temperature Totals Method	Error Using the Totals Method (days)	Forecast for Wax Ripening based on the New Method	Error Using the New Method (days)
1	2	3	4	5	6	7	8	9
Trubchevsk, Bryansky district	Spring barley, Viner	1965	30 VI	26 VII	29 VII	−3	28 VII	−2
"	"	1961	24 VI	18 VII	23 VII	−5	23 VII	−5
"	"	1960	18 VI	16 VII	18 VII	−2	17 VII	−1
"	"	1954	28 VI	26 VII	28 VII	−2	26 VII	0
"	"	1953	2 VII	4 VIII	1 VIII	+3	1 VIII	+3
"	"	1948	20 VI	25 VII	21 VII	+4	19 VII	+6

Average errors without distinction of positive and negative |5.0| |3.5|
Arithmetical average, M +3.1   +0.09
Standard deviation (from M), σ ±6.0   ±4.5
Errors in phenoforecasts, $M \pm \sigma$ 3.1 ± 6.0   0.09 ± 4.5

TABLE 3.6 Dates of wax ripeness of spring wheat calculated on the PTN, using the sprouting dates in Zavitinsk district of Amur region, in comparison to the actual dates over a number of years (from V. V. Golovin, 1959).

Year	Sprouting Date	Wax Ripeness Date	
		Actual	Calculated
1949	10 V	28 VII	31 VII
1950	8 V	28 VII	30 VII
1951	10 V	30 VII	31 VII
1952	16 V	4 VIII	1 VIII
1953	28 V	8 VIII	6 VIII
1954	29 V	30 VII	3 VIII
1955	18 V	2 VIII	2 VIII
1956	20 V	4 VIII	3 VIII
1957	12 V	2 VIII	1 VIII
1958	16 V	2 VIII	1 VIII

3.1c. Forecasts with Prolonged Foresight for Cultured and Wild Melliferous Plants, Medicinal Plants, Plant Allergens, Forest Berries, and Other Wild Plants.

When will common buckwheat flower in Ryazan if it was sown in 1970 on May 22? On Figure 1.96 we find the line of average temperatures with initial date May 22 (this is shown by the dashed line between the lines for May 15 and 25). The point of intersection of this line with phenocurve 1 for buckwheat, from sowing to flowering, has ordinate 36 days (see the horizontal dashed line). Therefore, the beginning of flowering occurs on June 27:

$$\text{May } 22 + 36 \text{ days} = \text{June } 27$$

Actually the buckwheat began to flower in 1970 on June 29. The error of the forecast is +2 days. This problem was solved in this way by T. N. Gavrilova (1972, Beekeeping Institute) on the basis of Podolsky's method.

If the main honey harvest on a farm is provided by common buckwheat, then the strongest bee families must be prepared exactly for June 27.

It is also important to know the sowing dates of cultured melliferous plants, which begin to flower when the flowering of wild melliferous plants is over. This problem can be solved by a retrospective (reverse) calculation, which will be discussed in more detail in the appropriate section.

Gavrilova used the Podolsky's method for cultured melliferous plants—common buckwheat and Phacelia (Fig. 1.96). V. A. Ryabov considerably extended the sphere of application of the new method in beekeeping. He constructed phenocurves for wild melliferous plants: for 11 species of herbs and 5 species of trees (Fig. 1.30 and others). Having superimposed these phenocurves onto the heat resources net of a Central Black-Earth Reserve (near Kursk), he calculated a large number of phenoforecasts with a foresight period of 40-70 days for a 15-year period, and came to the conclusion that the accuracy of these phenoforecasts was quite satisfactory.

On the basis of the PTN, Ryabov made phenoforecasting calendars for melliferous plants.

.   .   .

Phenoforecasting is of great importance for medicinal plants. The accumulation of the active elements of these plants is strictly related to very definite stages of development. Phenological forecasting helps to regulate the purveyance of medicinal raw materials in various natural climatic regions.

For example, Podolsky's method was used for the May lily, which is the raw material for preparations used in cases of cardiovascular diseases (1974). The phenocurves from Figures 1.25a and b were transferred onto the Kursk heat resources net. The average error in the forecasts over a 14-year period (1956-1970), without distinction of positive and negative, was 2.9 days for the beginning of flowering, and 3.1 days for the end of flowering.

On the basis of a nomogram we also made phenoforecasting calendars with temperature anomalies for the May lily (1977).

The very popular medicinal plant valerian was the subject of forecasting research by Ryabov. He constructed a number of pheno-curves that showed a rather close correlation between the temperature and a number of developmental stages (Fig. 1.26). Unfortunately, the correlation was not very close between the temperature and the last period—from seed ripening to the death of the plants—just when the underground part of valerian gives the most valuable raw materials. However, valerian is also gathered after flowering or after seed ripening, and here the forecasts are reliable: the errors in forecasts with prolonged foresight are ±4-5 days, and forecasts with shortened foresight ±2-3 days (see the next section for details).

In Figures 1.27 to 1.29 the reader will also see other medicinal plants.

Nevertheless, there is also a negative opinion: A. A. Juchko (1975) thinks that calculations should be abandoned in favor of observations. But what shall we do about forecasting? An observation states an accomplished fact, but not the future situation.

.   .   .

It is well known how severely some people suffer from allergies, and that the number of allergic diseases increases every year. Very often the allergens are wild or cultured plants. In the USSR, they include, for instance, wormwood (*Artemisia*), couch grass (*Agropyrum repens*), bluegrass (*Poa*), cocksfoot (*Dactylis glomerata*), saltbush (*Atriplex*), timothy grass (*Phleum*), fescue grass (*Festuca*), brome (*Bromus*), ambrosia (*Ambrosia*)—a dangerous quarantined plant that is "advancing" from the south to the north and has already reached the regions near Kursk, tall oat grass (*Arrhenatherum*), maize (*Zea*), rye (*Secale cereale*), hemp (*Cannabis*), sunflower (*Helianthus*), and other plants. All of the above and many other plants cause diseases during the flowering period.

A number of allergens are represented by their phenocurves in Chapter 1; in particular, see Figures 1.27 to 1.29.

It is clear that these phenocurves, if transferred onto the heat resources nets, will enable us to forecast the flowering dates of allergens in various regions. Such forecasts will give an opportunity to patients and doctors to prepare in advance for the outbreak of allergic diseases. Patients, for example, knowing their allergens, will plan the place and date of their holidays in order to avoid these allergens.

.   .   .

The first to use Podolsky's method for forecasting the developmental dates of wild berries was A. F. Cherkasov (1975). He constructed a PTN with phenocurves for a number of species of forest berries for the northern section of the ESU (Fig. 1.102; curves for other species of berries, after V. B. Gedikh, have also been introduced into this figure by the author of this book).

Podolsky's method was compared by Cherkasov with the method involving a simple augmentation of the average multiannual interval between stages of development, and then with the method based on totals of active and of effective temperatures. The methods were tested using a wide range of data (1923-1968).

The active temperatures method was unsuitable for forecasting: totals of active temperatures fluctuated sharply from year to year, in full agreement with the theory described here in Section 2.1.

The method of augmenting interdevelopmental stages also proved unsuitable: for example, in 1934 from the beginning of flowering to the beginning of ripening in the rowan tree, 86 days elapsed, whereas in 1944, 65 days elapsed; for the bilberry we have, in 1947, 64 days, and in 1948, only 39; and so on.

Of the other two remaining methods—the effective temperature totals method and the Podolsky's method—Cherkasov preferred the latter method, because of the low errors and scientific accuracy. For example, the effective temperature totals method proved to be unsuitable for phenocurves 4 and 5 (Fig. 1.102). These phenocurves have both optimal and above-optimal segments (see the theory described in Section 2.1).

For wild berries in Byelorussia Podolsky's method was applied by V. B. Gedikh (1976). It is very interesting that Gedikh calculated the spring frost zone apparently on his own (an example of such a zone is plotted by the author in Fig. 1.50). If the line of average temperatures intersects the phenocurve of the period, for example, from the time when the temperatures exceed 0° to the opening of the buds, in the zone of late spring frosts, then the buds will be "overtaken" by frost, and prospects for the harvest will not be good, according to Gedikh.

It is also interesting that Gedikh classifies phenocurves according to their steepness, which indicates the degree of the plant's reaction to temperature. This idea was developed—independently of Gedikh—by Ryabov. Only Gedikh's slip about a hyperbola gives cause for perplexity. A phenocurve over a wide range of temperatures is never a hyperbola, because there is an after-optimal branch of the phenocurve. Within a narrower range of temperatures, it is sometimes possible to express the phenological curve by *two* hyperbolas with different parameters, joined to each other. The artificial approximation of every type of phenological curve to a hyperbola is a deceptive habit.

A PTN was also used by M. N. Nazarova and S. I. Mashkin (1979) for analyzing phenocurves and for making phenoforecasts for five varieties of wild and cultured cherry (genus Cerasus Juss). The work was carried out in Voronezh (the central part of the ESU). For the analysis of phenocurves, see Section 3.4 about the biological "passport" system. In reference to phenoforecasts, the authors of the article reported that they achieved a high accuracy. For example, the average multiannual date for the first appearance of vegetative cone on the cherry trees is April 24. Therefore, the average multiannual flowering date, according to calculations made on a PTN, is May 12, and according to the actual multiannual observations it is May 11.

Nazarova and Mashkin note that exact phenological forecasts are important to plant introduction and in crop breeding.

. . .

The largest number of works on the application of Podolsky's method for studying wild plant ecology were written by Ryabov (1974, 1975, 1977-1979). He constructed more than 150 phenological curves for various developmental stages of 52 species of herbs, and 8 species of trees and shrubs. A number of phenocurves are given in this book (Figs. 1.26-1.30). Ryabov used abundant observed data from the Central Black-Earth Reserve* for classifying phenological curves according to

---

*Phenological observations were carried out by V. S. Zhmikhova and others; meteorological observations were made by the Reserve's meteorological station, under the direction of Ryabov.

the degree and sign (+, −) of the plant's reaction to heat. (The work was carried out under the guidance of the author of this book). At the same time the idea was expressed that movements of phenocurves on the field of the graph in the course of time could be a special biological indicator for physical changes in the environment. In this way, phenocurves would serve as an aid in the observation and control of the environment.

This will be discussed in the section about the biological passport system (Section 3.4); here we are interested in forecasting aspects. Ryabov showed, using mass data (about 1000 phenoforecasts), that in the majority of cases, the errors in phenoforecasts with prolonged foresight (30-50 days' forecasting period or more) do not exceed ± 4 days.

It is unfortunate that Ryabov did not compare the new method with other methods, especially with the temperature totals method.

This, however, was done by N. E. Buligin (1979). He showed, dealing with the small-leaved lime (*Tilia cordata*) in the Leningrad suburbs, that the average error in phenoforecasts of lime flowering, using Podolsky's method, is ± 5.3 days, and using the temperature totals method, the error is ± 7.3-7.5 days. These forecasts were made without taking into account the actual temperatures (i.e., they are pheno-climatical forecasts with prolonged foresight according to Podolsky's terminology) and in the extremely variable weather conditions of Leningrad.

Buligin also compared forecasts based on Podolsky's method with forecasts based on his own method. The average error with Buligin's method was smaller, namely ±3.7 days. This is one of two cases known to me (for the second one, in the case of the piorcer, *Carpocapsa pomonella L*, see Section 4.1b), when the results on PTN are of lower accuracy than other results. In this case, Buligin constructed a statistical graph of the relation between the *dates* of the initial developmental stage and the *dates* of the forecast developmental stage for a given plant. He accounted for his success by the fact that such a graph reflects, in latent form, the influence of not only temperatures, but also other "background" factors. However, our phenological curve also reflects the influence on the developmental rate of not only temperature, but also other factors in the region, data on that were used for constructing the phenocurve.

Thus the relative success of the forecasts by Buligin remains unexplained.

In addition the statistical relationship between two successive stages of development of a given plant, according to Buligin, is bound to the region for which it is obtained, and it cannot be transferred to another region where there are other temperatures. This is under-standable, since on Buligin's graph the temperature does not appear as such. However, using Podolsky's method, we can transfer phenocurves from one heat resources net to a net for any other region.

In another article (1977), Buligin compared Podolsky's method with the phenological intervals method (or logs, according to G. E. Schultz's terminology) in both his version and that of Shultz. As is known, the logs method was proposed by Shultz in 1936; this method consists of choosing, for the plant that interests us, another plant—a plant indicator—such that the indicator's developmental stages occur earlier than the developmental stages of the plant that interests us. If we know the multiannual average interval between the occurrence of a given developmental stage in the first plant and its occurrence in the plant indicator, it is possible to forecast the date when the developmental stage will occur in the first plant. This method was taken by Shultz, as the author understands it, to be the basis of "indicatory phenology."

Buligin, in turn, suggested relating the dates of the developmental stages of the plant that interests us (y) and of the plant indicator (x) by means of the statistical equation $y = ax + b$, instead of adding to x the multiannual average interval $(y - x)$.

This author does not understand how the forecasts based on Podolsky's method were compared with forecasts based on the logs method: these kinds of forecasts are entirely different. In Podolsky's method, the forecast is made from the date of the preceding developmental stage to the date of the next stage of *the same* plant. In the methods used by Shultz and Buligin, the forecast is made between the dates of the developmental stages in *different plants*, and often between stages of the same one. Therefore foresight periods of forecasts given by the two methods (Podolsky's and Shultz-Buligin's) are different.

In any event, the logs method suffers from the shortcoming mentioned above: neither the average multiannual log nor Buligin's equation can be transferred from one region to another.

Let us return to Ryabov. As was already mentioned, the average errors using Podolsky's method did not exceed, in most cases, ± 4 days. However, in some cases the errors turned out to be greater. Ryabov analyzed these cases and showed that they are related primarily to sharp temperature anomalies (e.g., in 1974). By applying the *short-term forecast* method (this method was developed by Podolsky in 1958 and later), large errors are reduced to ± 2-3 days. This gives sufficient accuracy for planning the dates for gathering the fruits and seeds of wild plants, and for carrying out forestry work.

Phenoforecasting calendars facilitate the practical application of phenoforecasts. Ryabov made such calendars for 30 develomental stages of species of wild trees, shrubs, and herbs.

But how does the phenoforecast method with shortened foresight give this increased accuracy? This question will be discussed next.

3.1d.   High-Accuracy Phenoforecasts with Shortened Foresight.

One sowing of cotton variety 108-F was carried out on April 1 in 1950 in Dushanbe. Mass sprouting occurred on May 4. However, the pheno-climatic forecast with prolonged foresight, calculated from April 1 using the nomogram for Dushanbe (Fig. 1.56), gave sprouting on April 24 (the line of average temperatures that corresponds to sowing on April 1 intersects phenological curve 11—for cotton variety 108-F from sowing to mass sprouting—at $n = 33$ days; therefore, the date of mass sprouting should be April 1 + 23 days = April 24). Thus the error of the forecast was 10 days.

This error could have been due to a number of causes, which can be divided into two groups: (1) not taking into account the temperatures in the given year; (2) nontemperature causes (the influence of other meteorological factors, of agricultural techniques, and even of geo-metrical errors when constructing the nets and the curves).

Therefore, we must first take into account the actual temperature trend in April 1950. If, parallel to the annual air temperature trend in Dushanbe (this trend was constructed on the basis of multiannual average data), we plot a temperature trend using the data for 1950, we will distinctly see that the temperature trend of April 1950 lies below the multiannual trend, but it remains more or less parallel to the latter. Therefore, in 1950 the air temperatures in April were lower than the norm and could indeed have caused the delay of sprouting in the fields.

Using the graph of the temperature trend in 1950 (it is not given here), or better still, using the meteorological table of 24-hour tempera-tures directly, let us calculate the actual line of average temperatures from April 1 in 1950. The average 24-hour temperature on April 1, 1950, was 15.9° C (we can obtain this datum only from the meteorological table, and not from the graph of the annual trend). The average temperature over the 5-day period (from April 1-5 inclusive) can be easily calculated using the meteorological table: $t_{av} = 10.3°$C. The average temperature from the same initial date of April 1 over a 10-day period (from April 1-10 inclusive) will be exactly equal to the average temperature over the first 10-day period of April: $t_{av,10} = 11.7°$ C. The average temperature from the same initial date, April 1, 1950, over a 20-day period (from April 1-20) is $t_{av,20} = 11.5°$ C. It is not necessary to total all the average 24-hour temperatures from the meteorological table, from April 1 to 20 inclusive; it is sufficient to divide by 2 the total of the average temperatures over the first and second 10-day periods of April, as taken from the meteorological table or from the graph:

$$\frac{11.7° C + 11.2° C.}{2} = 11.5° C$$

The average temperature from April 1, 1950, over 30 days is $t_{av} = 12.1°\,C$.

$$\left[\frac{11.7°\,C \;(\text{1st 10-day period}) + 11.2°\,C \;(\text{2nd 10-day period}) + 13.4°\,C \;(\text{3rd 10-day period})}{3} \right.$$

$$\left. =12.1°\,C \right] \qquad \text{and so on}$$

Next we plot all the results in Figure 1.56 in the form of the dashed broken line $fqhik$ (see Fig. 1.56, between approximately 11° and 14°C, and between 0 and 40 days). Point f with coordinates $t_{av} = 10.3°\,C$ and $n = 5$ days represents the average temperature from April 1 over 5 days; point q with coordinates $t_{av} = 11.7°\,C$ and $n = 10$ days represents the average temperature from April 1 over 10 days; and so on. (In Fig. 1.56, the initial section of the line from $t = 15.9°\,C$ is not plotted, so as not to overload the figure.)

The line $fqhik$ is the actual trend of average temperatures over various periods in 1950, with an initial date of April 1. As can be seen, this trend deviates considerably to the left of the multiannual line of average temperatures dated April 1, thus indicating the negative temperature anomaly in 1950.

The actual line of average temperatures, $fqhik$, related to the sowing date of April 1, intersects phenological curve 11, which represents the period from sowing to sprouting (with normal agricultural techniques), when $n = 33$ days. Hence the date of mass sprouting is April 1 + 33 days = May 4.

This calculated result coincided exactly with the date observed in nature. Therefore, the previous error of 10 days, which occurred in the phenoclimatic forecast with prolonged foresight, was due to the fact that the temperature factor was taken into account to an insufficient degree; nontemperature meteorological factors, agricultural techniques, and so on, in this particular case, did not affect the result.

The calculation described above, of the actual line of average temperatures from April 1, 1950, might have broken off, for instance, at point h which corresponds to April 21 (April 1 + 20 days), if we had not known the temperatures from April 22 onward. Not knowing these temperatures, we should have extrapolated the actual line fqh from point h onward. It would be most logical to make this extrapolation parallel to the multiannual line of average temperatures dated April 1, because peculiar as the temperature trend was in 1950, its tendency cannot differ constantly from the tendency of a normal temperature trend. This extrapolation is shown by the fine dotted line hr (Fig. 1.56).

Line hr intersects phenological curve 11 when $n = 30$ days; hence the date of mass sprouting, according to the phenoforecast with shortened foresight, will be April 1 + 30 days = May 1, whereas the actual date was May 4.

Therefore, the phenoforecast with shortened foresight reduced the error of the phenoclimatic forecast with prolonged foresight from 10 days to 3 days. However, the phenoclimatic forecast had a foresight period of 33 days (sowing on April 1; actual sprouting on May 4). The phenoforecast with shortened foresight had a foresight period of 14 days, because it was made not from the date of the developmental stage preceding the forecast stage, but from April 21, up to which date the actual temperatures are used.

The forecast with shortened foresight can be simplified considerably. First there is no need to calculate the entire trend of average temperatures based on actual dates; this was done here only for the purpose of visual clarity. It is sufficient to calculate only a single point h, which shows the actual average temperature over the 20-day period—from April 1 to 20 inclusive. It was already shown above that this average temperature is to be calculated from the average 10-day period temperatures, taken from the annual trend or directly from the meteorological tables:

$$t_{av,20} = \frac{t_{av} \text{ over the 1st 10 days of April} + t_{av} \text{ over the 2nd 10 days of April}}{2}$$

$$= \frac{11.7°\,C + 11.2°\,C}{2} = 11.5°\,C$$

Let us put onto the nomogram for Dushanbe (Fig. 1.56) a point with coordinates $t_{av} = 11.5°\,C$ and $n = 20$ days (this will be point h); from this point we draw a line parallel to the multiannual line dated April 1, and then we find the intersection of the line being drawn and the developmental curve, and so on.

For forecasting of this kind we use both the "biological inertia" of an organism's development (this was discussed in the previous paragraph) and the weather inertia. Moreover, even sharp weather changes cannot significantly alter the actual average temperature, which has already been formed during a part of the forecast period.

In the same way, that is, taking into account the actual temperatures over a part of the forecast period, almost all the forecasts with prolonged foresight available were recalculated as forecasts with shortened foresight (15-30 days) relating to the development of cotton variety 108-F in the Dushanbe region for 1947-1953.

The mean-root-square error of the forecasts with shortened foresight decreased (as could have been expected), but not to a large extent for plants in Middle Asia: $\sigma = \pm3.4$ days (as compared to ±3.7 for forecasts with prolonged foresight). The "guarantee" of these or smaller errors in the forecasts is 80%, that is, as good as for forecasts with prolonged foresight. The probable error with a 50% guarantee is $\sigma_w = \pm1.5$ days. Here we do not give the frequency spectrum of errors for forecasts with shortened foresight, but, like the spectrum for forecasts with prolonged foresight, it has a high-peak excess.

Even better results from the application of high-accuracy forecasts were obtained by V. K. Adzhbenov (1971) in Kazakhstan: the mean-root-square error of the forecasts decreased from ±3.9 to ±2 days for spring wheat.

However, phenoforecasts with shortened foresight taken on an importance of their own when solving ecological problems related, in particular, to plant protection against pests; moreover, such forecasts are important for regions where the weather and climate are more changeable than in Middle Asia.

It is possible to simplify even further the technique of forecasts with shortened foresight, so as to dispense with a special calculation of point h. It is sufficient to inquire at the local meteorological station about the temperature anomaly existing at the time that concerns us. Then we move the multiannual average temperature line needed for the calculations to an extent that corresponds to the value of this temperature anomaly.

Finally, it is possible to take into account actual temperatures without shortening the foresight period. We have only to use the actual temperature anomaly over the period preceding the forecast period: in our example, for instance, over the last 10-day period of March. Of course, the accuracy of such a forecast will be higher than that of a phenoclimatic forecast, but lower than the accuracy of a forecast with shortened foresight.

3.1e. Calculating the Dates of Plant Development under Field Conditions on the Basis of Temperatures Known Beforehand (phenosynoptical reliable forecast).

It is necessary to master this method of calculation for the purpose of solving a variety of operational problems concerning the volume and causes of the acceleration or deceleration of a plant's development in a given year, as well as for solving a number of problems in malariology.

This kind of calculation has been discussed previously. "Reliable forecasts" involve obtaining a certain line of average temperatures entirely on the basis of the actual data for a given year (see line $fqhik$ in Fig. 1.56). The intersection of such an actual line of average temperatures with the phenocurve that concerns us (in Fig. 1.56, with curve 11) gives the solution to the problem about the date of an organism's development.

However, here again it is quite obvious that there is no need to calculate the entire line of average temperatures ($fqhik$). It is sufficient to calculate only two points, i and k, in such a way that first, these points will be close to developmental phenocurve 11, and second, they will lie one on each side of curve 11. Thus the line joining these two points must intersect the phenocurve. The nearer to each other and to the phenocurve that points i and k are located, the less interpolation required, and the more exact the result.

The initial orientation as to where these points should lie is obtained by finding the ordinates of intersection of the corresponding multi-annual line of average temperatures with the biological curve. It is not important whether the first point is placed above or below the developmental phenocurve. If it is placed below, then for calculating the second point we have to take the longer period of time. If placed above, then we must take the shorter period.

In our experiments (in total, 188 experiments) there were 61 error-free forecasts using this method (phenosynoptical reliable forecasts), as compared to 40 error-free forecasts with prolonged foresight, and as compared to the figure of 30 given by the curve of normal frequency distribution corresponding to phenosynoptical reliable forecasts. The mean-root-square error of phenosynoptical reliable forecasts is $\sigma = \pm 2.6$ days, with a guarantee of more than 80%; the probable error is $\sigma_w = \pm 0.7$ days.

This kind of forecast is named "phenosynoptical reliable" because it is as if it were based on exact synoptical (weather) forecasts. Real weather forecasts will be discussed in the next section.

3.1f. Phenosynoptical Forecasts.

Let us consider phenoforecasts based on the use of regular long-range weather forecasts; the temperature will be given not in absolute values, but in the form of deviations from the norm, as is customary in long-range forecasting.

In this case, the average temperature over any period in a given year, as given by the weather forecast, will differ from the average temperature of the same period, as based on multiannual data, by exactly the value of the temperature deviation ($\Delta°C$) indicated in the weather forecast.

Various multiannual average temperatures are represented in relation to the length of the period n; this relationship, for each date from which the reckoning of n starts, is expressed by a line of average temperatures. If we draw a line parallel to the designated multiannual line, and at a distance of $\Delta°C$, then we obtain the supposed (i.e., as given by the long-range weather forecast) line of average temperatures (in relation to n) for a given year. The intersection of this line with a biological curve will give (following a simple calculation that has been used repeatedly before) the phenological forecast, taking into account the long-range synoptical forecast. For this purpose we do not need to start the line of average temperatures based on the long-range weather forecast directly from the axis of abscissae; it is sufficient to draw only a section of it, which will intersect the developmental curve concerning us.

For example, the mass flowering of cotton variety 108-F at the meteorological station in Dushanbe is recorded on July 14, 1952. When will the mass unfolding of the first seed bolls occur, if, according to the seasonal weather forecast, the temperature is supposed to be 1° C lower than the norm in July, near to the norm in August, and 1° C lower than the norm in September?

The interdevelopmental stage from flowering to the unfolding of the first bolls is approximately two months, that is, in our case, from approximately mid-July to mid-September. Therefore, we can assume that the average temperature of this entire period is 0.5° C lower than the norm. In fact, half of July plus half of September is the equivalent of one month when the temperature is 1° C lower than the norm; all of August has zero deviation; therefore, we have an average over the two months of 0.5° C lower than the multiannual average.

We mark with letters ab the section (nearest to biological curve 13) of the multiannual line of average temperatures from July 14 (see the place with approximate coordinates of 23° C and 70 days in Fig. 1.56). To the left of ab (i.e., on the side of the lower temperatures) we draw a parallel line cd at a horizontal distance from ab of 0.5° C according to the scale. The intersection of this parallel line with phenological curve 13 gives $n = 70$ days; hence the date of mass unfolding of the first bolls will be July 14 + 70 days = September 22. The actual date of mass unfolding of the first bolls was September 23.

It should be noted that such successful results were, in the author's experience, not always obtained with phenosynoptical forecasting. Unfortunately, long-range weather forecasts are not always successful in every region. Therefore, the application of long-range weather forecasts causes a deterioration in the results of phenoclimatic calculations with prolonged foresight. For example, the calculation of phenosynoptical forecasts and their errors for Tadzhikistan, according to all the available data for cotton in 1947-1953, showed that $\sigma = \pm 5.0$ days; such errors are guaranteed to occur in 70.5% of forecasts.

This author checked the monthly forecasts of temperatures over a 10-year period (1947-1956) in Tadzhikistan; 51% of them turned out to be right, but 49% were wrong.

Thus in Tadzhikistan, with its complicated orography, phenosynoptical forecasts proved to be the least successful, in comparison with the kinds of forecasts that have been discussed or will be discussed next.

3.1g.   Transitive Phenoforecasts.

Under this heading we refer to calculating the date of a particular developmental stage, using the preceding developmental stage, which has also been calculated in the same way (the term "transitivity," from the Latin *transitivus*, is adopted by this author from mathematics).

For example, it is necessary to define, in multiannual terms, when the mass unfolding of the first cotton boll variety 108-F in Dushanbe will occur, if it was sown on April 6. On the PTN for Dushanbe (Fig. 1.56) we can see that the line of average temperatures of April 6 intersects phenological developmental curve 11 for cotton variety 108-F for the period from sowing to mass sprouting at $n = 19$ days; this means that mass sprouting will occur on April 6 + 19 days = April 25.

Using this calculated date and phenocurve 12, we obtain in the same way the date of mass budding: April 25 + 41 days = June 5.

Using this date and developmental curve 8, we obtain the date of the next developmental stage, mass flowering: June 5 + 34 days = July 9.

Finally, using the calculated flowering date and phenocurve 13, we calculate the date of mass unfolding of the first bolls: July 9 + 63 days = September 10. September 10 is, then, the result of this transitive phenoforecast based on multiannual temperatures, and, like all the intermediate results, it coincides closely with the results of multiannual phenological observations.

Such calculations can be versatile in their application in bioclimatology and agroclimatology. They are also needed for long-term agricultural planning in new agricultural regions.

Sometimes it is necessary to make a transitive phenoforecast based on the actual temperatures in a given year (the actual temperatures relating to a part of the forecast period or to the entire period). Since we are dealing with actual temperatures, it is clear that elements of the transitive forecast will be carried out by means of the phenosynoptical reliable forecasting, which was discussed above.

Data for cotton accumulated from 1947 to 1953 were analyzed for the purpose of transitive forecasting based on actual temperatures. (However, when forecasting from the second to the last developmental stage to the last one, the calculation used was phenoclimatic with prolonged foresight, on the basis of multiannual temperatures.) This analysis showed the mean-root-square error of such forecasts to be $\sigma = \pm 3.9$ days; such errors or smaller ones are guaranteed in 76.5% of forecasts.

As can be seen, transitive phenoforecasts are not much less accurate than phenoclimatic forecasts with prolonged foresight, for which $\sigma = \pm 3.7$ days and the "guarantee" is 79.7%. The heat factor has such a great influence on the developmental dates of plants that it appears to be possible in practice, without prejudicing the accuracy, to use for the forecasts not the actual dates of those developmental stages taken as the initial ones for the calculations, but in many cases, the dates calcualted on the basis of actual temperatures. This is what L. N. Babushkin (1951, 1949) did: for the initial dates for phenoforecasts, he used not the observed dates, but the dates calculated on the basis of actual temperatures.

3.1h. Retrospective Calculations of the Dates of an Organism's Development.

In a number of cases it is necessary to know how to find the dates of preceding developmental stages on the basis of a given developmental stage. For example, such a process is necessary for solving the following questions: when is it necessary to sow in order that the interdevelopmental stage that interests us will have a given average temperature (Section 3.1i)?; what is the latest sowing date that will still allow time for a given crop to ripen (Section 3.1i)? The method used for such calculations must be mastered for the purpose of solving a number of ecological problems.

The problem of finding the date of the preceding developmental stage on the basis of the date of a given developmental stage can be solved in two ways: (1) by a calculation based either on multiannual temperatures, or (2) by a calculation based on the actual temperatures in the year concerned.

For example, it is necessary to find the date of mass flowering for cotton variety 5904-I in Kurgan-Tyube, if the mass unfolding of cotton bolls occurred on September 9 (all the temperatures are similar to multiannual temperatures).

On the PTN for Kurgan-Tyube, based on multiannual observations (Fig. 1.55), phenocurve 9 corresponds to the cotton's development from mass flowering to mass boll unfolding. We find the points where developmental curve 9 intersects the summer lines of average temperatures (when cotton flowers normally); we find the interdevelopmental stage periods n that correspond to these intersections; then we add the dates of the average temperature lines to the periods n and obtain the dates of boll unfolding. For example, the line of average temperatures dated June 25 gives, at the intersection with curve 9, for the date of boll unfolding June 25 + 68 days = September 1; the line dated July 5 gives July 5 + 69 days = September 12, and so on.

Having obtained the first dates of boll unfolding, we can see whether it is necessary to continue the search, or whether among the dates obtained we already have the given date of boll unfolding (September 9). For instance, on Figure 1.55, near phenocurve 9, we see that the marks for September 1 and 12 are those which interest us, because between them we have the given date of September 9. The mark for September 25 can be useful as a guide to tendencies, when interpolating, but the mark for October 8 is not needed.

Thus the date of boll unfolding, September 9, which was given in the problem, is located between the marks for September 1 and 12, at a distance from September 12 (it may be supposed) of one quarter of the geometric interval between these marks. However, taking into consideration the fact that the space between the next pair of marks (September 12 and 25) is greater, although the difference in time between the points is approximately the same, we must move the September 9 mark somewhat to the right. We mark it by means of a short line segment parallel to the line of average temperatures, and we see that it corresponds to the line of average temperatures dated July 2. That is the date of mass flowering, found on the basis of multiannual initial dates and corresponding to the ripening date of September 9: July 2 + 69 = September 9. Such a picture can be seen from the average multiannual data given in the *Agroclimatic Reference Book* of Tadzhikistan (1959, p. 100).

Now let us solve the same problem not on the basis of multiannual data, but in relation to a concrete year with concrete temperatures. Here it should be noted that it is advisable to convert these actual temperatures to the value of the deviations from the multiannual norm; this will greatly facilitate further calculations, and, moreover, the method of the solution will then prove to be analogous to the method of solving the problem using the long-range weather forecast.

Therefore, which flowering date is connected with the ripening date of September 9 for cotton in a given year, if the temperatures in July and August in this year are 2° higher than the multiannual norm?

Since the ripening date is the same (September 9), and the temperatures are higher than the norm, then the flowering, obviously, will occur not on July 2 (as with temperatures equal to the multiannual temperatures), but somewhat later (the cotton will ripen more quickly).

In accordance with the positive temperature anomaly of the given year, we transfer those lines of average temperatures, necessary for the solution, 2° further to the right (according to the scale) (Fig. 1.55). From the previous calculations it is known (although we can also manage without them) that the necessary temperature lines are those dated June 25, July 5, and possibly, July 15. We transfer only the lines of (June 25)′ and (July 5)′. As before, we find the ordinates n of the intersections of these new lines with biological curve 9; then we add these ordinates to the dates of the average temperature lines and obtain the dates for cotton ripening: (June 25)′ + 66 days = August 30; (July 5)′ + 67 days = September 10.

We mark the results (August 30 and September 10) by short line segments close to curve 9. Then, in the space between these results we look for the given ripening date of September 9. This is in the immediate proximity of September 10, and therefore, it relates to almost the same flowering dates as the ripening date of September 10, that is, July 5. A more exact interpolation indicates that cotton ripening on September 9 corresponds to flowering on July 4. *This is the solution of a retrospective problem using the actual temperatures in a given year.* As was expected, the same ripening date of September 9 can, in warm years, follow a flowering date later (July 4) than the multiannual one (July 2).

It should be noted that retrospective solutions can also be formed without the use of geometric constructions—by choosing a suitable date for the developmental stage that precedes the given one. The exact selection can be made successfully after two or three attempts.

Finally, the solving of retrospective problems can be maximally simplified by the use of phenological tables or calendars.

3.1i. Calculating Phenological Calendars; Solving Ecological Problems by Using Calendars and Nomograms, and the Theory of Sowing Dates.

For each organism and region, phenoforecasting calendars can be compiled, using the corresponding PTN. In the first column of such a calendar, the dates of the initial developmental stage are placed in consecutive order, and opposite them, in the next column, are noted the corresponding dates of the succeeding developmental stage as obtained from a nomogram. When doing this we can take into account temperature anomalies and nontemperature factors.

For example, on the PTN for Dushanbe (Fig. 1.56) it can be seen that the sowing date of March 26 corresponds to the following mass sprouting date: March 26 + 27 days = April 22, where $n = 27$ is an ordinate of the intersection of the average temperature line from the initial date of March 26 with phenocurve 11 (sowing to sprouting), assuming normal agricultural techniques. This is a "phenoclimatic" forecast with prolonged foresight. It is based entirely on multiannual temperatures, that is, it relates to zero temperature anomaly. In the second column of Table 3.7, headed $0°$ C, we read, opposite the sowing date of March 26 (26 III), mass sprouting date April 22 (22 IV).

If the temperatures differ from the multiannual temperatures by $+1°$ C, then in the column headed $+1°$, opposite the same sowing date of March 26, we read April 17 (17 IV). The latter date is obtained by transferring the multiannual meteorological line for March 26 (Fig. 1.56) $1°$ C further to the right, according to the scale. This can be easily done if the legs of a drawing guage are moved apart by a distance of $1°$ on the scale of the figure, then the left leg is positioned on the line of March 26, and the drawing guage is moved in this way (not changing the horizontal position of the guage's spread!) up to the moment when the right leg touches phenocurve 11. The point where it touches the curve will have the corresponding ordinate $n = 22$ days, which gives March 26 + 22 days = April 17.

If the temperature anomaly, as compared to the average multiannual temperature, is $-1°$, then the right-hand leg of the drawing guage is positioned on the meteorological line dated March 26. If the anomaly is $2°$, we increase the spread of the guage's legs, and so on.

Next the sowing date of March 31 can be taken as the initial date

TABLE 3.7. Phenoforecasting calendar (and a reference table) of mass sprouting of cotton variety 108-F in relation to sowing dates and to temperature anomalies, in fields with normal agricultural techniques and in fields with insufficient moisture and with soil crust, for Dushanbe, Tadzhikistan ($h = 803$ m).

	Normal Agricultural Techniques						
	Temperature Anomalies (°C)						
Sowing	0	+1	+2	+3	−1	−2	−3
11 III	25 IV	17 IV	10 IV	5 IV	—	—	—
26 III	22 IV	17 IV	14 IV	12 IV	26 IV	2 V	8 V
31 III	23 IV	19 IV	16 IV	14 IV	26 IV	2 V	7 V
5 IV	25 IV	21 IV	18 IV	17 IV	27 IV	2 V	7 V
10 IV	27 IV	24 IV	22 IV	20 IV	29 IV	2 V	7 V
15 IV	29 IV	27 IV	25 IV	24 IV	1 V	4 V	8 V
20 IV	3 V	30 IV	29 IV	28 IV	4 V	6 V	9 V
23 IV	5 V	3 V	1 V	30 IV	6 V	8 V	10 V
25 IV	6 V	4 V	3 V	2 V	7 V	9 V	12 V
28 IV	8 V	6 V	5 V	5 V	10 V	11 V	14 V
30 IV	9 V	8 V	7 V	7 V	11 V	13 V	15 V
5 V	13 V	12 V	12 V	11 V	15 V	17 V	18 V
10 V	18 V	17 V	17 V	16 V	19 V	20 V	22 V
15 V	22 V	22 V	21 V	21 V	23 V	24 V	26 V
20 V	26 V	26 V	26 V	26 V	27 V	28 V	30 V
26 V	31 V	31 V	31 V	31 V	1 VI	2 VI	3 VI

	Soil Crust and Insufficient Moisture						
	Temperature Anomalies (°C)						
Sowing	0	+1	+2	+3	−1	−2	−3
11 III	—	—	—	—	—	—	—
26 III	3 V	27 IV	23 IV	19 IV	9 V	18 V	27 V
31 III	3 V	28 IV	24 IV	21 IV	9 V	15 V	24 V
5 IV	3 V	29 IV	26 IV	22 IV	9 V	15 V	22 V
10 IV	4 V	1 V	28 IV	25 IV	9 V	14 V	20 V
15 IV	6 V	3 V	29 IV	27 IV	10 V	15 V	20 V
20 IV	8 V	5 V	2 V	30 IV	11 V	16 V	20 V
23 IV	10 V	7 V	4 V	3 V	13 V	17 V	21 V
25 IV	11 V	8 V	6 V	5 V	14 V	18 V	22 V
28 IV	12 V	10 V	8 V	7 V	15 V	20 V	22 V
30 IV	13 V	11 V	10 V	9 V	16 V	20 V	23 V
5 V	17 V	15 V	14 V	13 V	19 V	22 V	25 V
10 V	20 V	19 V	19 V	18 V	22 V	25 V	29 V
15 V	25 V	24 V	23 V	23 V	26 V	28 V	1 VI
20 V	29 V	28 V	28 V	27 V	30 V	1 VI	3 VI
26 V	3 VI	2 VI	2 VI	1 VI	3 VI	5 VI	6 VI

TABLE 3.7. (*Continued*)

For a Temperature Anomaly of 0°C Only and for
Normal Agricultural Techniques

Sowing	Sprouting	Sowing	Sprouting
31 V	5 VI	5 VIII	11 VIII
5 VI	11 VI	10 VIII	14 VIII
10 VI	16 VI	15 VIII	19 VIII
15 VI	21 VI	20 VIII	25 VIII
20 VI	26 VI	26 VIII	1 IX
25 VI	1 VII	31 VIII	6 IX
30 VI	8 VII	5 IX	11 IX
5 VII	17 VII	10 IX	17 IX
10 VII	21 VII	11 IX	19 IX
15 VII	26 VII	12 IX	20 IX
20 VII	29 VII	13 IX	21 IX
26 VII	3 VIII	14 IX	23 IX
31 VII	7 VIII	15 IX	24 IX

(interpolating this line on Fig. 1.56, between the lines of March 26 and April 5). The results are recorded in Table 3.7.

Data between the lines in Table 3.7 that correspond to the sowing dates of March 26 and 31 can be either interpolated or omitted when a calendar is compiled; those who use the calendar can easily make this interpolation themselves.

All of the above concerns normal agrotechnical conditions. For fields with insufficient moisture, and with soil crust during the period from sowing to sprouting, we use the second part of Table 3.7 (although in Fig. 1.56 there is no corresponding phenocurve, but this can be transferred from Fig. 1.2).

How is such a phenoforecasting calendar used?

Let us suppose that, in a particular year, cotton variety 108-F was sown in Dushanbe, in a field with normal agrotechnical conditions, on March 26 (this is not a usual sowing date, but a rather early one, when the agronomist's experience cannot help him in forecasting the future development of the cotton). We need to define when mass sprouting will occur. The temperature anomaly is not yet known; therefore, we assume, at first, that the temperature trend is identical to the multi-annual one. Then, in the column with temperature anomaly 0°, we find the date of mass sprouting to be April 22—opposite the sowing date of March 26.

The calculation described above is considered to be a preliminary phenoforecast, made as early as March 26. A more precise forecast can be made, for instance, 10 days later, having measured the actual average air temperature for the period from March 26 to April 4 inclusive, and compared it with the multiannual temperature over the same period. This will make it possible to obtain the temperature anomaly for the given year. Information about the temperature anomaly can also be obtained from the local meteorological station. Let us assume that the anomaly is $+2°$ C. Then in Table 3.7, opposite the date of March 26 and in the column designated $+2°$, we find the phenoforecast to be April 14. It is clear that the more days used for taking into account the actual temperature, the more exact is the phenoforecast, but the shorter is its foresight period. However, on March 26 it is already possible to take into account the temperature anomaly for the previous period, for example, from March 14 to 25.

Of course, if $+2°$ is given by the long-range weather forecast (for the end of March and beginning of April), and we trust this forecast, then the latter can be taken into account in calendars in the same way as above.

In a similar way we can calculate calendars for other developmental stages of cotton. The results give us the phenoprognostic calendars in Tables 3.8 to 3.10.

It should be noted that the range of cotton developmental dates in Tables 3.7 to 3.10 is wider than the range of dates observed in practice. This is so for the purpose of solving problems that arise in experimental work and in crop breeding.

In Table 3.10 phenological dates and the prefrost yield (in percent) are given. The latter is obtained from Figure 2.12, in accordance with, on the one hand, ripening dates (Table 3.10) and, on the other hand, the average multiannual date of the first autumn frost in the air of Dushanbe (November 2). As is known, the first autumn frosts stop the vegetation of cotton. For example, from ripening on September 14 until November 2, 49 days elapse; we find these 49 days on the axis of abscissae in Figure 2.12. From the 49-day point we trace a vertical line upward to the curve of the graph, and a horizontal line from the curve to the axis of ordinates, where we find 85%. We insert the latter figure in Table 3.10 near the ripening date of September 14.

Of course, we can compile such phenological calendars for any geographical region and for many organisms (plants, pests, poikilothermic carriers, and pathogen of illnesses) whose development depends on temperature or on any other factor with seasonal variability. For this purpose it is only necessary to have phenotemperature (or phenohumidity, etc.) nomograms.

Nevertheless, it is not always necessary to compile calendars for each region. If we know the *spatial* temperature "anomaly" of a given region in comparison to the "base" region for which we have a phenoforecasting calendar with *temporal* anomalies, we can use the latter. Let us assume that a phenological problem is to be solved for a region that is $+1°$ C warmer than the base region. Then, even if the

TABLE 3.8. Phenoforecasting calendar (and a reference table) for mass budding of cotton variety 108-F in relation to the dates of mass sprouting and to the temperature anomalies for Dushanbe, Tadzhikistan (h = 803 m).

Sprouting	Temperature Anomalies (°C)						
	0	+1	+2	+3	−1	−2	−3
5 IV	31 V	25 V	19 V	15 V	4 VI	11 VI	—
15 IV	1 VI	27 V	23 V	20 V	6 VI	11 VI	17 VI
16 IV	1 VI	27 V	24 V	21 V	6 VI	11 VI	18 VI
17 IV	2 VI	28 V	24 V	21 V	6 VI	11 VI	18 VI
18 IV	2 VI	28 V	25 V	22 V	6 VI	11 VI	18 VI
19 IV	3 VI	29 V	25 V	22 V	6 VI	11 VI	18 VI
20 IV	3 VI	29 V	26 V	23 V	7 VI	12 VI	18 VI
21 IV	4 VI	30 V	26 V	23 V	7 VI	12 VI	18 VI
22 IV	4 VI	30 V	27 V	24 V	7 VI	13 VI	18 VI
23 IV	5 VI	31 V	27 V	24 V	8 VI	13 VI	18 VI
24 IV	5 VI	31 V	28 V	25 V	8 VI	14 VI	18 VI
25 IV	6 VI	1 VI	28 V	25 V	9 VI	14 VI	18 VI
26 IV	6 VI	1 VI	29 V	26 V	9 VI	14 VI	19 VI
27 IV	6 VI	2 VI	29 V	27 V	9 VI	14 VI	19 VI
28 IV	7 VI	2 VI	30 V	27 V	10 VI	14 VI	19 VI
29 IV	7 VI	3 VI	31 V	28 V	10 VI	14 VI	19 VI
30 IV	7 VI	3 VI	31 V	29 V	11 VI	15 VI	20 VI
1 V	7 VI	4 VI	1 VI	30 V	11 VI	15 VI	20 VI
2 V	8 VI	4 VI	2 VI	30 V	12 VI	16 VI	20 VI
3 V	8 VI	5 VI	2 VI	31 V	12 VI	16 VI	21 VI
4 V	9 VI	5 VI	3 VI	1 VI	13 VI	17 VI	21 VI
5 V	10 VI	6 VI	4 VI	2 VI	13 VI	17 VI	22 VI
6 V	10 VI	7 VI	4 VI	3 VI	13 VI	18 VI	22 VI
7 V	10 VI	7 VI	5 VI	3 VI	14 VI	18 VI	22 VI
8 V	11 VI	8 VI	6 VI	4 VI	14 VI	18 VI	23 VI
9 V	11 VI	9 VI	6 VI	5 VI	15 VI	18 VI	23 VI
10 V	12 VI	10 VI	7 VI	6 VI	15 VI	19 VI	23 VI
11 V	13 VI	10 VI	8 VI	7 VI	16 VI	19 VI	24 VI
12 V	14 VI	11 VI	9 VI	7 VI	16 VI	20 VI	24 VI
13 V	14 VI	12 VI	9 VI	8 VI	17 VI	20 VI	24 VI
14 V	15 VI	13 VI	10 VI	9 VI	17 VI	21 VI	25 VI
15 V	16 VI	13 VI	11 VI	10 VI	18 VI	21 VI	25 VI
16 V	16 VI	14 VI	12 VI	11 VI	18 VI	22 VI	25 VI
17 V	17 VI	15 VI	13 VI	12 VI	19 VI	22 VI	26 VI
18 V	18 VI	16 VI	13 VI	12 VI	19 VI	23 VI	26 VI
19 V	18 VI	16 VI	14 VI	13 VI	20 VI	23 VI	26 VI
20 V	19 VI	17 VI	15 VI	14 VI	21 VI	24 VI	27 VI
21 V	20 VI	18 VI	16 VI	15 VI	21 VI	24 VI	27 VI
22 V	20 VI	19 VI	17 VI	16 VI	22 VI	25 VI	28 VI
23 V	21 VI	20 VI	18 VI	17 VI	23 VI	26 VI	28 VI
24 V	22 VI	20 VI	19 VI	19 VI	23 VI	26 VI	29 VI
25 V	22 VI	21 VI	20 VI	20 VI	24 VI	27 VI	30 VI

TABLE 3.8. *(Continued)*

Sprouting	\multicolumn{7}{c}{Temperature Anomalies (°C)}						
	0	+1	+2	+3	−1	−2	−3
26 V	23 VI	22 VI	21 VI	21 VI	25 VI	28 VI	30 VI
27 V	24 VI	23 VI	22 VI	22 VI	26 VI	28 VI	1 VII
28 V	25 VI	24 VI	23 VI	23 VI	26 VI	29 VI	2 VII
29 V	25 VI	25 VI	24 VI	24 VI	27 VI	30 VI	2 VII
30 V	26 VI	25 VI	25 VI	25 VI	28 VI	30 VI	3 VII
31 V	27 VI	26 VI	26 VI	26 VI	29 VI	1 VII	4 VII
1 VI	28 VI	27 VI	27 VI	27 VI	29 VI	2 VII	4 VII
2 VI	29 VI	28 VI	28 VI	29 VI	30 VI	2 VII	5 VII
3 VI	29 VI	29 VI	29 VI	30 VI	1 VII	3 VII	6 VII
4 VI	30 VI	30 VI	30 VI	1 VII	1 VII	4 VII	6 VII
5 VI	1 VII	1 VII	1 VII	2 VII	2 VII	4 VII	7 VII
6 VI	2 VII	2 VII	2 VII	3 VII	3 VII	5 VII	8 VII
7 VI	3 VII	3 VII	3 VII	5 VII	4 VII	6 VII	8 VII
8 VI	4 VII	4 VII	4 VII	6 VII	5 VII	7 VII	8 VII
9 VI	5 VII	5 VII	5 VII	7 VII	6 VII	7 VII	10 VII
10 VI	6 VII	6 VII	6 VII	8 VII	7 VII	8 VII	10 VII
11 VI	7 VII	7 VII	7 VII	10 VII	8 VII	9 VII	11 VII
12 VI	8 VII	8 VII	9 VII	11 VII	8 VII	10 VII	12 VII
13 VI	9 VII	9 VII	10 VII	13 VII	9 VII	10 VII	13 VII
14 VI	10 VII	10 VII	12 VII	15 VII	10 VII	11 VII	13 VII
15 VI	11 VII	11 VII	13 VII	16 VII	11 VII	12 VII	14 VII
16 VI	12 VII	12 VII	14 VII	17 VII	12 VII	13 VII	15 VII

\multicolumn{4}{c}{For Temperature Anomaly 0° C}			
Sprouting	Budding	Sprouting	Budding
20 VI	16 VII	31 VII	27 VIII
25 VI	22 VII	5 VIII	2 IX
30 VI	27 VII	10 VIII	8 IX
5 VII	1 VIII	15 VIII	15 IX
10 VII	5 VIII	20 VIII	26 IX
15 VII	10 VIII	24 VIII	8 X
20 VII	15 VIII	26 VIII	22 X
26 VII	22 VIII	27 VIII	2 XI

temporal anomaly of the temperature is zero in the given region, we use the +1° column in the calendar of the base region. If the temporal anomaly is also +1°, then we use the +2° column. Here the +2° consists of a +1° spatial anomaly and a +1° temporal anomaly. If, on the contrary, the temporal anomaly is −1°, then we use the column for zero anomaly in the base calendar, since +1° + (−1°) = 0°; and so on.

To use phenological calendars one does not need any special

TABLE 3.9. Phenoforecasting calendar (and a reference table) for mass flowering of cotton variety 108-F in relation to the dates of mass budding and to the temperature anomalies for Dushanbe, Tadzhikistan (h = 803 m).

Budding	Temperature Anomalies (°C)						
	0	+1	+2	+3	−1	−2	−3
25 V	29 VI	27 VI	25 VI	24 VI	1 VII	3 VII	5 VII
26 V	30 VI	28 VI	26 VI	25 VI	2 VII	4 VII	6 VII
27 V	1 VII	29 VI	27 VI	26 VI	3 VII	5 VII	7 VII
28 V	2 VII	30 VI	28 VI	27 VI	4 VII	6 VII	8 VII
29 V	2 VII	30 VI	28 VI	27 VI	4 VII	6 VII	8 VII
30 V	3 VII	1 VII	29 VI	28 VI	5 VII	7 VII	9 VII
31 V	4 VII	2 VII	30 VI	29 VI	6 VII	8 VII	10 VII
1 VI	5 VII	3 VII	1 VII	30 VI	7 VII	9 VII	11 VII
2 VI	6 VII	4 VII	2 VII	1 VII	7 VII	9 VII	11 VII
3 VI	6 VII	4 VII	2 VII	1 VII	8 VII	10 VII	12 VII
4 VI	7 VII	5 VII	3 VII	2 VII	9 VII	11 VII	13 VII
5 VI	8 VII	6 VII	4 VII	3 VII	9 VII	11 VII	13 VII
6 VI	9 VII	7 VII	5 VII	4 VII	10 VII	12 VII	14 VII
7 VI	9 VII	8 VII	6 VII	5 VII	11 VII	13 VII	15 VII
8 VI	10 VII	8 VII	6 VII	5 VII	12 VII	14 VII	16 VII
9 VI	11 VII	9 VII	7 VII	6 VII	12 VII	14 VII	16 VII
10 VI	12 VII	10 VII	8 VII	7 VII	13 VII	15 VII	17 VII
11 VI	13 VII	11 VII	9 VII	8 VII	14 VII	16 VII	18 VII
12 VI	13 VII	12 VII	10 VII	9 VII	15 VII	17 VII	19 VII
13 VI	14 VII	12 VII	11 VII	10 VII	15 VII	17 VII	19 VII
14 VI	14 VII	13 VII	11 VII	10 VII	16 VII	18 VII	20 VII
15 VI	15 VII	14 VII	12 VII	11 VII	17 VII	19 VII	21 VII
16 VI	16 VII	15 VII	13 VII	12 VII	18 VII	20 VII	22 VII
17 VI	17 VII	16 VII	14 VII	13 VII	19 VII	21 VII	23 VII
18 VI	18 VII	16 VII	15 VII	No data	19 VII	21 VII	24 VII
19 VI	19 VII	17 VII	16 VII	"	20 VII	22 VII	24 VII
20 VI	20 VII	18 VII	17 VII	"	21 VII	23 VII	25 VII
21 VI	21 VII	19 VII	18 VII	"	22 VII	24 VII	26 VII
22 VI	22 VII	20 VII	19 VII	"	23 VII	25 VII	27 VII
23 VI	22 VII	21 VII	19 VII	"	24 VII	26 VII	28 VII
24 VI	23 VII	22 VII	20 VII	"	25 VII	27 VII	29 VII
25 VI	24 VII	23 VII	21 VII	"	26 VII	28 VII	30 VII
26 VI	25 VII	24 VII	22 VII	"	27 VII	29 VII	31 VII
27 VI	26 VII	25 VII	23 VII	"	28 VII	29 VII	1 VIII
28 VI	27 VII	25 VII	24 VII	"	29 VII	30 VII	2 VIII
29 VI	28 VII	26 VII	25 VII	"	30 VII	31 VII	3 VIII
30 VI	29 VII	27 VII	26 VII	"	31 VII	1 VIII	4 VIII
1 VII	30 VII	28 VII	27 VII	"	1 VIII	2 VIII	5 VIII
2 VII	31 VII	29 VII	28 VII	"	1 VIII	3 VIII	6 VIII
3 VII	1 VIII	30 VII	29 VII	"	2 VIII	5 VIII	7 VIII
4 VII	2 VIII	31 VII	30 VII	"	3 VIII	6 VIII	8 VIII
5 VII	3 VIII	1 VIII	31 VII	"	4 VIII	7 VIII	9 VIII

TABLE 3.9.   *(Continued)*

Budding	Temperature Anomalies (°C)						
	0	+1	+2	+3	−1	−2	−3
6 VII	4 VIII	2 VIII	1 VIII	″	5 VIII	8 VIII	10 VIII
7 VII	5 VIII	3 VIII	2 VIII	″	6 VIII	9 VIII	12 VIII
8 VII	6 VIII	5 VIII	3 VIII	″	8 VIII	11 VIII	13 VIII
9 VII	7 VIII	6 VIII	4 VIII	″	9 VIII	12 VIII	14 VIII
10 VII	8 VIII	7 VIII	5 VIII	″	10 VIII	13 VIII	15 VIII

For Temperature Anomaly 0°C

Budding	Flowering	Budding	Flowering
12 VII	9 VIII	4 VIII	9 IX
15 VII	13 VIII	5 VIII	10 IX
20 VII	20 VIII	6 VIII	12 IX
26 VII	27 VIII	7 VIII	13 IX
28 VII	30 VIII	8 VIII	15 IX
30 VII	2 IX	9 VIII	16 IX
31 VII	3 IX	10 VIII	18 IX
2 VIII	6 IX	11 VIII	19 IX

training. With their help (sometimes together with PTNs) problems can be solved as follows:

1. *Definition of Multiannual Average Dates of Phenological Phenomena in the Organism's Life in a Given Region, as Given in a Normal Phenological Calendar.* Here the phrase "multiannual average dates" is used intentionally, because the PTN, from which the above phenological calendars were derived, is based on multiannual data (and it can be concrete for a given year only if when working with a nomogram we use the actual initial dates for a given year, rather than the multiannual dates, and if we use the actual temperatures).

In Table 3.23 we give the results of transitive phenological extracts from the zero columns of calendars for cotton variety 108-F sown on April 13. It is not mere chance that these data from our calendars are close to the results of multiannual phenological observations after sowing on April 13 in the Dushanbe region; these results were published in the *Agroclimatic Reference Book* of Tadzhikistan (1959), and are given in Table 3.23. Nevertheless, it is obvious that in order to compose phenological calendars with the help of a nomogram for a given region, there is no need for multiannual phenological observations to be made in the very same region. This is because the biological curves on the nomogram could be constructed on the basis of phenological observations in other regions.

TABLE 3.10. Phenological calendar (and a reference table) for mass boll unfolding of cotton variety 108-F and its prefrost yield in relation to the dates of mass flowering at normal temperatures for Dushanbe, Tadzhikistan ($h = 803$ m).

Flowering	Ripening	Prefrost Yield (in %)	Flowering	Ripening	Prefrost Yield (in %)
21 VI	25 VIII	95	16 VII	22 IX	78
22 VI	26 VIII	95	17 VII	24 IX	76
23 VI	27 VIII	95	18 VII	25 IX	75
24 VI	28 VIII	94	19 VII	26 IX	74
25 VI	29 VIII	94	20 VII	27 IX	73
26 VI	30 VIII	93	21 VII	28 IX	71
27 VI	1 IX	93	22 VII	29 IX	70
28 VI	2 IX	92	23 VII	30 IX	69
29 VI	3 IX	92	24 VII	1 X	68
30 VI	4 IX	91	25 VII	3 X	65
1 VII	5 IX	91	26 VII	5 X	63
2 VII	6 IX	90	27 VII	7 X	60
3 VII	8 IX	89	28 VII	8 X	57
4 VII	9 IX	88	29 VII	10 X	53
5 VII	10 IX	88	30 VII	11 X	50
6 VII	11 IX	87	31 VII	12 X	46
7 VII	12 IX	86	1 VIII	14 X	42
8 VII	13 IX	86	2 VIII	15 X	41
9 VII	14 IX	85	3 VIII	16 X	40
10 VII	16 IX	83	4 VIII	18 X	35
11 VII	17 IX	82	5 VIII	21 X	30
12 VII	18 IX	82	6 VIII	23 X	25
13 VII	19 IX	81	7 VIII	26 X	20
14 VII	20 IX	80	8 VIII	30 X	14
15 VII	21 IX	79	9 VIII	2 XI	8

Note. The opening of seed bolls is determined according to the method used by the NTV. The boll is considered to be open if the segments of the boll are fully opened (in contrast to the method used by the Hydrometeorological Service, whereby the boll is considered to be open if there is an opening of only 0.5-1 cm between the segments. As a result, ripening is observed by the Hydrometeorological Service 7 days or more earlier than by the NTV, after regular sowing dates; see Figs. 1.1, 1.2, and others).

Moreover, in the *Agroclimatic Reference Book* the calendar for the development of cotton variety 108-F was compiled on the basis of multiannual phenological observations in Dushanbe, and it relates to the average multiannual sowing date of April 13 only. Phenological calendars based on PTN enable us to judge the average multiannual developmental dates and the prefrost yield of cotton following any given sowing date. For example, a late resowing of cotton in May (this sometimes occurs in the Hissar Valley, because of damage from hail) gives multiannual developmental dates and yield, according to the phenoprognostic calendars given above, as follows: sowing on May 15, mass sprouting on May 22, mass budding on June 20 (in zero column of Table 3.8, opposite the date of May 22), mass flowering on July 20, ripening on September 27, and a prefrost yield of 73% of the entire yield.

Of course, such calendars or reference books, once compiled, can be used every year. The agricultural regulations which are often published in the USSR for agronomists could be supplied with such calendars.

2. *Phenoforecasts with Prolonged Foresight Can Be Made More Easily than with PTNs.* For example, in 1958 at the Dushanbe agricultural station, cotton variety 108-F was sown on April 28 in a field with normal agrotechnical conditions. When can we expect mass sprouting?

In Table 3.7 we find the actual sowing date of April 28, and opposite this we see the date of mass sprouting is May 8. In fact, mass sprouting in 1958 occurred exactly on May 8.

When can we expect mass budding? In Table 3.8 we find the actual sprouting date of May 8, and opposite this we see the budding date is June 11. Unfortunately, the NTV does not observe the budding stage of cotton, but, as can be seen, the budding date can easily be found with the help of the phenological calendar. In the same way, in Tables 3.9 and 3.10, we find the dates of mass flowering and ripening and the prefrost yield.

3. *Determining Early Sowing Dates.* It is very clear that there is no sense in sowing any crop at temperatures lower than the biological threshold of development, that is, the temperature below which the plant must stop its development, not according to the mathematical extrapolation of the line $1/n = \varphi(t)$ to the axis of abscissae, where $1/n = 0$, but in terms of the actual conditions in the field and in experiments.

For cotton variety 108-F during the period from sowing to sprouting, such a biological temperature threshold is, apparently, close to $9.8°$ C. For instance, according to our field experiments, sowing on March 11, 1950, was followed by sprouting only 56 days later, while the average temperature over the entire period from sowing to sprouting was $11.4°$ C. In Figure 3.5 an experimental curve of the developmental rate $1/n = \varphi(t)$ is given, which is constructed on the basis of the experiment mentioned above, and on the basis of a number of other experiments in which development also took place at temperatures as low as those mentioned above (close to $11.4°$ C).

Extrapolating this curve to the axis of abscissae, drawn at a distance of only $1.5°$ C, indicates a threshold of $9.8°$ C. On this curve, it can be

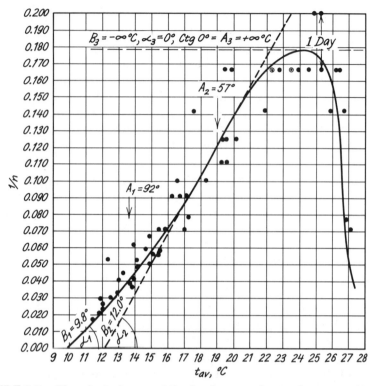

FIGURE 3.5.   Phenological curve of the developmental rate of cotton variety 108-F during the period from sowing to mass sprouting, in relation to the air temperature, under normal agrotechnic conditions (using data from field experiments carried out by the author and others at the agrometerological station of Dushanbe in 1948-1954).

seen how changeable are the thresholds and totals of effective temperatures.

A 24-hour average air temperature of 9.8° C (and a very similar 24-hour average temperature at the depth of location of the cotton seeds in spring in the conditions of Tadzhikistan) can usually be observed in the Dushanbe region, on about March 25 (in Fig. 1.56 see the temperature on the axis of abscissae where the line of average temperatures dated March 25 begins). Therefore, the end of March is among the early dates for cotton sowing in Dushanbe.

However, sometimes cotton has been sown at even earlier dates—in very warm springs, and in the hope of an intensive daytime warming up of the soil. Calculating for such cases, the phenological calendars for sprouting (Table 3.7) begin with a sowing date of March 11. However, it is interesting that sowing on this date gives, according to the multiannual average data (i.e., with zero anomaly), mass sprouting on April 25, and a sowing date of March 26 gives sprouting on April 22, that is, not later and not even on the same date as sowing on March 11, as might be expected, but earlier. Thus the calculations on the nomogram corroborate what is known from practice: seeds put into moist, but not thoroughly warmed soil, lose their sprouting energy. In addition, such early sowing often causes rotting of the seeds and thinning out of the sprouts.

Therefore, sowing is considered to be early when it is carried out in the following circumstances: after the 24-hour average temperature of air and soil (at the depth of location of the seeds) in a given year reaches stability at levels higher than the lower biological threshold for the development of a given crop during the period from sowing to sprouting. Nevertheless, the earliest sowings are not the most advisable, as will be shown in the next problem.

4. Calculating Optimum Sowing Dates.   Let us find, with the help of phenological calendars (Tables 3.7-3.10), the developmental dates and the prefrost yield of cotton variety 108-F in the Hissar valley (Dushanbe) when the sowing date is March 25, and April 5, 15, and 25 (Table 3.11).

As can be seen, the first three sowing dates give almost identical results. This happens because of a known phenological principle, which can be correctly explained with the help of the PTN: later sowings, under the influence of higher temperatures, develop more quickly than early sowings.

If the difference between the developmental dates and the yield of crops sown during the period from March 25 to April 15 is so small, is it worth sowing the cotton too early, that is, exposing it to the danger of rotting, hail, and heavy showers? Maybe it is better to postpone cotton sowing in the Dushanbe region until the middle of April? However, to postpone sowing until after the middle of April is not desirable: sowing on April 25, as compared with sowin on April 15, results in a delay in the ripening of not 0 to 1 day, but 5 days, as well as a decrease in the prefrost yield of not 0 to 1%, but 4%.

Therefore, there is a range of sowing dates within the limits of which

TABLE 3.11.   Average multiannual developmental dates and prefrost yield of cotton vari/ ty 108-F in the Hissar Valley in Tadzhikistan after different sowing dates (according to phenological calendars).

| Sowing Date | Date of | | | | Prefrost Yield (in %) |
	Sprouting	Budding	Flowering	Ripening	
25 III	22 IV	4 VI	7 VII	12 IX	86
5 IV	25 IV	5 VI	8 VII	13 IX	86
15 IV	29 IV	7 VI	9 VII	14 IX	85
25 IV	6 V	11 VI	13 VII	19 IX	81

the ripening dates and the yield change very little. In the majority of cases, we must look within this range for the optimum dates, according to the soil moisture conditions, unfavorable meteorological phenomena, and even farming conditions.

Thus simple calculations with phenological calendars show that, in the majority of years, the optimum sowing dates for cotton variety 108-F in the Dushanbe region are in the middle of April. The same is indicated by the data given in the *Agroclimatic Reference Book* for Tadzhikistan (1959).

However, as a result of various factors, including sowing too early, in practice we have sometimes had to resow or to carry out a supplementary sowing of cotton in May and even in June. What are the limits for the maximally late dates, after which cotton has no time to ripen before the autumn frost, or after which it will produce a lot of unopened bolls, instead of top-quality fiber?

It thus proves necessary to calculate, for given climatic conditions, the maximally latest sowing dates which are able to give a certain quantity of prefrost fiber.

5. *Calculating the Maximally Latest Sowing Dates.* The problem can be presented as follows: to which sowing date does mass unfolding of the first bolls on the eve of the autumn frosts correspond?

Let us solve this problem on the PTN for Dushanbe (Fig. 1.56). On this nomogram we see that phenocurve 14, giving the development of cotton variety 108-F from flowering to ripening, enters the shaded area of the autumn frost zone simultaneously with the line of average temperatures dated August 9. This means that ripening can occur at the time of the autumn frosts only if flowering occurred on August 9. In fact, the intersection of the line of average temperatures from August 9 with phenocurve 14 gives $n = 85$ days; hence ripening will occur on August 9 + 85 days = November 2. This is the multiannual average date of the first autumn frost in the air of Dushanbe.

Now, on the basis of the flowering date of August 9, it is necessary to find the mass budding date; this procedure continues up to sowing. As we already know, this can be done using the method of a retrospective calculation (see Section 3.1h).

The retrospective problem can be solved much more easily with the use of phenological calendars. On the basis of the ripening date of November 2, we find in Table 3.10 the mass flowering date of August 9; using this date we find in Table 3.9 the budding date of July 12 (with a temperature anomaly of $0°$); using the date of July 12 we obtain from Table 3.8 the mass sprouting date of June 16; finally, using the sprouting date we find the sowing date of June 10 (Table 3.7). Therefore, cotton sown later than June 10 (and, really, after the end of May) will not give even a beginning of mass biological ripening before the autumn frosts.

For example, according to experiments in the fields of the agro-meteorological station at Dushanbe, sowing on May 15, 1947, May 17, 1950, and May 15, 1951 was followed by mass unfolding of the first bolls on October 5, and 6, respectively, that is, almost a month later than the norm, and sowing on May 15, 1953 was followed by unfolding of the first bolls only on October 20. Sowing on May 25, 1949 and May 25, 1954 were not followed by mass unfolding of the first bolls at all.

What percentage of the yield can we expect to be prefrost—that is, of the highest quality—if mass unfolding of the first bolls occurred on the eve of the autumn frost (on November 2)?

In Table 3.10 we find that the figure will be only 8% of the entire yield (and on Fig. 2.12 we see that this 8% corresponds to the zero period from the beginning of mass ripening to the cessation of vegetation, and is produced by the very beginning of ripening). Therefore, resowings at the end of May and in June in the Hissar Valley in Tadzhikistan are of little efficiency in normal years.

In Figures 2.6 to 2.14, which show the accumulation of yield and which are based on actual data, it can be seen that not only for cotton variety 108-F, but also for other varieties, the prefrost yield at the moment of the mass unfolding of the *first* bolls does not exceed 20-30%. Thus the mass unfolding of the first bolls does not give a clear view of the cotton yield. This is why it seems advisable to change somewhat the method for determining the ripening dates of cotton as established by the NTV and by the Hydrometeorological Service of the USSR.

Let us note the interesting properties of the graphs of yield accumulation. In Figure 2.6 (see the line prolonged to the axis of abscissae) we see the volume of the prefrost yield that could be gathered 5, 8, and 10 days before the mass unfolding of the first bolls: −5 days corresponds to 8%, and −10 days corresponds to 0%. We see how many days before the beginning of mass ripening the unfolding of the very first bolls began.

Therefore, the very first unfolding of bolls in the field, in fine-fiber cotton variety 5904-I, begins approximately 10 days before the beginning of mass unfolding. Analogous or somewhat longer periods can also be seen for other varieties.

Such a method enables us to determine the prefrost yield if the autumn frost comes unexpectedly before the mass unfolding of the first bolls, and when separate harvesting of the bolls that opened before the frost has not yet been carried out (first we find on a nomogram, on the basis of the flowering date, the date of ripening).

In conclusion, it should be noted that a good long-range weather forecast enables us to make all the calculations of sowing dates and yield mentioned above, not in multiannual terms, but in concrete terms for each year. It is also possible to use for this purpose a number of actual temperatures from the current year. This can be done directly on

PTNs, or by using the calendars with temperature anomalies mentioned above.

6. *Problems Involving a Combination of the Developmental Stage and Temperature. The Planning of Experiments.* When is it necessary to sow cotton variety 108-F in the Dushanbe region, so that the interdevelopmental stage from sprouting to budding proceeds at the lowest possible average temperature for this period under the given climatic conditions? Solving such problems is of considerable importance in the selection of fast-ripening and cold-resistant varieties of cotton by means of "training" it during the budding stage at lower temperatures. The problem can also be related to the selection of dates for the summer sowing of potatoes in the south. Such solutions are of interest for experimental purposes, for example, for filling in "blank spaces" on phenological graphs with the help of experiments.

Obviously, here we can discuss two sowing seasons: spring and autumn. Let us first consider the autumn season. The lowest possible temperature for the interdevelopmental stage will obviously be the temperature that will give budding as late as possible, on the eve of the usual date of the autumn frosts—for instance, at the beginning of November 2. On the phenological calendar (Table 3.8) we find that this budding date corresponds to mass sprouting on August 27. This means that the cotton should be sown on August 22 (see Table 3.7). All of the above applies under conditions of zero temperature anomaly.

In the spring version of the experiment, in order to obtain the lowest possible temperature we have to find the dates of the earliest sprouting. However, the earliest sprouting does not always correspond to the earliest sowing date. This is clearly reflected in Table 3.7—the earliest sprouting will occur not after sowing on March 11, but after sowing on March 20 to 26. Therefore, cotton sowing on March 26 will give mass sprouting on April 22 and mass budding on June 4 (Table 3.8); the interdevelopmental stage from sprouting to budding will proceed at low temperatures.

In general, the autumn version of the experiment gives lower temperatures than the spring version. We can verify this only with the help of a PTN, which is able, in contrast to calendars, to solve such a problem in a general form.

In general form, problems concerning combinations of the interdevelopmental stage and temperature can be expressed as follows: when is it necessary to sow a given crop, so that a given interdevelopmental stage will coincide with a given average temperature over this period? And, as a concrete example, when is it necessary to sow cotton variety 108-F so that the developmental stage from sprouting to budding will take place, in the conditions of Dushanbe, at $t_{av} = 17.2°C$?

On the PTN for Dushanbe (Fig. 1.56), we trace a vertical line from the mark 17.2°C on the axis of abscissae to the intersection with phenocurve 12, giving the development of cotton variety 108-F from sprouting to budding. We see that through this intersection run two lines of average temperatures—from August 26 and from April 3. This means that the given combination of developmental stage and temperature can occur twice a year—with mass sprouting on August 26 (when budding will be on October 22) and on April 3. Directly on the PTN, or on the phenological calendar (Table 3.7), we find that sprouting on August 26 corresponds to sowing on August 21, and it is totally impossible to obtain mass sprouting as early as April 3 in the natural conditions of Dushanbe.

Therefore, in the majority of years the developmental period from sprouting to budding will take place with an average temperature for this period of $17.2°$ C only in the case of autumn sowings carried out on approximately August 21.

In Figure 1.3, at the upper end of phenocurve 1, we see the experimental point that corresponds to this calculation. This point relates to the experimental sowing made on August 17, 1953, at the agrometeorological station of Dushanbe. Mass sprouting was observed on August 23 and mass budding on October 24, and the entire interdevelopmental stage took place at $t_{av} = 17.2°$ C.

It is absolutely clear that since the entire period from August 23 to October 24 has an average temperature of $17.2°$ C, the second part of this period, particularly at the end of October, had (with successful cotton development) average temperatures much lower than $17.2°$ C.

The development of cotton from sprouting to budding at average temperatures between $17°$ and $19°$ C was observed by the author both in the fields of Dushanbe's agrometeorological station and in the fields of surrounding farms in different years, up to 10 times (Fig. 1.3). In a number of cases, the theoretical calculation was made first, and afterward, on the basis of the calculated data, a field experiment was conducted.

For example, in 1954 it was calculated on the PTN for Dushanbe that a t_{av} of $19°$ C over the period from mass sprouting to mass budding could be achieved (at zero temperature anomaly) by sowing on August 18 and March 23 to 26 (through the intersection of the vertical line from $19°$ C with developmental curve 12, in Fig. 1.56, run the lines of average temperatures corresponding to sprouting on August 23 and on April 22; hence with the help of a retrospective calculation on the PTN or on a calendar, we obtain the corresponding sowing dates of August 18 and March 23 to 26, respectively.) According to the result of these calculations, on August 18, 1954 cotton variety 108-F was sown in a field belonging to the Dushanbe agrometeorological station. Mass sprouting was observed on August 27 (the error in the calculated date of August 23 is 4 days); budding was observed on October 6 (the error in the forecast based on the initial date of August 23 is 1 day); the average temperature over the period from August 27 to October 6, calculated on the basis of 24-hour averge temperatures, turned out to be 19.0° C—exactly in accordance with the theoretical calculation. Moreover, a temperature only slightly higher than this—namely, 19.8° C—is the usual average temperature for the period from sprouting to budding for the majority of commercial sowings of cotton variety 108-F in the Dushanbe region.

In the period from budding to flowering, average temperatures of 18° to 19° C were also often observed.

However, the possibility of such low temperatures during the period of budding was formerly theoretically disputed, in particular by T. D. Lisenko.

All the solutions described here were made on a multiannual basis. However, it is clear that they can also be made for a given year, either by using a good long-term weather forecast or by drawing the lines of average temperatures necessary for the calculation, with the help of initial line segments based on actual temperatures for a part of the period.

7. *Solving Operational Problems. Estimating the Development of Agricultural Crops in the Current Season.* When assessing the state of particular crop, agronomists are often guided by a comparison of the present state of the plants with their state in the previous year, or with the average multiannual state. However, the sowing dates in a particular year on a given farm can differ considerably from the previous year's dates or from the multiannual dates for the region. Therefore, it seems essential to calculate the multiannual or the previous year's dates of development for any given sowing date. By using phenological calendars or nomograms (PTNs), it is possible to do this.

Let us return to the example discussed above. In one field in the Dushanbe region, cotton variety 108-F was sown on April 1, 1950. Mass sprouting occurred on May 4. The agronomist's problems are whether the cotton is behind in its development in comparison with the multiannual norms which characterize normal weather and normal agrotechnical conditions; if it is behind, then by how many days?; why is it behind?—because of the weather (temperature conditions) or agrotechnical errors (in the agronomist's opinion, of course, he has made no errors).

On a phenological calendar (Table 3.7), or directly on a PTN (Fig. 1.56), the agronomist can find out very quickly the average multi-annual date of mass sprouting for cotton corresponding to sowing on April 1. This is April 24 (phenoclimatic forecast with prolonged foresight). Since sprouting in 1950 did not, in fact, occur until May 4, the agronomist's apprehension proved to be not unfounded—sprouting is 10 days behind.

To find out the causes of this development lag, it is sufficient to calculate the mass sprouting date entirely on the basis of the actual temperatures in 1950 (i.e., to give a phenosynoptical reliable forecast). This calculation cannot be done precisely using the calendar: a nomogram is necessary, and, specifically, we need to use the small area directly around the developmental phenocurve.

We shall not repeat in detail the calculation on the PTN, because this has been described in Sections 3.1d and 3.1e. The segment of the actual (for 1950) line of average temperatures from April 1, ik (Fig. 1.56), intersects phenocurve 11 for cotton development from sowing to sprouting at $n = 33$ days; hence mass sprouting in crops sown on April 1 should appear (given the actual temperatures in 1950, and assuming normal agrotechnical conditions) on April 1 + 33 days = May 4, that is, the date when they actually appeared in 1950.

Therefore, the lag in development of cotton in 1950 in comparison to multiannual norms occurred only because of the low temperature conditions in the spring of 1950. Nevertheless, it is necessary to give the crops more intensive care in order to decrease this 10-day lag in development.

Thus an equation can be constructed:

$$\epsilon_i = \epsilon_t + \epsilon_c \tag{15}$$

where the integral error ϵ_i of the phenoclimatic forecast (+10 days) consists, in our case, entirely of the temperature error ϵ_t; however, the complex error ϵ_c, which includes agrotechnical errors, the effect of not taking into account the influence of nontemperature meteorological factors, and errors in geometric constructions, is (ϵ_c) zero.

It is absolutely clear that in order to compare the conditions of cotton development in 1950 with the conditions in 1949, it is necessary to make an analogous calculation relating to sprouting for crops sown on April 1, on the basis of the actual temperatures in 1949 and on the basis of the same PTN and phenocurve 11 for normal agrotechnical conditions. This calculation must be made because it may happen that there was no sowing on April 1, 1949; however, the calculation based on actual temperatures is, as usual, so exact that it may be considered to be the actual date for a field with normal agrotechnical conditions.

Let us take another example from practice. Cotton was sown on March 26, 1951 and mass sprouting actually occurred on May 1. The questions asked are whether the cotton's development is behind or ahead of the multiannual norm at the stage from sowing to sprouting in 1951, and by how much and why?

The data in Table 3.7 show that given normal weather and agrotechnical conditions, mass sprouting should have occurred on April 22. The cotton's development is thus 9 days behind in 1951. Therefore +9 days is the integral error of the phenoforecast (ϵ_i).

The calculation based on the actual temperatures in 1951 gives a sprouting date close to the multiannual date, but differing from the actual date by +12 days (i.e., according to the calculation sprouting was to occur 12 days earlier). This will be the complex error ϵ_c of the phenoclimatic forecast (if we exclude from the latter the temperature error ϵ_t by using a phenosynoptical, trustworthy forecast, as was, in fact, done). Inserting these values into Equation (15), we obtain $+9 = \epsilon_t + 12$, hence $\epsilon_t = -12 + 9 = -3$ days.

Thus, the 9-day lag in cotton development in 1951 as compared with the multiannual norm occurred not because of the temperature factor (which in 1951 could have caused the development to be 3 days ahead of the developmental norms, i.e., $\epsilon_t = -3$), but as a result of the 12-day complex error. Since such nontemperature factors as length of daylight, cloudiness, haze, and wind cannot significantly influence seeds that are already in the soil, then, in all probability, the 12-day complex error relates entirely to agrotechnical errors. In fact, in the April 1951 Observer's Notebook there is a note concerning "soil crust."

The mass unfolding of the first bolls of cotton sown on April 27, 1949 occurred 6 days later than was forecast by the phenoclimatic calculation (with prolonged foresight), and 3 days later than forecast by the calculation based on the actual temperatures in 1949 (both calculations used the one and the same starting date—the actual flowering date). Therefore, the delay in cotton development during the stage from flowering to ripening in 1949, in comparison with the multiannual development—that is, with the phenoclimatic calculation, had two causes: 3 days of the developmental delay were due to the lower (compared with the multiannual) temperatures in 1949, and 3 days were due to the influence of nontemperature factors. In point of fact, the Observer's Notebook for 1949 contains important notes to the effect that in July and August there were long, dry, hazy periods, and the fruit-bearing elements of the plants began to fall. Because of the decrease in the atmosphere's transparency to solar radiation, this period was also notable for slightly lower temperatures, together with changed values of PAR (photosynthetically active radiation).

Such analyses of the causes of delay or acceleration in the development of agricultural crops enable agronomists to respond promptly and correctly to concrete situations.

The entire phenoforecasting complex described above can also be used for calculating operational problems. In particular, when sowing a specific crop (especially if sowing is not carried out on the optimal dates), the agronomist can check beforehand the probable dates of all developmental stages. As is already known, these can be calculated by transitive phenoforecasts, or, simpler still, defined on the basis of phenological calendars.

The idea of phenoforecasting calendars, which were published by this author for the first time in 1957 for cotton, proved to be a fruitful one. Now, many authors, on the basis of PTN, compile phenoforecasting calendars for various organisms of the plant and animal world.

Some of them are given here in an abbreviated form in Tables 3.12 to 3.21 (the tables shown here refer to plants).

TABLE 3.12. Phenoforecasting calendar (and reference table) for the development of spring wheat variety Saratovskaya-29 under normal agrotechnical conditions for Uritzk, in the northern part of Kazakhstan (from V. K. Adzhbenov, 1973).

Sowing	Mass Sprouting	Beginning of Tillering	Full Heading	Wax Ripeness
25 IV	9 V	24 V	28 VI	1 VIII
28 IV	12 V	26 V	30 VI	3 VIII
1 V	14 V	27 V	2 VII	5 VIII
4 V	17 V	31 V	4 VII	7 VIII
7 V	19 V	2 VI	6 VII	9 VIII
10 V	22 V	4 VI	8 VII	11 VIII
13 V	24 V	6 VI	10 VII	13 VIII
16 V	27 V	8 VI	12 VII	16 VIII
19 V	29 V	11 VI	14 VII	18 VIII
22 V	2 VI	14 VI	17 VII	22 VIII
25 V	5 VI	17 VI	19 VII	25 VIII
28 V	7 VI	19 VI	21 VII	27 VIII
31 V	9 VI	21 VI	23 VII	30 VIII
3 VI	12 VI	24 VI	26 VII	3 IX
6 VI	15 VI	26 VI	28 VII	8 IX
9 VI	18 VI	29 VI	31 VII	13 IX

TABLE 3.13. Phenoforecasting calendar for mass heading of spring wheat variety Kharkovskaya-46 for Kursk district, Russian, * in relation to the date of mass sprouting and to temperature anomalies.

	Temperature Anomalies (°C)					
Sprouting	0	+1	+2	+3	−1	−2
25 IV	14 IV	10 VI	8 VI	7 VI	—	—
30 IV	15 VI	13 VI	12 VI	12 VI	20 VI	—
5 V	19 VI	18 VI	17 VI	19 VI	21 VI	26 VI
10 V	23 VI	22 VI	23 VI	26 VI	25 VI	29 VI
15 V	27 VI	27 VI	29 VI	—	28 VI	1 VII

*The Russian part of the USSR.

TABLE 3.14. Phenoforecasting calendar for wax ripeness of spring wheat variety Kharkovskaya-46 for Kursk district, Russia, in relation to the date of mass heading.

Heading	Temperature Anomalies (°C)					
	0	+1	+2	+3	−1	−2
5 VI	16 VII	13 VII	10 VII	6 VII	20 VII	—
10 VI	21 VII	17 VII	14 VII	10 VII	24 VII	—
15 VI	24 VII	21 VII	17 VII	14 VII	28 VII	31 VII
20 VI	28 VII	25 VII	21 VII	18 VII	1 VIII	4 VIII
25 VI	1 VIII	29 VII	25 VII	22 VII	5 VIII	9 VIII
30 VI	6 VIII	2 VIII	29 VII	26 VII	10 VIII	14 VIII

TABLE 3.15. Phenoforecasting calendar for mass heading of spring barley variety Valtitzky for Kursk district, Russia, in relation to the dates of mass sprouting and to temperature anomalies.

Sprouting	Temperature Anomalies (°C)						
	0	+1	+2	+3	−1	−2	−3
25 IV	11 VI	9 VI	7 VI	6 VI	12 VI	14 VI	—
30 IV	15 VI	13 VI	11 VI	9 VI	16 VI	18 VI	20 VI
5 V	18 VI	16 VI	15 VI	13 VI	20 VI	22 VI	23 VI
10 V	22 VI	19 VI	18 VI	17 VI	14 VI	26 VI	28 VI

TABLE 3.16. Phenoforecasting calendar for mass wax ripeness of spring barley variety Valtitzky for Kursk district, Russia, in relation to the date of mass heading and to temperature anomalies.

Heading	Temperature Anomalies (°C)					
	0	+1	+2	+3	−1	−2
1 VI	4 VII	2 VII	1 VII	28 VI	6 VII	—
5 VI	7 VII	5 VII	3 VII	2 VII	9 VII	—
10 VI	11 VII	9 VII	8 VII	6 VII	13 VII	15 VII
15 VI	16 VII	14 VII	13 VII	12 VII	18 VII	20 VII
20 VI	20 VII	18 VII	16 VII	15 VII	22 VII	24 VII
22 VI	22 VII	20 VII	18 VII	17 VII	24 VII	26 VII

246

TABLE 3.17. Phenoforecasting calendar for the mass appearance of panicles in oats variety Lgovsky-1026 for Kursk district, Russia, in relation to the date of mass sprouting and to temperature anomalies.

Sprouting	Temperature Anomalies (°C)						
	0	+1	+2	+3	−1	−2	−3
15 IV	19 VI	14 VI	10 VI	7 VI	—	—	—
20 IV	20 VI	15 VI	12 VI	9 VI	24 VI	—	—
25 IV	21 VI	17 VI	14 VI	11 VI	25 VI	30 VI	—
30 IV	23 VI	20 VI	16 VI	13 VI	28 VI	2 VII	—
5 V	26 VI	22 VI	19 VI	16 VI	30 VI	4 VII	9 VII
10 V	29 VI	25 VI	22 VI	20 VI	3 VII	7 VII	11 VII
15 V	2 VII	29 VI	26 VI	24 VI	6 VII	9 VII	14 VII

Note. Tables 3.13 to 3.17 are taken from A. S. Podolsky and O. S. Yermakov (1973).

TABLE 3.18. Phenoforecasting calendar for the beginning of flowering of lily of the valley for Kursk, in relation to the date of the beginning of vegetation and to temperature anomalies.

Beginning of Vegetation	Temperature Anomalies (°C)						
	0	+1	+2	+3	−1	−2	−3
5 IV	9 V	6 V	3 V	1 V	—	—	—
15 IV	11 V	8 V	6 V	4 V	13 V	16 V	20 V
25 IV	14 V	13 V	12 V	10 V	17 V	19 V	22 V
5 V	21 V	20 V	19 V	18 V	23 V	24 V	26 V
15 V	30 V	29 V	28 V	27 V	31 V	1 VI	2 VI
26 V	9 VI	8 VI	7 VI	—	10 VI	11 VI	12 VI
5 VI	18 VI	17 VI	—	—	19 VI	20 VI	21 VI
15 VI	28 VI	—	—	—	28 VI	29 VI	30 VI
25 VI	7 VII	—	—	—	8 VII	9 VII	9 VII

3.2. ELEMENTS OF MATHEMATICAL ECOLOGY IN CONNECTION WITH AGROCLIMATIC PROBLEMS

3.2a. Ripening of Agricultural Crops, and Crops Sown after Harvest.

If we superimpose the phenocurves of plants in which we are interested ("biological passports," obtained from observations in various regions) onto the heat resources net of a new region, where the behavior of these organisms is not yet known, we can carry out a number of bioclimatic calculations without resorting to numerous new experiments.

TABLE 3.19. Phenoforecasting calendar for the end of flowering of lily of the valley for Kursk, in relation to the date of the beginning of flowering and to temperature anomalies.

Beginning of Flowering	Temperature Anomalies (°C)						
	0	+1	+2	+3	−1	−2	−3
20 IV	24 V	22 V	19 V	15 V	—	—	—
30 IV	29 V	26 V	24 V	21 V	1 VI	5 VI	—
10 V	2 VI	31 V	30 V	28 V	5 VI	9 VI	14 VI
20 V	9 VI	7 VI	6 VI	6 VI	12 VI	15 VI	19 VI
31 V	18 VI	18 VI	17 VI	17 VI	21 VI	23 VI	26 VI
10 VI	28 VI	27 VI	27 VI	27 VI	29 VI	1 VII	4 VII
20 VI	7 VII	7 VII	6 VII	—	8 VII	10 VII	12 VII
30 VI	17 VII	16 VII	—	—	18 VII	19 VII	21 VII
10 VII	26 VII	26 VII	—	—	27 VII	28 VII	29 VII

Note. Tables 3.18 and 3.19 are taken from A. S. Podolsky and L. V. Shatunova (1977).

TABLE 3.20. Phenoforecasting calendar of the beginning of flowering of *Valeriana rossica* for Kursk, in relation to the date of the beginning of vegetation and to temperature anomalies (from V. A. Ryabov, 1978, abridged).

Beginning of Vegetation	Temperature Anomalies (°C)						
	0	+1	+2	+3	−1	−2	−3
5 IV	20 V	17 V	13 V	10 V	23 V	27 V	31 V
10 IV	20 V	18 V	15 V	12 V	24 V	27 V	31 V
15 IV	21 V	19 V	16 V	14 V	25 V	28 V	1 VI
20 IV	23 V	21 V	19 V	17 V	26 V	30 V	2 VI
25 IV	26 V	24 V	23 V	21 V	29 V	31 V	3 VI
27 IV	27 V	25 V	24 V	23 V	30 V	1 VI	4 VI
30 IV	29 V	28 V	26 V	26 V	31 V	3 VI	5 VI
2 V	31 V	29 V	28 V	28 V	2 VI	4 VI	6 VI
5 V	2 VI	31 V	31 V	31 V	4 VI	5 VI	8 VI
7 V	4 VI	3 VI	2 VI	2 VI	5 VI	7 VI	9 VI
10 V	7 VI	6 VI	5 VI	—	7 VI	9 VI	11 VI
12 V	9 VI	8 VI	8 VI	—	9 VI	10 VI	12 VI
15 V	11 VI	11 VI	10 VI	—	11 VI	12 VI	14 VI
17 V	13 VI	13 VI	12 VI	—	12 VI	14 VI	16 VI
20 V	16 VI	15 VI	15 VI	—	15 VI	16 VI	18 VI
22 V	18 VI	18 VI	18 VI	—	17 VI	18 VI	20 VI
25 V	21 VI	20 VI	20 VI	—	20 VI	21 VI	22 VI
28 V	24 VI	23 VI	23 VI	—	23 VI	23 VI	24 VI
31 V	26 VI	26 VI	26 VI	—	26 VI	27 VI	28 VI
2 VI	28 VI	28 VI	—	—	28 VI	29 VI	30 VI
5 VI	1 VII	1 VII	—	—	1 VII	2 VII	3 VII

TABLE 3.21. Phenoforecasting calendar for the beginning of flowering of *Senecio czernjaevii* for Kursk, in relation to dates of budding and to temperature anomalies (from V. A. Ryabov, 1978).

Budding	Temperature Anomalies (°C)						
	0	+1	+2	+3	−1	−2	−3
31 III	6 V	3 V	30 IV	27 IV	10 V	—	—
10 IV	8 V	6 V	3 V	1 V	11 V	13 V	15 V
15 IV	9 V	7 V	5 V	2 V	12 V	15 V	17 V
20 IV	11 V	8 V	6 V	4 V	13 V	16 V	18 V
25 IV	13 V	11 V	9 V	8 V	15 V	18 V	20 V
30 IV	16 V	14 V	13 V	12 V	17 V	20 V	22 V
5 V	19 V	18 V	17 V	16 V	20 V	22 V	25 V
10 V	23 V	22 V	21 V	21 V	24 V	26 V	28 V
15 V	27 V	26 V	26 V	26 V	29 V	30 V	1 VI
20 V	31 V	31 V	31 V	31 V	3 VI	4 VI	7 VI
25 V	5 VI	5 VI	5 VI	6 VI	6 VI	7 VI	8 VI
31 V	11 VI	11 VI	12 VI	—	11 VI	12 VI	13 VI
5 VI	15 VI	16 VI	17 VI	—	16 VI	16 VI	17 VI
10 VI	21 VI	22 VI	—	—	21 VI	21 VI	22 VI
15 VI	26 VI	27 VI	—	—	26 VI	26 VI	27 VI
20 VI	1 VII	3 VII	—	—	1 VII	1 VII	1 VII
25 VI	7 VII	8 VII	—	—	6 VII	6 VII	6 VII
30 VI	12 VII	13 VII	—	—	11 VII	11 VII	11 VII

In this way we can estimate, more accurately and more reliably than with the temperature totals method, the unexploited agroclimatic resources of various regions in terms of new crops and varieties, in terms of crops sown after harvest or after mowing, and for the purpose of using the earth year-round in the south, and so on. *By using the mathematical phenology method, it is possible to transfer a number of plant-growing problems from an experimental basis to a calculatory basis, as is already done in the exact and the engineering sciences.*

Let us assume that we do not know how many hay harvests the lucerne variety Tashkentskaya 721 can give, with irrigation, in the conditions of Kurgan-Tyube, in averge multiannual terms. We find the date when the 24-hour average air temperature exceeds the known threshold of 5°C: on the horizontal axis of the PTN for Kurgan-Tyube (Fig. 1.55) we see that, near the mark for 5°C, the line of average temperatures begins approximately from February 12. We define this date as the time when lucerne begins to grow in the spring. Then we trace the line of average temperatures from February 12 to the intersection with phenocurve 10 for lucerne's development, and we see that $n = 82$ days; therefore, the beginning of the first flowering of lucerne, on the average, will occur on February 12 + 82 days = May 5.

Since the crop is to be mowed at the beginning of flowering, May 5 will be the date of the first mowing.

Using the line of average temperatures dated May 5, we find in the same way the date of the second mowing—May 5 + 35 days = June 9; then the third mowing—June 9 + 34 days = July 13; the fourth mowing—July 13 + 34 days = August 16; and the fifth mowing—August 16 + 35 days = September 20.

A sixth mowing of the highest quality and quantity is not possible in the majority of years (if five optimal mowings have already been made), because the line of average temperatures from September 20 does not intersect developmental curve 10. The date of the last mowing which could still be followed by another flowering is indicated by the line of average temperatures from September 14, tangential to curve 10. Flowering would have occurred in this case if autumn frosts had not interferred; however, the point of contact between curve 10 and the line from September 14 occurs well inside the autumn frosts zone (the beginning of this zone is shown on Fig. 1.55 by a shaded area). For this reason, the last mowing that can be followed by another prefrost flowering of lucerne is not on the 14th, but on September 7 to 10, because the line of average temperatures from September 7 intersects developmental curve 10 exactly at the entrance to the autumn frosts zone. At this intersection $n = 51$ days, and therefore, flowering will occur on September 7 + 51 days = October 28, that is, on the eve of the first autumn frosts in Kurgan-Tyube.

Thus in the majority of years lucerne variety Tashkentskaya 721 is able to give, under the conditions of Kurgan-Tyube, and when mowing is carried out on time, five optimum-value harvests and a sixth of less value. This has been confirmed to be accurate by observations.

The reliability of calculations concerning the developmental dates of agricultural crops in new regions can also be judged by the exact coincidence of the results calculating the developmental dates of wheat sown in winter in Kurgan-Tyube with the results of multiannual observations of this crop by the Kurgan-Tyube meteorological station (Table 3.22). These calculations are based on phenocurves from the

TABLE 3.22. Developmental dates of wheat sown in autumn, variety Surkhak 5688, according to calculations on the PTN for Kurgan-Tyube, and according to average multiannual actual dates from the *Agroclimatic Reference Book* of Tadzhikistan (1959).

Data	Sowing	Mass Sprouting	Beginning of Tillering	Mass Stalk Shooting	Mass Heading	Wax Ripeness
Calculated	5 XII	26 I	7 III	7 IV	1 V	2 VI
Obtained from observations	5 XII	No data	No data	8 IV	4 V	4 VI

NTV; the curves were constructed *without data* for Kurgan-Tyube, because the Kurgan-Tyube agricultural station does not carry out observations on cereals.

As can be seen, if in Kurgan-Tyube wheat had never been sown, or if phenological observations on wheat had never been made, it would be possible to calculate all the multiannual developmental dates quite exactly, using the PTN for Kurgan-Tyube with the temperature curves of development based on observations of wheat in other regions.

In any event, the proposed method enables us to dispense with multiannual phenological observations in a new region. In the *Agroclimatic Reference Books* for Tadzhikistan there are no data about the sprouting and tillering of cereals sown in the autumn, because crops in these stages are often covered by snow, and it is difficult for the meteorological stations to observe them. This lack of data can be easily remedied by means of calculations (Table 3.22). It is necessary to bear in mind that for regions where autumn-sown crops are snow covered during the winter period of rest, we must also construct phenological developmental curves for the periods from the spring resumption of vegetation (24-hour average temperatures exceed $+4°$ - $+5°$ C) to tillering or stalk shooting, and sometimes to mass sprouting. Which of these curves is to be used, and when, will be evident from the PTNs themselves. It would be even better to use a special phenological curve from tillering to stalk shooting. When constructing such a curve on the basis of field experiments, it is necessary to exclude the period of winter rest, taking a single t_{av} over the autumn (tillering) and spring (stalk-shooting) period. However, then the average temperatures for the autumn lines of the heat resources net are also to be taken without winter temperatures, and the autumn lines of the net should be constructed without taking into account the period of winter rest.

When making subsequent forecasting calculations, we have to add the winter period, which was not taken into account, to the ordinate of the intersection of such an autumn line of average temperatures with the phenocurve. It should be mentioned here that the same principle is used for hibernating poikilothermic animals. This makes it possible to take into account the degree of development of a living organism before the winter rest begins (A. S. Podolsky, and I. M. Suslov).

Table 3.23 gives an analogous comparison for another region of Tadzhikistan and for another crop. Here, as before, we find agreement between the calculation and the actual multiannual data.

With a PTN, it is easy to find multiannual average temperatures over interdevelopmental stages. For example, the abscissa of the intersection of the average temperature line from April 13 with phenological curve 11 designates the average temperature over the period from sowing to sprouting in cotton; this temperature is $15.4°$ C for Dushanbe.

TABLE 3.23. Developmental dates for cotton variety 108-F according to calculations on the PTN for Dushanbe, and according to an average of over 16 years of actual data from the *Agroclimatic Reference Book* for Tadzhikistan (1959).

Data	Sowing	Mass Sprouting	Mass Budding	Mass Flowering	Mass Unfolding of the First Bolls	Length of Vegetation Period (in days)
Calculated	13 IV	28 IV	7 VI	9 VII	9 IX[a]	149
Obtained from observations (Hydro-meteorology Division)	13 IV	1 V	12 VI	11 VII	9 IX	149

[a]According to developmental phenocurve 13 in Figure 1.56, the curve corresponds to the observation method of the Hydrometeorological Service.

For the period from sprouting to budding we obtain 19.8°C; from budding to flowering, 24.9°C; from flowering to ripening, 23.8°C, according to the phenocurve from the NTV, and 24.1°C, according to the curve from the Hydrometeorological Service.

The next agroclimatic problem that we will solve concerns a plant that is in fact not yet sown in Tadzhikistan. This is a valuable crop that prefers a short day length—soya. We wish to determine whether it is possible to cultivate for grain the soya variety Easy-Cook, with irrigation, under the temperature conditions of the Hissar Valley. (This is the most humid valley in Tadzhikistan. It has more than 600 mm of precipitation per year, and the average relative humidity of the air in summer is approximately 50% at a meteorological ground and 80% or higher in the cotton crops). Which dates are the most advisable for sowing soya?

As is known, soya is to be sown at an average temperature of 12°C. In the conditions of Dushanbe, such a temperature occurs, according to average multiannual data, on April 3. This can be determined on the basis of the annual temperature trend or by a PTN: from the point marked 12°C on the PTN's axis of abscissae begins the line of average temperatures from April 3.

On the PTN for Dushanbe (Fig. 1.56), we trace the line of average temperatures dated April 3 up to the intersection with phenocurve 15 for soya development from sowing to flowering; we find that this period is $n = 119$ days, and therefore the flowering will occur on April 3 + 119 days = July 31. Then we trace the line of average temperatures dated July 31 up to the intersection with phenological curve 16 for soya development from flowering to ripening; at this intersection $n = 63$ days. Soya will ripen on October 2 (July 31 + 63 days).

Since the first autumn frosts in the Dushanbe region do not occur until November 2, the yield will ripen, as can be seen, a month before the frosts.

Thus, given the temperature conditions in the Hissar Valley, soya can be cultivated quite well. However, is there a more expedient date for sowing than April 3, and is soya a crop suitable for sowing after the harvest of another crop? We see that phenocurve 16 from flowering to ripening, having been extrapolated to the left, enters the autumn frosts zone simultaneously with the line of average temperatures from September 16. This means, as we already know, that the last flowering date to be followed by ripening before the frosts will be September 16.

By means of a retrospective calculation (described above), on the basis of the phenological curve of development from sowing to flowering (curve 15), we find that flowering on September 16 corresponds to sowing on June 10.

Therefore, if soya is sown on June 10, flowering will occur on September 16, and ripening will occur on the eve of the first autumn frost (November 1-2).

Thus in the climatic conditions of the Hissar Valley, there is quite a wide range of dates suitable for sowing soya—from April 3 to June 10. Apparently, late sowing is profitable, because it makes it possible to use one and the same field for sowing, first, any crop which ripens early and, after this, soya.

For example, in the autumn of the year preceding soya sowing—for instance, on November 15—it is possible to sow (for grain) barley variety Khordjau 18 or Persicum 64. Then the developmental stages of barley will proceed as follows: full sprouting on December 25; beginning of tillering on February 15; full stalk shooting on April 4; full heading on May 3; full wax ripeness on June 1. Thus 10 days will remain before the sowing of the soya. These dates can be calculated, if we draw onto tracing paper the corresponding phenological curve for barley from Figures 1.17, 1.19, 1.31, and 1.32, and superimpose the tracing paper onto the meteorological net of the Dushanbe nomogram, that is, onto Figure 1.56. It should be remembered that the phenological curves and the net must be drawn to the same scale.

Instead of barley, we can sow, in late autumn and even in spring, peas (there are varieties of peas with a vegetation period of only 60 days), as well as legume and cereal mixtures for forage—for example, barley with peas. Finally, one more version is interesting: planting a number of vegetable crops before sowing soya. For example, in Tadzhikistan it is already frequent practice to sow potatoes just before the onset of winter.

The feasibility of sowing soya in June has been proved by experiments near Tashkent (L. N. Babushkin, 1951).

As can be seen, a PTN makes it possible to solve problems about feasible combinations of agricultural crops for a given natural climatic region, that is, to solve problems concerning crop succession involving the concentration of crops timewise in one and the same field (Podolsky, 1957, 1958, 1967, 1974).

It is interesting to note that a postponement of the dates of soya sowing by more than two months (from April 3 to June 10) delays the ripening dates by only one month (from October 2 to November 2). This is a manifestation of a phenological law (which was mentioned above): the postponement of sowing dates entails a considerable increase in the average temperatures over developmental periods, and therefore a shortening of the stage from sowing to flowering. As a result, the flowering dates (July 31 and September 16) differ to a lesser extent than the sowing dates. The dates of the next developmental stage are also brought closer together because of the fact that, in the period from flowering to ripening, the development of soya at high temperatures proceeds not more quickly, as might be expected, but more slowly. This can be seen clearly on experimental phenocurve 16, which represents soya development from flowering to ripening on Figure 1.56; at high temperatures this curve has only an after-optimal branch.

Identical calculations were also carried out in the Moscow suburbs. For example, T. N. Gavrilova (1972) successfully used Podolsky's method for forecasting the developmental dates of melliferous plants. The intention is to use the mathematical method for calculating the "green conveyor," that is, calculating the sowing dates for cultured melliferous plants which should flower when wild melliferous plants finish blossoming.

The problem of a "green conveyor" arises in the *canning industry*. A concentrated succession of vegetables would be calculated here in the manner described above.

However, let us return to the subject of cotton. The Yavan Valley is one of the most fruitful regions of Tadzhikistan, but until 1968 it was not irrigated; therefore, cotton was not sown there. Nevertheless, using the new method of bioclimatic estimations, it was possible to make prior calculations.

Let us assume that we wish to find out whether fine-fiber cotton (*Gossypium barbadense*) variety 504-V will ripen in the Yavan region. It is known that even the variety 108-F (*G. hirsutum*) is not to be sown if the 24-hour average temperature of the soil has not reached at least the biological temperature threshold of 10°C, and that it is better to sow when the 24-hour average temperature of air and soil (at a depth of 5 cm) is 12°-14°C.

In Yavan a temperature of 14°C occurs on April 5 (see in Fig. 1.50, on the PTN for Yavan, the line of average temperatures from the mark 14° C on the axis of abscissae). Therefore, we assume that April 5 is one of the sowing dates.

At the intersection of the average temperature line dated April 5 with the phenocurve giving the development of cotton variety 504-V from sowing to mass sprouting (this curve is absent from the Yavan nomogram, as are all biological curves for cotton variety 504-V; these curves must be transferred from Fig. 1.31), we obtain the multiannual date of mass sprouting: April 5 + 15 days = April 20.

At the intersection of the average temperature line from April 20 with the phenocurve giving the development from sprouting to budding, we obtain the multiannual average date of budding: April 20 + 45 days = June 4.

In the same way we find the dates of mass flowering and ripening, July 1 and September 3, respectively.

Since the first autumn frosts in the air of Yavan do not occur until November 17, according to average multiannual data, then cotton variety 504-V will not only ripen in Yavan (this can be seen at the point of intersection of the average temperature line dated July 1 with the developmental curve from flowering to ripening; the intersection occurs at some distance from the shaded area of the first autumn frosts zone), but will give a 92% prefrost yield. The amount of the prefrost yield is obtained from Figure 2.7, since we know that the period between the calculated date of ripening (September 3) and the first autumn frosts (November 17) is 74 days.

If we want to obtain a prefrost yield from cotton variety 504-V of not less than 70%, then the period between frost and ripening must be not less than 42 days (Fig. 2.7), when the ripening must occur not later than November 17 − 42 days = October 6. With the help of a retrospective calculation, we find that sowing must be carried out in this case not later than May 18.

3.2b. Concrete Agroclimatic Division into Regions, and Mapping

The direct solution of equations for the heat requirements of crops (as well as humidity and light requirements) and the region's resources can give the answers to a number of detailed questions concerning the advised dates of sowing in a given region, the multiannual average dates of the occurrence of various stages, ripening dates, the volume of yield, concentrated one-year successions of crops in the same field, the optimal placing of different varieties, and so on. These questions cannot be answered by the other methods available (artificial temperature totals).

By the term "concrete agroclimatic division into regions," we mean such placing of crops that answers to questions not only about "strategy," but also about the "tactics" of plant growing in a given region and zone.

A concrete agroclimatic division into regions with the example of the Yavan and Dangara Valleys in Tadshikistan was published in 1963 (Podolsky, 1963).

In these valleys it had been proposed to supply with water, for irrigation purposes, vast tracts of arable land, including many thousands of hectares for cotton. Of course, beforehand it was necessary to estimate not only the soil and hydroland improvement resources, but the agroclimatic potential of these regions for the ripening of the most valuable *fine-fiber* (*Gossypium barbadence*) cotton varieties. Cotton had never been sown in these regions.

For the irrigation of the Yavan and Dangara Valleys, the intent was to use water from the Vakhsh River and energy from the Nurek hydroelectric power station. In order to bring water into the Yavan Valley it was necessary to dig a tunnel 7-km long through rocks, and in the case of the Dangara Valley a tunnel 13-km long. It was clear that this would entail great expenditure, and before the financing could be granted, it was necessary to know whether fine-fiber cotton varieties would grow in these valleys. As is known, areas for cultivating this high-quality raw material are very limited in the USSR.

There were two possible methods of investigation: the purely experimental method, involving experiments in these valleys, and the method using semitheoretical calculations based on many experiments in other regions. The purely empirical way would have taken too long; moreover, if the experiments were limited to only one or two years, then this method would not have been reliable, since these one or two years could have been abnormal.

On the other hand, the method involving a quite exact "engineering" calculation in agrobiology made it possible during a very short time to obtain details about the ripening of different cotton varieties in the Yavan and Dangara Valleys (including new varieties which were still being tested in other regions at that time), and also to obtain details about cotton pests.

The potential of various cotton varieties for ripening was estimated by forecasting, on PTNs, the developmental and ripening dates, as well as forecasting the prefrost yield for various sowing dates. This was done by superimposing onto the Yavan heat resources net the phenological curves for different cotton varieties; these curves were obtained *from observations in other regions of Tadzhikistan.* One and the same nomogram was used for the different mesoclimatic regions of Yavan, by means of moving the average temperature lines to the extent of the temperature anomaly of a particular mesoclimatic region in comparison to the base region for which the nomogram was constructed. The northwestern part of the Yavan Valley was taken as the base region (Fig. 1.50).

In Figure 3.6 some results of the forecasting calculations are represented. As can be seen, in the warmest, northwestern part of the Yavan Valley, before the extensive provision of water in this area, all cotton varieties should have been able to ripen (i.e., irrigated cotton) from the fast-ripening variety S-4727 (*G. hirsutum*) to the late-ripening variety of fine-fiber cotton 2I3, which was bred at the selection station of Iolotan, in Turkmenistan, from Egyptian varieties of *G. barbadense* (variety 2I3 is given below only for comparison). All the varieties ripen with a wide range of sowing dates—from March 25 to May 15—and

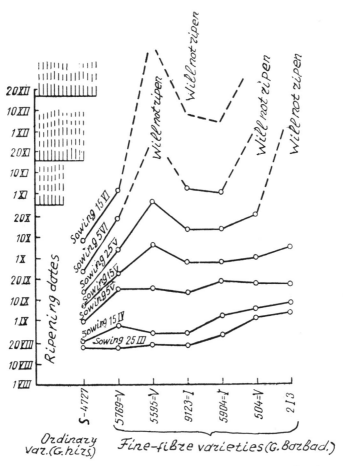

FIGURE 3.6. Calculated dates of mass boll unfolding for cotton of various varieties, after different sowing dates, in the northwestern part of the Yavan Valley, under the temperature conditions that existed before it was supplied with water. The shaded areas show the dates of the first autumn frosts in the air: the lower area—early occurrence of frosts; middle area—normal occurrence; top area—late occurrence.

variety S-4727 ripens with a much later sowing date. With still later sowings, some varieties do not ripen at all: first 2I3 (sowing on May 25); then 504-V and 5595-V (sowing on June 5), and 5904-I (sowing on June 15). This can be seen clearly in Figure 3.7, where the results of forecasting calculations (made on the basis of Figs. 2.6-2.14) for prefrost yield are represented. We also give data for other regions in Figure 3.8 and 3.9.

Thus the calculations showed that, in the Yavan and Dangara Valleys, fine-fiber cotton varieties can indeed ripen, and that these valleys can considerably widen the limited areas available for fine-fiber cotton in the USSR.

Moreover, the Yavan Valley is so warm and favorable that, according to the calculations, it is possible to obtain a harvest of wheat and cotton from the same field and in the same year. As an example, if, in the northwestern part of the Yavan Valley, wheat variety Surkhak 5688 is sown on December 26, then, according to the PTN (Fig. 1.50), mass sprouting will occur on February 3, the beginning of tillering on March 12, stalk shooting on April 12, mass heading on May 7, and wax ripeness on June 9. After harvesting the winter wheat, a fast-ripening cotton variety S-4727 (*G. hirsutum*) can be sown on June 15 to 20 in the same field, and can be harvested with a 70 to 75% prefrost yield. (The reader should repeat all these calculations for practice. For wheat the phenocurves are shown in Fig. 1.50, and for cotton the reader should

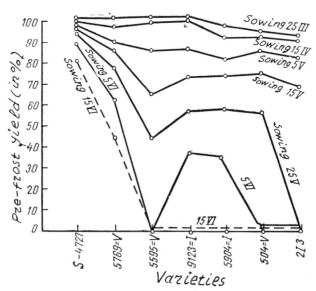

FIGURE 3.7. Results of calculations of prefrost yield (as a percentage of the total yield) for various cotton varieties, after different sowing dates, in the northwestern part of the Yavan Valley before it was supplied with water.

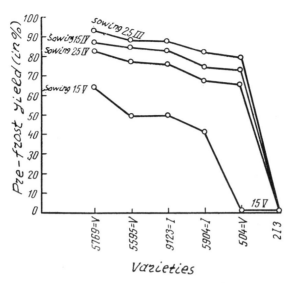

FIGURE 3.8. Results of calculations of the prefrost yield (as a percentage of the total yield) for various cotton varieties, after different sowing dates, in the middle section of the Yavan Valley after the valley was supplied with water. The middle section of the valley is approximately 2°—3°C cooler than the northwestern part.

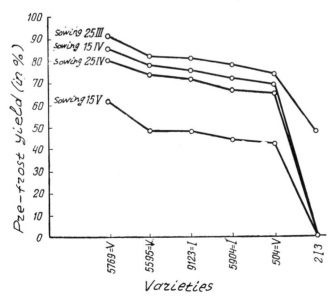

FIGURE 3.9. Results of calculations of the prefrost yield (as a percentage of the total yield) for various cotton varieties in the conditions of Dangara.

transfer the phenocurves onto the PTN from the corresponding figures.)

It should be noted that in periodicals and journals in the USSR the question of producing cotton and cereals in the same field in Middle Asia is very often discussed.

The above calculations were tested by a sicentist, Sh. Azamov, from the Tadzhikistan Institute of Agricultural Research. He became interested in the unusually late date of sowing for cotton variety S-4727. Azamov carried out experiments on plots of land in the northwestern part of the Yavan Valley (water for irrigation was brought in water-carts, because the experiments were done before Yavan was supplied with water).

Here is an extract from the field notebook:
Place where experiments were done—*Collective farm "Yavan."*
Sowing of cotton variety S-4727—*28 June 1962.*
Preliminary light irrigation—*30 June.*
Beginning of sprouting—*5 July.*
Second preliminary light irrigation—*9-10 July.*
First irrigation—*29 July.*
Beginning of budding—*1 August.*
Beginning of flowering—*23 August.*
Developed bolls of full value, 5-6 per bush—*30 September.*
Beginning of boll unfolding—*10 November.*
Mass boll unfolding—*12 November 1962.*

As can be seen, even when sown on June 28, the coarse-fiber cotton S-4727 has time to ripen. Of course it will ripen if sown on June 15 to 20 (after harvesting the wheat).

If we calculate the ripening of cotton S-4727 on the PTN, taking into account the actual date of flowering (August 23) in the experiment by Azamov, as well as taking into account the actual temperatures in 1962, then we can obtain the date of November 9 for mass boll unfolding: the difference between the calculation and the experiment is only +3 days.

Thus the forecasting calculations proved correct. The Yavan Valley now gives rich yields of cotton. However, as in other regions, it is proving difficult to introduce the production of high-quality, fine-fiber cotton in spite of the intervention of the press, and even of L. I. Brezhnev himself (see the newspaper *Izvestiya*, December 7, 1979). The cause for this is the striving for the gross yield of cotton, and the gross yield is a little higher in coarse-fiber varieties or in fine-fiber varieties of lower fiber quality.

For example, in Figures 3.6 to 3.9 we see that, according to calculations, with early, normal, and even delayed (but not late!) sowing dates, fine-fiber cotton variety 504-V ripens only 4 to 5 days later than the variety 5904-I. Moreover, the percentage of the prefrost yield differs little between these varieties. The facts bear out the calculations entirely: according to G. N. Borisov (1958), the difference in ripening is 6 to 7 days. According to this author's calculations, based on purely factual data from the Tadzhikistan NTV from 1955 to 1961 inclusive, the average percentage of the prefrost yield for variety 5904-I is 85.7, and for variety 504-V it is 81.8, that is, only 3.9% less. However, 504-V gives first-class fibers, whereas 5904-I gives third-class fibers, although variety 5904-I is also referred to as fine-fiber (*G. barbadense*) in the USSR. These small differences, however, were sufficient enough to cause the chairman of collective farms to prefer variety 5904-I. Only on occasions when all the sowing dates of cotton 5904-I had been missed, did they quickly "fill the gap" with variety 504-V. However, this latter variety does not "like" late sowing (see Figs. 3.6 to 3.9). Of course, in these cases variety 504-V gave low yields. In this way, 504-V, which was a good variety, bred by a talented crop breeder, V. P. Krasitchkov, was discredited.

Fortunately, Krasitchkov and his colleagues soon bred cotton varieties even better than 504-V: for example, 5595-V, which will be discussed below.

However, let us return to the subject of agroclimatic division into regions.

The fastest ripening of the coarse-fiber varieties is S-4727, and the fastest among the fine-fiber varieties, as can be seen in Figures 3.6 and 3.7, is 5769-V. Even after sowing on May 15, variety 5769-V gives a prefrost yield of 85%; variety S-4727 gives almost the same yield (more exactly, 80%), even after sowing close to June 15.

However, all other varieties give a good percentage of prefrost yield—80 and more—if they are sown before May 5 and in the warmest northwestern part of the Yavan Valley.

In other mesoclimatic regions of this valley, as well as in Dangara, it is cooler; therefore, the dates of ripening shift correspondingly.

After being supplied with water, the valleys will become somewhat cooler in summer, owing to the output heat of evaporation and transpiration of water, and owing to the absorption and dispersion of some components of solar radiation by water vapor. The qualitative structure of the solar radiation will also change somewhat. Calculations show that, in this case, for all the main varieties, an 80% and more prefrost yield can be obtained only by sowing not later than the end of April or the beginning of May, even in the warmest northwestern part of the Yavan Valley. However, the normal dates for sowing are at the end of March and the beginning of April.

The variety with the highest potential for the Yavan and Dangara Valleys proves to be the fine-fiber variety 5595-V.

Bioclimatic calculations similar to those given above were made by this author for cotton variety Acala S.J.-2 (*Gossypium hirsutum L.*) in

Israel. This was done by transferring the phenocurves from Figures 1.4 and 2.15 onto the heat resources nets for Israel (Fig. 1.111-1.128). The results of the calculations will not be published until they have been checked by planned field experiments. However, it can be assumed that, in view of the warm winter-spring period (in the majority of years), cotton can be sown in Israel at earlier dates than those usual for this area. The advantage of early and very early sowing is that budding and flowering will proceed, in that case, at relatively low temperatures, and therefore less of the fruit-bearing elements will fall. In addition, with these sowing dates advantage can be taken of the rain, thus reducing the cost of production. There are shortcomings, however: a prolonged vegetation period is required, and the assimilative apparatus is developed to a less considerable extent. The final word must rest with field experiments and economic calculations.

Calculations have also shown that in a number of regions in Israel the fine-fiber cotton *G. barbadense* (including Egyptian varieties) can be cultivated.

It is clear that similar calculations can also be made for cereals. Rich data on this subject are available from O. S. Yermakov (USSR).

. . .

It is absolutely clear that an agroclimatic estimation of a not-too-large region (such as Yavan) can be given without drawing the isolines of particular agroclimatic elements onto a geographical map. Where large territories are concerned, however, agroclimatic mapping methods considerably facilitate agroclimatic division into districts.

It seems that for mapping on the basis of Podolsky's method, we need, on the one hand, a climatic map of isotherms for each month or season, and on the other hand, PTNs for a few base points in each climatic zone as limited, for example, by two successive isotherms.

On the nomogram we calculate the multiannual date of a particular developmental stage for the organism that interests us, in the way described above. This date, of course, will be the actual date for the point to which the nomograms relate. However, if we move, on the nomogram, the average temperature lines necessary for the calculations, to the extent of the spatial temperature anomaly, we can obtain developmental dates not only for the base point, but also for neighboring points. The temperature anomaly for any point, in comparison with the base point, can be easily obtained using an isotherm map for that month or season to which the calculation on the nomogram relates. In this way we obtain a net of the dates that interest us, on the geographical map of a region. All that remains is to draw, an *isophene* between these dates, which will be a direct expression, and not an indirect one (by means, e.g., of temperature totals), of the vital activity of an organism.

Of course, we can similarly construct an agroclimatic map giving the optimum sowing dates; the dates of ripening and the volume of yield with different sowing dates; the number of harvests from one field in one year in the south; and other information concerning detailed agroclimatical division into regions.

As can be seen, such an agroclimatic map could become a program for action in the field of plant growing in each region.

In these cases, when the rate of an organism's development is related not only to temperatures, but also to nontemperature factors, we know the procedure (Section 2.2): onto the net of temperature resources we superimpose two to three phenological curves for one and the same organism, but for different values of the nontemperature factor. Then, when doing agroclimatic mapping, we use the base PTN with the phenological curve that corresponds to the state of the nontemperature factor in the given region of a large territory.

Of course, a heat resources net, once constructed, can be used for very different plants and poikilothermic organisms, as well as for agroclimatic and phenoprognostical purposes.

The method of bioclimatic mapping on the basis of Podolsky's method will be described in more detail using the example of poikilothermic organisms (Section 4.2b).

3.3 SOME ELEMENTS OF THE THEORY OF TESTING CROP VARIETIES

In Figures 3.6 and 3.7, it can be seen that the variety 5595-V (*G. barbadense*) does not respond well to late sowing in all parts of the Yavan and Dangara Valleys; when sown late, the ripening of this variety is progressively delayed, and the prefrost yield is reduced in comparison to other varieties.

All of this is explained by the phenocurve for this variety's development from sprouting to flowering (see curve 4 in Fig. 1.18). If the data are sufficient for constructing the curve, then we can draw the conclusion that the optimum temperature for this period is only 25°C, that is, much lower than for other varieties (it is convenient to compare the varieties shown in Fig. 1.50). Therefore, the increase of 24-hour average temperatures above this level (25°C) results not in a decrease, but in a renewed increase in the length of the interdevelopmental stage. Cotton that is sown late will be at the stage from sprouting to flowering during the highest temperatures (and will be at the ripening stage during the low temperatures). Hence the interdevelopmental stages become longer, especially in the warm, northwestern part of the Yavan Valley. Therefore, variety 5595-V is good for relatively cool regions (such as the middle section of the Yavan Valley and Dangara), and it is good starting material for further breeding of cold-resistant fine-fiber varieties: the critical stage of budding occurs in the middle of its "cold-requiring" stage, from sprouting to flowering. It is important to point out that one of the "parents" of variety 5595-V is a relatively cold-resistant Soviet cotton variety.

It should be noted that when cotton has a developmental curve of a kind similar to what variety 5595-V has in the period from sprouting to flowering (with a descending and an ascending branch), the temperature totals method cannot be used, because it contradicts the law of the optimum.

Variety 504-V (*G. barbadense*) also develops poorly after late sowing, but the causes are different: 5595-V develops poorly because with a late sowing date, the interdevelopmental stage from sprouting to flowering coincides with high temperatures; but with variety 504-V the reason is mainly that after late sowing, the period from *flowering to boll unfolding* coincides with relatively low temperatures, and the whole vegetation period becomes shorter. Hence the conclusions are diametrically opposed: the growing region for variety 5595-V can extend to the north of Tadzhikistan, but the region for variety 504-V is mainly in the south, although both varieties seemed to react identically to the delay in sowing.

Variety 5769-V (*G. barbadense*) is notable because its developmental curves are almost horizontal (Fig. 1.31). Therefore, the interdevelopmental stages of this variety depend on temperature to a lesser degree than those of other varieties. This means that the very fast-ripening variety 5769-V can be cultivated in quite a wide range of areas.

These examples are sufficient to show how important it is to classify varieties according to their ripening rate, taking into account the temperature conditions of the environment. It will then be clear why a particular variety, in a given region and with a given sowing date, turned out to be early-ripening in comparison to another given variety in a particular year, whereas a year later, or in another region or with a different sowing date, the same variety turned out to be late-ripening. For example, with sowing dates of March 25 to May 5, varieties 5769-V and 5595-V prove to develop almost identically (Figs. 3.6 and 3.7), whereas with later sowing dates, variety 5595-V is later-ripening, and the later it is sown the later-ripening is this variety in comparison to variety 5769-V. Varieties 5904-I and 504-V (both *G. barbadense*) in the northwestern part of the Yavan Valley, before it was supplied with water, differed little when sown before May 15, but variety 504-V, when sown later, had a considerably reduced prefrost yield in comparison with variety 5904-I (Fig. 3.7).

Thus the terms "early-ripening variety" or "late-ripening variety" are often senseless without an indication as to the region and the sowing date concerned, or more exactly, without indication as to the environmental temperatures to which these estimations relate.

In this connection, it would be very important if centers for the testing of crop varieties, as well as selectionists, when drawing up a "passport" system for varieties according to their speed of ripening, would construct phenocurves for various varieties on the basis of data over several years and from several places, or, even better, if they would construct an entire PTN. This would make it possible to assess the varieties under an entire range of temperature conditions and of sowing dates (the latter could be given on the PTN in a wide range). A PTN would enable selectionists and even agrometeorologists to see these important nuances in the varieties' behavior, which would be difficult to detect using any other method, and which could otherwise be found only as a result of very long, laborious, special experiments.

In addition, it should be remembered that Podolsky's method of phenological forecasts enables selectionists to make prior calculations of those cotton sowing dates which will be followed under given climatic conditions, by budding at low average temperatures over the period from sprouting to budding (Section 3.1i).

Figures 3.6 to 3.9 enable us to make other additions and amendments to comparative estimations of the rate of ripening for different varieties. For example, G. N. Borisov (1958) notes that variety 504-V ripens almost simultaneously with variety 2I3 (*G. barbadense*, Egyptian). In fact, in Figure 3.6 it can be seen that in the hot, northwestern part of the Yavan Valley, before the valley was irrigated, varieties sown before May 5 did not differ much from each other. If they are sown late, when ripening coincides with low temperatures, the difference between varieties 504-V and 2I3 increases very much, both in the ripening dates and in the volume of prefrost yield. This difference increases even more in relatively cold regions. In Dangara, for example, the difference in ripening dates between cotton varieties 504-V and 2I3 after sowing on March 25 increases up to 27 days (variety 2I3 ripens later); when they are sown even later, the variety 504-V ripens, whereas variety 2I3 does not ripen (the graph of ripening dates for this case is not given here, but this can be judged from Fig. 3.9).

Thus divergence is noted between the varieties in their ripening dates at low temperatures. In fact, phenological curves for various varieties in the period from flowering to boll unfolding, in the majority of cases, diverge sharply at low temperatures and converge at high temperatures (compare curves 20, 19, 16, and 6 in Fig. 1.31).

It should be noted that this applies not only to plants, but to other biological organisms. This is proved by Figures 1.55 to 1.57, 1.62, 1.74, 1.99, and many others. In these figures the phenological curves converge into a cluster at high temperatures and diverge at low temperatures.*

---

*This author ventures a small comparison: the "character" of a plant or of a cold-blooded organism can be judged better in unfavorble conditions than in favorable ones, just as a "friend in need is a friend indeed."

In temperature conditions of the northeastern part of the Yavan Valley, before it was supplied with water, variety 2I3 and other varieties would have ripened, according to the calculations, and would have given a 77% prefrost yield, even after sowing on April 25; whereas after the provision of water, variety 2I3 probably would not have ripened here at all, since the extensive irrigation most probably would have resulted in this region's becoming colder. However, such cooling is favorable to variety 5595-V, which does not require excessively high temperatures in the period from sprouting to flowering. With sowing on May 15, before the region was supplied with water, this variety ripened (according to calculations) on October 2, and gave a 70% prefrost yield; whereas after the provision of water, it ripened on September 27 with a prefrost yield of 76%.

In most cases, when valleys are supplied with water and are well irrigated, cotton ripens 4 to 5 days later, and the difference between varieties increases (according to calculations).

According to G. N. Borisov, variety 5904-I is somewhat faster in ripening than variety 9123-I—*G. barbadense* (in his book he gives the vegetation period for 5904-I as 130-150 days, i.e., an average of 140 days; for variety 9123-I he gives 142 days). However, in the conditions of Tadzhikistan, in the majority of cases, variety 5904-I, on the contrary, ripens 5 to 8 days later than 9123-I. Most figurs, like Figure 3.6, show a similarity in ripening dates (there may be some genetic relationship) between the fine-fiber varieties (*G. barbadense*) 5769-V, 5595-V, and 9123-I. This group of varieties differs comparatively sharply from the fine-fiber varieties (*G. barbadense*) 5904-I, 504-V, and 2I3.

Of course, the difference in ripening dates between varieties is related by means of temperatures to sowing dates, to the region, and to irrigation arrangements, and it can range from zero ad infinitum (when one of the varieties does not ripen). However, in the majority of cases, the difference in ripening dates between the "extreme" varieties among the fine-fiber varieties is 17 to 22 days, and the difference in the volume of prefrost yield is approximately 20% (with the help of Fig. 3.6, it is easy to calculate the length of the vegetation period for different varieties and sowing dates).

It is also difficult to indicate the precise difference in the ripening dates of a given variety in different regions of the Yavan Valley and of Dangara: this depends on the variety, but in most cases, ripening occurs in different locations and regions with a difference of 5 to 8 days for the same variety, corresponding to temperature differences of 1° to 2°C.

It is interesting to compare the graphs of yield accumulation (Figs. 2.6-2.14). These were constructed without any relation to one another; nevertheless, the prefrost yields, for example, on the 60th day after the mass unfolding of the first bolls, proved to be almost identical for all fine-fiber varieties (with the exception of 5769-V), without any relation to the type of branching—that is, both for branched varieties and for zero branching. For example, 5595-V accumulates an 83% prefrost yield on the 60th day after the mass unfolding of the first bolls, 9123-I—82%, 5904-I—82%, 504-V—84%, and 2I3—81%.

There is also almost no difference in the prefrost yield among different coarse-fiber varieties (*G. hirsutum*): 108-F on the 60th day gives 92%, 149-F—94%, and C-4727—92%.

However, there is a great difference between fine-fiber (*G. barbadense*) and coarse-fiber (*G. hirsutium*) varieties: fine-fiber varieties accumulate on the 60th day an average 82% prefrost yield, whereas coarse-fiber varieties accumulate a 93% prefrost yield. There is also a considerable difference between the shapes of the curves representing the yield accumulation: coarse-fiber varieties, initially accumulate the yield more quickly than fine-fiber varieties, and then they accumulate it more slowly (Fig. 3.10).

There is some difference in the case of fine-fiber variety (*G. barb.*) 5769-V: on the 60th day it gives a prefrost yield of 88%—that is, more than all fine-fiber varieties, but less than all coarse-fiber varieties—and its curve of yield accumulation has an intermediate shape (Fig. 3.10). Apparently, this is the nature of variety 5769-V, which is descended from a fine-fiber (*G. barb.*) and a coarse-fiber (*G. hirs.*) variety.

When conclusions made independently by different authors in different regions and at different times coincide, it means that these conclusions are correct: ". . . the ripening of cotton is related, obviously, to different rates of boll unfolding, and the capacity for rapid and simultaneous boll ripening has its own genetic basis" (Ter-Avanesyan, 1954).

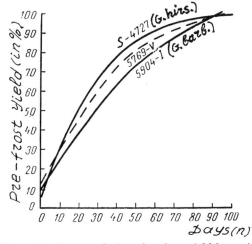

FIGURE 3.10. Dynamics of accumulation of prefrost yield for various cotton varieties (curves are transferred from Figs. 2.6, 2.11, and 2.13).

However, it may be that in this case there is not only a genetic factor, but also an ecological one: a relative delay in the boll unfolding of coarse-fiber varieties in late autumn (at $n > 40$ days, Fig. 3.10) can also be related to low temperature and higher humidity in autumn in the relatively cool regions of Tadzhikistan where coarse-fiber varieties are cultivated.

Variety 504-V gives a prefrost yield on the 60th day that is a little larger than that of 5904-I, but variety 5904-I ripens somewhat earlier and has a longer period of yield accumulation. Therefore, the final results (yield in %), up to the moment when vegetation stops, prove to be slightly higher for variety 5904-I.

The USSR State Committee for Testing Agricultural Crops called these calculations "a good beginning to working out a theory of crop testing and to the application of mathematics for studying problems in plant growing" (1964). However, it failed to introduce these methods in the practice of crop testing.

3.4. SOME ELEMENTS OF THE BIOLOGICAL "PASSPORT" SYSTEM

Let us pay attention to several important comparisons. If curves 5 (Fig. 1.17) and 8 (Fig. 1.18), constructed by A. R. Mustafayev, are correct, and if we compare them with our curves 4 and 3, respectively, we can see an astonishing coincidence between the phenocurves for winter wheat varieties Shark (Azerbaidzhan) and Surkhak 5688 (Tadzhikistan). An astonishingly exact coincidence was also shown between phenocurves 2 and 5 for spring barley (Fig. 1.21) in Tadzhikistan and in the European section of the USSR, respectively. It seems natural to question whether in all cases, when necessary, the varieties undergo considerable selective alteration upon being transferred from one region to another, and whether such a comparison of phenological curves can be used by crop breeders and by centers for crop testing as an indicator of the originality of the proposed new variety.

Of course, for cultured plants and their varieties, which have recently been transferred into new climatical conditions, the time factor has not yet played its part in the formation of a steady phenotype or ecotype. Otherwise, there would be no coincidence of curves such those noted above, because native plants display the obvious specificity of their phenotype to a give climatic region in comparison with the same plants native to another region. When comparing, for example, phenocurve 4 for the Turkmenistan mulberry tree with curve 5 for the Azerbaidzhan mulberry tree (Fig. 1.16), we see that the Turkmenistan mulberry requires more heat than the Azerbaidzhan mulberry, as should be the case, since Turkmenistan is hotter than Azerbaidzhan. When comparing phenocurve 6 (Fig. 1.18), or curve 7, for the Turkmenistan apricot with curve 9 for the Azerbaidzhan apricot, we see again the greater heat requirements of the Turkmenistan plant. Moreover, this difference between phenotypes naturally increases at low temperatures, as can be seen on the phenocurves.* Such "splitting" of phenotypes takes a very long period of time.

It is interesting to compare temperature curves for the development of spring wheat of the southern variety Irodi 1006 (Figs. 1.21, 1.22, 1.32) with the curves for spring wheat that has undergone a long acclimatization in the severe conditions of the Far East (Fig. 1.108). It is clear that the Far Eastern wheat ripens faster than Irodi 1006 at indentical temperature and that the relative early-ripeness of the Far Eastern wheat increases with decreasing temperatures.

The same conclusion follows from a comparison of the phenocurves for spring wheat Kharkovskaya 46 in Kazakhstan (a province bordering on Middle Asia) with the phenocurves for the same variety in Central Russia (curves 4a and 4c in Fig. 1.5).

The phenocurves for cherries *Cerasus pennsylvanica* and *Mahaleb* were compared by M. N. Nazrova and S. I. Mashkin (1979). These species were introduced into the botanical gardens of Voronezh University (central Russia), where they were observed. *Mahaleb* proved to have higher heat requirements than *Cerasus pennsylvanica* in various interdevelopmental stages. Both these species, as well as others (*Cerasus tomentosa, Cerasus japonica*), show a convergence of phenocurves at high temperatures and a divergence at low temperatures.

The same is true with regard to wild berries: see phenocurves 11 III for Byelorussia and 13 III for Kostroma, which is much farther north (Fig. 1.27).

A similar phenomenon can be seen in cold-blooded organisms. To see this, it is sufficient to compare the phenocurves of northern and southern ticks (Fig. 3.11). Probably *Ixodes ricinus* ticks are very plastic and have adapted well to the conditions of the severe north. According to some data, they are of Mediterranean origin (Pomerantzev, 1948).

The southern population of the piorcer (*Carpocapsa pomenella*) has higher heat requirements than the northern population. This can be seen by comparing phenocurve 1 in Figure 1.36 and phenocurve 1 in Figure 1.37.

In what order (and why) is the capacity for plasticity in plants, cold-blooded organisms, and microorganisms arranged? How is this property related to the level of an organism's organization?

---

*In connection with all of the above, it should be noted that the historical development of fruit trees in a particular region obviously relates to the spontaneous settling of seeds.

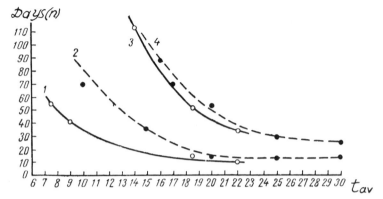

FIGURE 3.11. Temperature curves for the development of pasture ticks from the family *Ixodidae*. Solid lines: northern ticks *I. ricinus*, according to the laboratory experiments by E. M. Kheisin (1954) in Karelia; dashed lines: southern ticks *Hyalomma detritum*, according to the laboratory experiments by Galuzo (1947) in Kazakhstan; curves 1 and 2: the length of the period (*n*) between the time when the females fall from the host and the beginning of egg laying, in relation to the average temperature; curves 3 and 4: the same as curves 1 and 2, but from egg to larva.

The phenocurve for a given genotype is seen to shift in its entirety either up or down the *n* axis when the genotype undergoes more or less prolonged phenotypical or ecotypical changes; whereas the optimal temperature on this curve is apparently a more constant indicator, relative to the conditions of the genotype's formation.

In connection with this, the reader's attention is drawn to the fact that a number of phenological curves which are given in Chapter 1 have descending and ascending branches, that is, display optima of temperatures and development (curve 5, Fig. 1.1; curve 1, Fig. 1.2; curve 2, Fig. 1.4; curve 2, Fig. 1.7; curve 1, Fig. 1.8; curve 1, Fig. 1.9; curve 2, Fig. 1.17; curves 3-5, Fig. 1.18; curve 2, Fig. 1.20; curves 1 and 4, Fig. 1.22; curve 1, Fig. 1.24; curves 2 and 7, Fig. 1.27; curve 7, Fig. 1.29; curves 2, 4, and 10, Fig. 1.30; curve 2, Fig. 1.36; Fig. 1.43; curve 2, Fig. 1.45, and others). This type of phenocurves (more precisely, the ascending sections in the curves) is seldom obtained in the field with the usual sowing dates *in the middle and the northern sections* of the USSR (and even if it is obtained, it occurs in special experiments with late sowing). The cause of this is that the provenance of many cultured plants is further south than the northern and middle sections of the USSR; in the south they find optimal temperatures for their development, and also, apparently, in the southern mountains.

Therefore, the shape and position of a developmental curve for a plant (as well as for a cold-blooded organism) on a coordinates net can indicate the place of the organism's origin.

For example, the optimal temperature for flax development from sowing to sprouting is about 15°C (Fig. 1.24, curve 1), and for cotton variety 108-F from sowing to sprouting it is 25°C (Fig. 1.2, curve 1). Therefore, it is clear that the provenance of cotton is further south and at a lower altitude above sea level than the provenance of flax. We can also say that the developmental curves of cotton variety 108-F in subsequent developmental stages do not display any optimum, according to the experiments with sowing at different times which were made at the agrometeorological station of Dushanbe (Figs. 1.2 and 1.3); therefore, the two parent varieties of cotton variety 108-F, as well as those of other varieties, came from regions where it is warmer than in Dushanbe. On the other hand, the places of origin of wheat varieties Surkhak 5688 or Shark are on latitudes close to Tadzhikistan, because the initial developmental stages, and also the later stages (Fig. 1.18, curve 2) and the last stage (Fig. 1.18, curves 3 and 8), find optimal temperatures here. We should remember that "The Tadzhik agricultural and horticultural flora is a part of one of the most important world centers of origin for cultured crops" (N. I. Vavilov, selected works, Volume 5, Moscow-Leningrad, 1965, p. 565).

It should be noted that such analyses of curves aid the selection of pairs for crossing distant forms of plants.

Curve 5 for the cotton bollworm (Fig. 1.1) displays optimal temperature conditions for the organism's development in Tadzhikistan. Therefore, the cotton bollworm is of Tadzhikistan origin, or at least it has acclimatized well to the region.

It sometimes happens that totally different organisms give curves that coincide astonishingly well. For example, very important conclusions follow from the coincidence of the temperature curves giving the development of the malarial plasmodium and of the yellow fever virus in Figure 1.38 (for more details see Section 6.8).

The phenological curves for such geohelminths as *Haemonchus contortus* and *Bunostomum trigonocephalum* (Fig. 1.57) coincide, however, these are a closely related species.

Another interesting phenomenon is the similarity between the curve representing the distribution of the intensity of the mosquito's egg layings from cycle to cycle (gonotrophic), and the curve representing the sexual productivity of ticks (Fig. 6.6). The reader's attention is also drawn to the astonishing similarity between the shape of the curves representing the egg layings of mosquitoes and of ticks in relation to temperatures (Fig. 6.7). Moreover, it is interesting to note the fact that a number of cold-blooded organisms live longer at low temperatures: for mosquitoes, see Figure 1.40; for the oncospheres of the cattle tapeworm, see Figures 1.59 to 1.61, and Figures 6.1 and 6.2; for fleas, see Figure 1.63; for the virus of foot-and-mouth disease, see Figures 1.93, 6.16 to 6.18. Let us review all these figures on a single graph (Fig. 3.12).

The comparison and analysis of phenological curves is in itself a

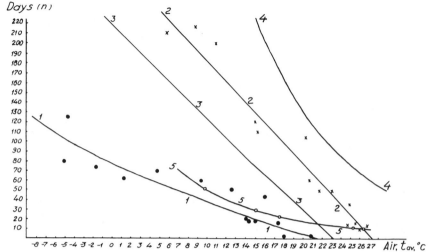

FIGURE 3.12. Phenological curves for the lifetime of organisms of different organizational levels, related to the air temperatures. Curve 1: duration of activity of the A_{22}-type virus of foot-and-mouth disease on straw substrate in shade. (The curve is drawn through the black, experimental points.) (After A. A. Slepov and A. S. Podolsky); curve 2: oncospheres of beef tapeworm (*Taeniarhynchus saginatus*) in overshaded grass. (The line is drawn through the crosses obtained from experiments.) (After R. I. Babayeva and A. S. Podolsky.); curve 3: the same as curve 2, but for a sunlit, bare soil surface; curve 4: souslik and marmot fleas (*Aphaniptera*) (after observations by I. G. Ioff); curve 5: malaria mosquito (*Anopheles superpictus*) (The curve is drawn through the outlined circles.) (After A. S. Podolsky, from observations by A. Ya. Storozheva and E. S. Kalmikov.)

large and original topic for further research, in light of N. I. Vavilov's theory about the centers of origin of cultured plants; there is room for further research into the problems of the genotype and the phenotype, the similarity between various organisms, and the transference of Vavilov's law about homologous series from the world of plants to the world of animals, and so on.

It follows from all of the above that it is risky to transfer phenological curves into sharply different climatic zones, especially if there is no unity of varieties and species. *Nevertheless, in more or less similar climatic conditions one and the same phenological curves for a given organism can be plotted on different heat resources nets (nomogram).*

For example, phenological curves for peaches and wheat were plotted onto the net for Dushanbe (Tadzhikistan), as constructed on the basis of observations in Uzbekistan (and taken from L. N. Babushkin, 1951); nevertheless, the forecasts based on these PTNs proved to be quite accurate.

In Figure 1.2 we see a convergence between cotton curves 1 and 2 based on data from relatively distant regions—Tadzhikistan and

Azerbaidzhan; a good coincidence with the data from Uzbekistan can also be seen.

The developmental curve for the cotton bollworm constructed by this author on the basis of observations in Tadzhikistan lies, in the graph's field, very close to the analogous curve constructed by the Institute for Plant Protection Research on the basis of observations in Uzbekistan. Even the analogous curves for Azerbaidzhan lie quite close to the curves for Tadzhikistan (Fig. 1.1). E. S. Budrik (1972), when using Podolsky's method of phenoforecasts in agricultural aviation (where an exact forecast is especially necessary), successfully superimposed the Tadzhikistan curves for the cotton bollworm onto the Azerbaidzhan heat resources net.

The phenocurves for the Colorado potato beetle, constructed on the basis of Kursk data (central Russia), lie close to the phenocurves constructed by L. I. Arapova on the basis of data gathered from south of Byelorussia (Fig. 1.41).

The same can be seen in Figures 1.46 and 1.47, taken from A. S. Danilevsky (1961).

Let us compare the experimental curves 3 and 3a that concern the class of parasitic fungi (Fig. 1.36). As can be seen, they coincide very well over a considerable temperature interval, in spite of the difference in geographical regions.

O. S. Yermakov showed over a wide range of geographical areas a remarkable degree of identity in phenocurves for the reproductive period in the development of cereals. In the prereproductive period, the effect of photoperiodicity becomes apparent: the phenocurves constructed for plants, which prefer long days on the basis of data from northern locations, are located below the analogous curves relating to the south; this indicates accelerated development in the north.* However, day length does not change very sharply with latitude; thus curves for the preproductive period can also be transferred to a certain extent.

Yermakov also showed that there is comparatively little relationship between phenocurves and the photoperiod in the case of cold-blooded agricultural pests. An exception is the piorcer, which displays considerable plasticity, as expressed by various phenocurves in the northern and southern regions of the USSR. This can be verified by comparing phenocurve 1 in Figure 1.36 with the phenocurve in Figure 1.37. Not without reason, piorcer has the widest area of distribution on the Earth; this will be mentioned again in the section concerning bioclimatic mapping (Section 4.2).

The investigations by K. A. Timiryazev, G. T. Selaninov, and A. A.

---

*Here it can also be seen that the northern populations of plants "become accustomed" to managing with less heat.

Shigolev are also indicative of the constancy of the heat characteristic of the development of plants in different geographical regions.

Thus plants and poikilothermic organisms that have been very adaptable over historical periods—when migrating through countries and continents—display a certain constancy (conservatism) within the bounds of comparatively small areas (especially if a species or a variety has not undergone centuries of acclimatization in the place to which this species' phenocurve is transferred).

It has already been mentioned that in the course of many decades of experimental work a vast amount of phenological data, often unused, has been collected in the archives of experimental stations, crop testing stations, research institutes, and meteorological stations. It is therefore perfectly possible to begin making a "passport" system for plants, varieties, and poikilothermic organisms by plotting phenotemperature curves, and to compile a catalog and atlas of such curves.

As a matter of fact, the first 300 phenological curves for such an atlas are given in this book. A similar quantity of curves has been drawn, but not yet published, by O. S. Yermakov, V. A. Ryabov, V. V. Denisov, and E. A. Sadomov. If we add to these 600 phenocurves a number of curves "scattered" by a large number of authors in various publications, then this will be a good beginning for a catalog and atlas of phenotemperature "passports" for various organisms.

Such an atlas, without a doubt, would help in studying the biological, genetic, and ecological peculiarities and principles of the development of plants, of different varieties, and of cold-blooded organisms. In the end, such an atlas would help in the selection and testing of crop varieties, and in parasitology.

The first steps in this direction are already being taken. First, it is necessary to systematize the phenocurves. Some phenocurves indicate a strong reaction to temperature on the part of organisms, whereas others indicate a weak reaction; in still others, the sign (plus or minus) of this reaction changes (i.e., when temperatures increase, the development decelerates, instead of accelerating).

This reaction of an organism to heat can be expressed at each point of a phenocurve in an analytical manner, by the first derivative of the developmental period (n) with respect to the temperature (t): dn/dt. However, since the developmental rate can be influenced not only by temperature, but also by other factors, then when we discuss only one factor t (argument), it is logical to write the organism's reaction to heat in the form of a partial derivative: $\partial n/\partial t$.

It is clear that each point on a phenocurve has its own $\partial n/\partial t$. However, for each phenocurve it is possible to calculate a certain average value, $(\partial n/\partial t)_{av}$, over the interval from the beginning to the end of the curve. This average value can be obtained approximately with respect to the endpoints of the interval, if we assume conditionally that in this place the function $n = f(t)$ is linear. In other words, $(\partial n/\partial t)_{av}$ is replaced by the value $\pm \Delta n/\Delta t$ (minus for a descending phenocurve and plus for an ascending phenocurve).

The higher the value $|\Delta n/\Delta t|$, the steeper, generally speaking, is the phenocurve, and the more pronounced is the organism's reaction to heat.

Although the intervals that are used for calculations of $\Delta n/\Delta t$ prove to be unequal for different phenocurves, nevertheless, each organism has its own heat characteristics, and thus its own "passport."

For example, Ryabov (1979) divided 25 meadow-steppe plant species, which form cenoses in the Central Black-Earth Reservation in the USSR, into the following groups:

1.   Species that are the most susceptible to changes in temperature, such as hellebore (*Veratrum*), awnless brome grass (*Bromus inermis*), and others; in all there are eight species. For these species:

$$\frac{\Delta n}{\Delta t} = -\frac{\text{5-8 days}}{1°C}$$

2.   Species of intermediate sensitivity, such as pot marjoram (*Origanum vulgare*), feather grass (*Stipa pinnata*); in all there are 14 species. For these species:

$$\frac{\Delta n}{\Delta t} = -\frac{\text{3-5 days}}{1°C}$$

3.   Species of low sensitivity, such as lychnis (*Lychnis*) and forget-me-not (*Myosotis*). For these species:

$$\frac{\Delta n}{\Delta t} < \frac{\text{3 days}}{1°C}$$

These will be brief analytical "passports," which form a good supplement to the graphic passports of the corresponding species, that is, phenological curves.

Apparently the idea of characterizing species through phenological curves is a sound idea. It occurred not only to this author and to Ryabov, but also to V. B. Gedikh (1976).

He also writes that the shape of phenological curves enables us to draw conclusions about the biology and ecology of plants. For example, Gedikh thinks that the relatively marked steepness of the phenological curves for wild berries (cranberry, bilberry, bog whortleberry), during the period from the time when the 24-hour average temperatures exceed

0° C to the opening of the buds, and from flowering to fruit ripening, indicates high heat requirements during these periods.

In contrast, the declivity of phenocurves for the period from bud opening to flowering indicates the endogenic character of the factors that cause the developmental rates during this period.

M. N. Nazarova and S. I. Mashkin (1979) gave the ecological characteristic of the cherry species, based on their phenological curves (see Section 3.4).

Of course, it is possible to classify organisms on the basis of analytical equations of phenological curves (as opposed to graphs). However, this is less convenient and less reliable. It is less convenient because it is relatively laborious to calculate the equations; it is less reliable because empirical phenocurves are approximated here, usually to a few types of standard lines: straight line, hyperbola, and parabola. However, an empirical curve usually has more variety than these standards: its various sections can be expressed by different equations, and this fact greatly complicates the classification and the "passport" system drawn up on the basis of equations.

CHAPTER FOUR

Phenological Forecasts and Bioclimatic Estimations for Entomologists and Phytopathologists

4.1. OPERATIONAL FORECASTING FOR DEVELOPMENTAL DATES OF POIKILOTHERMIC AGRICULTURAL PESTS

4.1a. A Little History. The Temperature Totals Method in Parasitology, in General, and Some Critical Notes on It.

Although the founder of the temperature totals method both in plant growing and in entomology was the French physicist and biologist Réaumur, who used this method for plants and insects in the first half of the eighteenth century, the use of the method developed in plant growing much earlier than in entomology. For example, the basic equation of the effective temperature totals method,

$$n = \frac{A}{t_{\mathrm{av}} - B}$$

which was proposed for plants by Gasparen (1844) and Babinet (1851), was not repeated for insects by H. Bluck until 1923, and then by F. S. Bodenheimer in 1924.

True, it should be noted that the idea of totals of squared temperatures, which began to be used relatively recently in entomology (A. Rodd, 1952), is also not entirely original: the equation, where the length of the interdevelopmental stage of a plant is inversely proportional to the squared average temperature of this period as counted from $0°C$ (and therefore, the product of multiplying the period by the squared temperature gives a constant total of squared temperatures), was introduced into plant growing at the end of the nineteenth century by the meteorologist Sonee (see E. G. Loske, 1913).

Nevertheless, although the temperature totals method was used in entomology later than in plant growing, entomologists and parasitologists have made better progress than plant growers in the understanding and analysis of the defects in this method.

For example, Carradetty, using the hyperbola equation mentioned above, which was adapted by Bodenheimer for describing malaria mosquito development, obtained results that differed greatly from actual developmental dates (V. N. Beklemishev, 1944).

Academician Beklemishev, as well as Schelford, obtained, using actual data, a phenological curve (but not a straight line!) for the developmental rate of *Anopheles maculipennis* (this curve has a shape very similar to the curve for cotton shown in Fig. 3.5, as well as to the scheme given by B. B. Wiggles-Worth, 1965). Beklemishev's curve, in full accordance with the mathematical theory described in Section 2.1, shows that the totals and thresholds of effective temperatures are manifestly inconstant along the phenological curve. Therefore, he concludes that Bodenheimer's hyperbola correctly describes the relationship between temperature and the duration of the development of *A. maculipennis* only at temperatures between 14° and 25°C. Moreover, he writes that in those days (1944) there was no generally accepted theory about the relationship between temperature and developmental rate. Moreover, he continues, it is very likely that the forces and factors influencing the developmental rate over different sections of the temperature scale are very different. There is also no theoretical equation that would express the relationship that interests us. All of the available equations—hyperbola, catenary curve, and so on— enable us merely to construct interpolated curves from which, as a result of defining the duration of development at several different temperatures, we can judge the duration of development at any intermediate temperatures (all of the above is paraphrased from Beklemishev).

Beklemishev also showed that accustoming mosquitoes to high or low temperatures, we can considerably shift the thresholds of activity.

Facts given by other entomologists also indicate that the lower thresholds of development change in relation to the temperature conditions in which the organism lives. For example, it is known that wintering larvae of the beet webworm (*Loxostege sticticalis*) have a lower developmental threshold of 10°C in spring, whereas larvae of the summer generations have a threshold of 17° to 18°C. The same is true of the pupae of the cotton bollworm, and of the pine butterfly, and this phenomenon is generally widespread among insects having several generations in a year (Kozhanchikov, 1961).

The works of N. Chumachenko (1954) also indicate the totals of effective temperatures change in relation to the intensity of the metabolic processes during the development of insects, and that this total increases in most cases from high temperatures to low temperatures within the bounds of one and the same interdevelopmental stage—in full accordance with the conclusions obtained above for plants (Fig. 3.5). For instance, Chumachenko obtained the totals of squared 24-hour average temperatures required for the full developmental cycle of *Laphygma exigua Hb.* in sugar-beet fields; one of these totals was 17,643° C, when *Laphygma exigua Hb.* developed within the temperature limits of 30° to 26.8° C, and the other total was 20,470° C, when *Laphygma exigua Hb.* developed within the temperature limits of 25.2° to 20° C.

Dj. D. Kudina gives in her work (1971) the fluctuations, in various years, of effective temperature totals which are accumulated up to the actual beginning of pupation of the European corn borer (*Ostrinia nubilalis*) in the Ukraine: in the Poltava district the fluctuation is 101° to 396° C, in the Sumy district it is 172° to 261° C, and in the Chernovtcy district it is 160° to 278° C. Kudina also obtained considerable fluctuations over the period up until the beginning of the flight of butterflies and she writes that with such fluctuations of totals of effective temperatures in various years, it is hardly possible to use these totals for forecasting individual developmental stages of the European corn borer.

E. M. Kheisin (1954) found considerable inconstancy in the temperature totals that are necessary for the development of the ticks *Ixodes ricinus*. To this list could be added a considerable number of other examples.

From these fluctuations errors result in phenoforecasts. A number of authors note a systematical increase of phenoforecasting errors for insects from generation to generation (sometimes comparatively large errors are found, not only for the latest generations, but also for the earliest generation, because of 24-hour fluctuations in temperatures). L. N. Gaplevskaya (1960) writes, on the basis of experiments in 1957 and 1958 in Tadzhikistan, that errors in phenoforecasts calculated for the cotton bollworm by the temperature totals method increase up to the last generation. The causes of this are discussed in Section 2.1; they lie in the fact that temperature totals lose their biological sense at optimal and above-optimal temperatures.

N. Ya. Sokolenko (1956) used the method of the Middle Asian branch of the USSR Plant Protection Institute. This method states that the flight of spring butterflies of the piorcer *Carpocapsa pomonella* occurs after the accumulation of effective air temperature totals of 190°C, with a lower threshold of 10°C; and that the flight of the subsequent generations occurs after the accumulation of totals of squared active air temperatures of 24,000°C. Sokolenko obtained the following results by using this method (Table 4.1).

The data in Table 4.1 show that during each of the three years in the Hissar Valley in Tadzhikistan, the errors in phenoforecasts decrease slightly from the wintering generation to the first generation. They increase very sharply as we move to the second generation (from 1-3 days to 10-19 days). For the warmer Vakhsh Valley we can see a constant increase in the extent of the errors from generation to generation (+1, +7, +16 days). The cause is the same. It is directly related to the fact that in 1953, on the actual flight date of butterflies of the first generation—June 16—the total of squared active temperatures was 23,688°C (see the line "observations"), whereas on the actual flight date of the second generation—August 6 (beginning from June 16)— the total was 38,296°C, almost twice as much, despite the principle of the temperature totals method; in 1954, the totals were 26,404° and 31,528°C, respectively, and in 1955 they were 27,298° and 31,197°C (all of this is for Dushanbe).

As a result, there is an increase in the differences between the calculation, based on the constant total of 24,000°C, and the actual state of affairs (it should be noted that Sokolenko obtained these large errors in spite of the fact that the calculation of 24,000°C was based entirely on actual temperatures).

Insects, apparently, are somewhat more adaptable than plants to environmental conditions: they more often find a developmental optimum, especially in the south, and especially in the case of summer generations; this is why the temperature totals method needs a substitute in the field of parasitology even more than in plant growing.

Nevertheless, this author is not opposed to using different approaches to the problem. Science and practice need methods involving different degrees of exactness and usefulness. The work of B. V. Dobrovolsky (1960, 1961), for example, in which he defines the average multiannual developmental dates of agricultural insect pests on the basis of actual data, may also be useful. The large amount of cartographic data compiled by N. K. Shipitzina (1957) (malaria mosquito) on the basis of the temperature totals method is also useful. The series of malariological calculations carried out by Sh. D. Moshkovsky and N. N. Dukhanina using temperature totals were also necessary. Of great interest are works carried out outside the USSR, including the works mentioned previously. To them F. S. Bodenheimer's works on poikilotherms should be added, as well as works by a number of other authors—K. Fedra (1980), P. W. Wellings (1981), J. R. Willard (1972), and others.

TABLE 4.1. Dates of flight of piorcer butterflies, both calculated (using constant totals of temperatures accumulated according to the actual data) and observed in Tadzhikistan (according to Sokolenko).

		Wintering Generation				First Generation				Second Generation			
			Total of Effective Temperatures (in °C)	Difference Between Observations and Calculations			Total of Squared Active Temperatures (in °C)	Difference Between Observations and Calculations			Total of Squared Active Temperatures (in °C)	Difference Between Observations and Calculations	
Year	Data	Date		In Days	In Temperature Total	Date		In Days	In Temperature Total	Date		In Days	In Temperature Total
1	2	3	4	5	6	7	8	9	10	11	12	13	14
						Hissar Valley (Dushanbe)							
1953	Calculated	29 IV	195	—	—	17 VI	24400	—	—	18 VII	24430	—	—
	Observed	25 IV	152	-4	-43	16 VI	23688	-1	-712	6 VIII	38296	+19	+13866
1954	Calculated	27 IV	194	—	—	15 VI	24636	—	—	22 VII	24418	—	—
	Observed	17 IV	121	-10	-73	18 VI	26404	+3	+1768	5 VIII	31528	+14	+7110
1955	Calculated	27 IV	190	—	—	20 VI	24141	—	—	23 VII	24194	—	—
	Observed	22 IV	147	-5	-43	22 VI	27298	+2	+3157	2 VIII	31197	+10	+7003
						Vakhsh Valley							
1955	Calculated	11 IV	191	—	—	7 VI	24166	—	—	10 VII	24280	—	—
	Observed	12 IV	End of flight	+1	—	14 VI	33825	+7	+9659	26 VII	31162	+16	+6882

4.1b. Short-Term Forecasts with Prolonged Foresight for Developmental Dates of Agricultural Pests (Phenoclimatic Forecasts)*

It is known that forecasts of the developmental dates of pests and diseases of agricultural crops are necessary, first of all, in order to determine the appropriate dates for expensive mass treatment with pesticides and fungicides, especially when chemical poisons have limited periods of toxic effect. The wide application of technical means of plant protection (e.g., agricultural aviation) requires a more exact knowledge of the optimal dates for their application. Moreover, useful insects may perish together with the pests if we do not first forecast the developmental dates of both.

The intensive development of biological methods in pest and disease control in recent years requires forecasts not only for the parallel developmental dates of a pair of poikilothermic organisms, but often also of the third organism—the plant. For example, in order to effectively use egg eaters to annihilate the pest *Eurygaster integriceps*, it is necessary to know the phenology of the egg eaters, of the pest, and of the wheat upon which the pest feeds.

Forecasts of the developmental dates of pests are also necessary for choosing the right time for the application of agrotechnical pest-control measures (for instance, treatment of fallow or treatment of space between the rows for the annihilation of pupae). Finally, in recent years entomologists have begun to take more and more interest in the phenology of useful insects.

Phenoclimatic forecasting of the developmental dates of cold-blooded organisms does not differ in its form from analogous forecasting for plants.

For example, we require a forecast for the beginning of the first mass egg laying of the cotton bollworm (*Chloridea obsoleta* F.) for 1959 in Kurgan-Tyube: it is known that the actual date of this pest's spring awakening in 1959 was April 6 (the date of the cotton bollworm's awakening is usually assumed to be the time when the 24-hour average air temperatures regularly exceed 15°C, unless there are direct observations of the spring awakening).

---

*According to the terminology accepted by the USSR Plant Protection Service short-term phenoforecasts are precalculations with a shorter period of foresight than seasonal forecasts for the distribution of insects, or than forecasts for the following year. That is why phenoclimatic forecasts with prolonged foresight, which were discussed previously in relation to plants, and will be discussed now, come under the heading of short-term forecasts. The foresight period of short-term entomological forecasts, as made by the Plant Protection Service on the basis of temperature totals, is only 5 to 10 days, whereas the foresight period of our phenoclimatic forecasts is more than 30 to 40 days. Therefore, the latter, although they are relatively short-term forecasts, definitely have prolonged foresight.

As before, we find on the PTN for Kurgan-Tyube (Fig. 1.55) the line of average temperatures relating to April 6 (this will be a little to the right of the line dated April 5). We trace this line up to the intersection with phenocurve 11, representing the cotton bollworm's development during the period from the moment when the 24-hour average temperatures exceed 15°C to the first egg laying (remember that the phenological curve is transferred onto the PTN from Fig. 1.1, where it was constructed on the basis of actual dates over different years in different regions). From the intersection we move the pencil in a strictly horizontal direction to the left, to the axis of ordinates, where we find that the intersection corresponds to $n = 64$ days. This will be the period between the awakening and the first egg laying. Thus the first egg laying should occur in 1959 on April 6 + 64 days = June 9.

In actual fact, the beginning of the first mass egg laying of the cotton bollworm was observed in 1959 in Kurgan-Tyube on June 13. If we take into account that the forecast was made 70 days ahead, then its error of +4 days is quite small.

As was already mentioned, there can be difficulties in defining the date when the 24-hour temperatures regularly exceed 15°C, although it has been agreed to take this date as being the time after which the 24-hour average temperatures either never fall below 15°C, or do so only on rare occasions. Therefore, there may sometimes be room for doubt. For example: should we take the date mentioned above of April 6 as the date when the 24-hour average temperatures regularly exceed 15°C, or should we take, for example, April 15?

It turns out that even such a big difference in the initial forecasting date in the spring does not cause any considerable change in the result of the forecast. In fact, the line of average temperatures from April 15 intersects phenocurve 11 at $n = 56$ days, hence the date of the first mass egg layings proves to be close to the date calculated above: April 15 + 56 days = June 10.

The cause here is the phenological principle (apparent especially in spring) that individuals that begin to develop later at high temperatures overtake the earlier individuals in their development. To this law a special chapter will be devoted (Chapter 8).

On the basis of the first actual date of mass egg laying for the cotton bollworm in Kurgan-Tyube—June 13, 1959—we must calculate the forecast for the second egg laying.

We find in Figure 1.55 the line of average temperatures dated June 13, and we trace it up to the intersection with phenocurve 12 for the cotton bollworm's development between egg layings. At this intersection $n = 30$ days. Thus the forecast for the beginning of the second mass egg laying will be June 13 + 30 days = July 13.

In the field conditions of 1959, egg laying was observed on July 13. As can be seen, the forecast proved to have no error, in spite of its one-month period of foresight and in spite of the fact that between the actual date of the first egg laying and the date of the second egg laying, only the multiannual temperature trend (the heat resources net) was taken into consideration. This is obviously the influence of the actual initial date of the forecast, which is an integral (total) of the concrete influences of place and time. In the same way, we find in Figure 1.55 the date of the third egg laying—August 13. This forecast also has no error.

Finally, on the basis of the actual date of the beginning of the third mass egg laying of the cotton bollworm—August 13, 1959—it is necessary to define whether the beginning of the fourth egg laying in Kurgan-Tyube will occur and when.

Even if we prolong phenocurve 12 (Fig. 1.55), we can see that the average temperatures line of August 13 at best will only touch curve 12. This means that such a late third egg laying as August 13 is the latest that can still be followed by a fourth egg laying, and in this case, in fact, the fourth egg laying would occur very late.

However, this would all be correct only if the temperatures in 1959, beginning from August 13, were equal to the multiannual temperatures, because the heat resources net shown in Figure 1.55 is constructed on the basis of multiannual temperatures.

But these temperatures in 1959 could prove to be different from the multiannual temperatures. In Figure 1.55 it can be seen that the slightest movement upward or downward of the average temperature line dated August 13 considerably changes the ordinate of intersection of this line with developmental curve 12, because the latter rises up very steeply in its left-hand, low-temperature end.

Thus it is clear that if, using such steep phenological curves, it is possible to obtain good forecasts in spring (when average temperature lines intersect phenocurves at wide angles), then in autumn (when average temperature lines intersect phenological curves at narrow angles) it is necessary to determine the course of the average temperature lines using partial actual data. In short, in these cases phenoclimatic forecasts must have not prolonged foresight, but shortened foresight (Section 4.1c).

By analogy with the above, let us calculate phenoclimatic forecasts with prolonged foresight for the piorcer (*Carpocapsa Pomonella* L.) in 1959 for another region—Dushanbe—using Figure 1.56 (PTN) together with biological curves 17 and 18. The mass flight of the wintering generation is forecast for March 19 + 42 days = April 30 (here, March 19 is the date when the averge air temperatures regularly exceed 10°C). The actual date of the mass flight of butterflies from the wintering reserve in 1959 in Dushanbe was April 27. The date of the flight of the first generation is forecast as April 27 + 59 days = June 25 (using curve 18). In fact, the mass flight of butterflies of the first generation was observed in 1959 in Dushanbe on June 23. The flight of the second generation is forecast for June 23 + 46 days = August 8; in fact, it took place on August 10.

Finally, it is necessary to find out whether a flight of butterflies of a third generation will occur. The multiannual line of average temperatures from August 10 does not intersect phenocurve 18. From the theory of PTNs, we know that this means it will be impossible for butterflies of a third generation to fly, because of insufficient heat.

However, since here (in the second half of summer) the angles of intersection of the average temperature lines with the phenocurve are narrow, such a phenoclimatic forecast with prolonged foresight must be checked by means of a forecast with shortened foresight using the actual temperatures over part of the forecast period. In this particular case, such a forecast did not change the result—the flight of a third generation is again forecast as impossible.

And in fact such a flight did not occur. However, in Figure 1.56 it can be seen that if the year of 1959 had been anomalously warm to a sufficient extent, and if, owing to this, the line of average temperatures dated August 10 had shifted to the right by 1.5 to 2°C, then a third generation would have developed. In this case, however, the piorcer should have been looked for, not on apples (which would have been gathered long before), but on late varieties of quince. A third generation is most likely to occur in the valleys of southern Tadzhikistan in hot years.

We give in their entirety all the results of forecasting in 1959 for the developmental dates of the cotton bollworm and the piorcer, using Podolsky's method, as published by the USSR Plant Protection Institute—VIZR (Polyakov, 1960), and by the Tadzhikistan Ministry of Agriculture (Stativkin, 1959) (Tables 4.2 and 4.3).

As can be seen, the forecast dates coincided in the majority of cases with actual dates not only in the Hissar Valley, but also in the warmer Vakhsh Valley. Moreover, there is no increase in the values of the errors in the case of the last generation.

Sokolenko (1956) compared, for other years (1953-1955), the forecasts for the developmental dates of the piorcer in Tadzhikistan: on the one hand, forecasts made using the method that combines effective temperature totals with totals of squared active temperatures; on the other hand, the forecasts made using Podolsky's method of phenological forecasting. Sokolenko found a divergence between the actual dates and the results of calculations made on the basis of constant temperature totals; this divergence was as much as 19 days. With such a divergence, he notes that it is impossible to use the totals method for the purpose of forecasting in the conditions of depressive temperatures (i.e., above-optimal and optimal temperatures) that occur during the second half of the summer. On the other hand, Sokolenko says that the accuracy of forecasts made using Podolsky's method in conditions of

TABLE 4.2. Comparison of piorcer's developmental dates, calculated on the basis of the new mathematical phenology method, with the same dates from actual data for Tadzhikistan in 1959 (after V. G. Stativkin, Editor).

	Date of the Mass Flight of Butterflies					
	Wintering Generation		First Generation		Second Generation	
Region	According to the Forecast	Actual	According to the Forecast	Actual	According to the Forecast	Actual
Dushanbe (Hissar Valley)	30 IV	27 IV	25 VI	23 VI	8 VIII	10 VIII
Kurgan-Tyube (Vakhsh Valley)	21 IV	15 IV	9 VI	10 VI	27 VII	25 VII

depressive temperatures is ±3 to 5 days. He therefore concludes, after checking the new method in zones of depressive temperatures, that is *can be used for forecasting the developmental dates of the piorcer in orchards in Tadzhikistan.*

It should be noted that when no combination of temperature totals had been able to achieve the desired results, phenoforecasting for the piorcer in the USSR was offically removed from the "arsenal" used against this pest. However, as can be seen, there does exist a method that ensures the necessary accuracy.

The same was also stated by I. T. Korol (1969), who used Podolsky's method during a number of years for forecasting the dates of the piorcer's developmental stages in Byelorussia in relation to the application of microbiological preparations against the piorcer.

There have also been successful phenoforecasts for the piorcer in the CBEZ (O. S. Yermakov and A. S. Podolsky, 1973). In the same way as Korol, we used the phenocurve obtained not from data for Tadzkikistan, but from phenological observations in the ESU (see, for instance, Fig. 1.37).

However, there is also a partly negative result. When I. T. Kulakova (1972) used Podolsky's method for the piorcer in the Gorkovsky district of the USSR (to the east of Moscow), phenoclimatic forecasts with prolonged foresight proved to have little success. The cause is, as Kulakova says, that during the years of testing the temperatures differed greatly from the multiannual temperatures. The use of fore-

TABLE 4.3. Comparison of the dates of the beginning of mass egg laying in the cotton bollworm, as calculated using the mathematical phenology method, with the same dates according to actual data in various regions of Tadzhikistan in 1959 (after I. Ya. Polyakov, Editor; V. G. Stativkin, Editor).

	Dushanbe		Kulyab		Kurgan-Tyube	
Generation	According to the Forecast	According to Field Observations	According to the Forecast	According to Field Observations	According to the Forecast	According to Field Observations
First	20 VI	22 VI	11 VI	10 VI	9 VI	13 VI
Second	26 VII	27 VII	11 VII	5 VII	13 VII	13 VII
Third	—	27 VIII	5 VIII	4 VIII	12 VIII	13 VIII

Note. Forecast of the third egg laying in Dushanbe and the fourth egg laying in Kurgan-Tyube are calculated below in Section 4.1c using forecasts with shortened foresight.

casts with shortened foresight diminished the errors from 1 to 20 days (?!) to 0 to 5 days.

The Podolsky's method of phenoforecasting also came into use in Georgia (1976).

However, let us return to the cotton bollworm. The data of Table 4.3 show that the forecast dates and the actual dates of the cotton bollworm's development also coincide during the period of active piorcer control in Tadzhikistan. In the neighboring republic of Uzbekistan, however, the temperature totals method gave, during the same developmental period, errors of ± 10 to 13 days. Moreover, author's method gave forecasts 30 to 60 days before the event, whereas the temperature totals method provided forecasts only 5 to 10 days before the event. A review published by the Tadzhikistan Ministry of Agriculture states that in 1959 early notice was given of the beginning, and the mass appearance, of individual developmental stages of the cotton bollworm. All districts were informed in advance about the beginning of mass egg laying for each generation. The forecasts were made on the basis of PTN's, which proved to be entirely correct.

Almost identical relationships between the errors of forecasts for the egg layings of the cotton bollworm as made using the totals method and Podolsky's method were obtained in special testings made over previous years (Gaplevskaya, 1960).

However, the evaluation of the proposed method for cold-blooded organisms would not be complete if we had no idea about the extent of the fluctuations in the natural elements forecasted. It turned out that the duration of comparable interdevelopmental stages of the cotton bollworm changed by 20 to 25 days between different years and 20 to 25 days between different regions in Tadzhikistan. Moreover, the difference in the lengths of interdevelopmental stages for different generations in one and the same year and place can be as great as 50 days. In different years and in different places, the length of interdevelopmental stages differs by even more than 50 days: it is sufficient to compare, in Figure 1.1, the highest points of line 4 with the lowest points (α and β) of curve 5. Of course, in those cases, when, for example, the fourth egg laying of the cotton bollworm occurred in one year, and not in another year, the difference between years is infinite (from the mathematical point of view).

As can be seen, the exactness of forecasts based on average (over a number of years) lengths of interdevelopmental stages, without any relation to temperatures and dates, would in Tadzhikistan be at best ± 10 to 12 days.

In another neighboring republic—Kazakhstan—the method involving prolonged foresight was used for the pest *Apamea anceps Shiff.— H. sordida Bkh.* The mean-square error σ proved to be ± 4.6 days with an 80% guarantee* (V. K. Adzhbenov, 1971, 1972), as compared with ± 3.9 days for spring wheat. We note that the rate of success in forecasts for cold-blooded organisms is somewhat lower than for plants.

The urgency of the problem of the spreading of the fall webworm (*Hyphantria cunea Dr*) is well known. Comparatively recently this omnivorous and dangerous, quarantined pest was not found in the USSR. Now the development of the fall webworm in the southern regions of the USSR has become disturbing, and it should be eliminated.

For solving this problem, mathematical forecasts of the developmental dates of this pest are of great importance, especially if we take into consideration its small numbers, which makes it difficult to find this pest in nature (I. A. Churayev, 1962).

In Figure 1.48 phenocurves for the fall webworm are given, and in Figure 1.71 a PTN for Eisk, incorporating these phenocurves, is given. The system for calculating phenoforecasts for the fall webworm does not differ in principle from the calculations described above. If, for example, the actual date of the caterpillar's appearance is June 18, then pupation is to be expected on June 18 + 33 days = July 21, where $n = 33$ days is the ordinate of intersection of phenocurve 2 with the line of average temperatures for June 18 (Fig. 1.71). Immediately below we give the forecast results (Table 4.4).

As can be seen in Table 4.4, the errors in phenoforecasts concerning the fall webworm are in the range of 1 to 3 days. The same results for this pest were obtained by A. S. Kharlamov for Odessa in 1970 and 1971.

Not long ago the Colorado potato beetle (*Leptinotarsa decemlineata*)—the most dangerous of potato pests—was, like the fall webworm, a quarantine object in the USSR. Now the Colorado potato beetle is no longer a quarantine object, because of its massive spreading in many regions of the European section of the USSR (ESU).

According to published, independent data[†] by Krasnovskaya and Vorotintzeva (1967), Britzky (1963), Arapova (1974), and Svikle (1967), results from testing the new method were obtained respectively in Moldavia, Ukraine, Byelorussia, and Latvia (see Table 4.5).

In these investigations the above authors used the corresponding PTNs given in Chapter 1. They gave transitive phenoforecasts with foresight periods up to 100 days and more. It should be remembered that a transitive forecast is a calculation made so that the date of each

---

*As was mentioned in Chapter 3, this means that in 80% of forecasts such errors, or smaller ones, are guaranteed.

[†]That is, data that were not used in constructing the phenocurves.

TABLE 4.4. Accuracy of forecasts for developmental dates of the fall webworm (*Hyphantria cunea Dr.*), on the basis of Podolsky's method, in the Eisk district of the Krasnodar region in 1969 (according to A. E. Stadnikov).

Generation and Developmental Stage	Date of Occurrence of Developmental Stages		Error of the Forecast (in days)
	According to the Mathematical Forecast	According to the Actual Date	
1	2	3	4
1. The first egg layings	—	5 VI	The actual date of spring re-generation of pupae is not observed; there-fore, there is no forecast
2. Birth of the first larvae of the first generation	16 VI	18 VI	+2
3. Beginning of the pupation of the larvae of the first generation	21 VII	24 VII	+3
4. Flight of butter-flies of the first generation	11 VIII	10 VIII	−1
5. Beginning of birth of larvae of the second generation	21 VIII	22 VIII	+1
6. Beginning of pupation of larvae of the second generation	16 X	There are no observations	—

succeeding developmental stage is obtained on the basis of the stage calculated before, and not on the basis of actual data. In a number of cases amendments were made in a forecast with prolonged foresight owing to the difference between the multiannual temperatures and the actual temperatures in one year or another (the method that involved taking into account the actual temperatures over a part of the forecast period, or before this period, was already described above for plants, and will also be discussed in the next section for cold-blooded organisms).

Forecasting of the second and third generations of the Colorado potato beetle was carried out taking into account photoperiodicity, as described in Section 2.3.

Podolsky's principle—as it was called by Byelorussian scientists and practical workers*—has been used by them, in particular, for forecasting the optimal dates for pest control measures in cruciferous crops: cabbage flies, cabbage moth, *Plutella maculipennis Curt.*, and cabbage white butterfly. Inopportune treatment against these pests causes a 30% harvesting waste in cabbages—the key vegetable crop in Byelorussia.

On the basis of observations carried out from 1967 to 1973, N. N. Kharchenko et al. (1973, 1978, 1979) constructed corresponding pheno-curves, and superimposed them onto the heat resources nets for different regions of Byelorussia (e.g., see Fig. 1.76).

The results of testing the method over a number of years in specialized vegetable-raising farms in Byelorussia showed that it is possible to determine, more than a month ahead, the dates of the most harmful and the dates of the most vulnerable developmental stages of cabbage flies and moths with an accuracy of ± 2 to 3 days. For instance, in 1977 the error of the forecast was -1 day, in 1976 there was no error, in 1970 the error was $+1$ day, and so forth. It is also possible to determine the dates of planting for a crop for the purpose of avoiding the harmful periods for this crop and for making retrospective analyses of the pests's development.

Kharchenko, as well as other authors, tried to use the method of phenoindicators and totals of effective temperatures, but unsuccessfully. On the basis of PTNs, a collection of handy phenoforecasting calendars was published for pests of crucifers in different parts of Byelorussia (1979).

In the Ukraine, A. D. Sheludko (1971) used Podolsky's method for controlling the weevil (*Tanymecus dilaticollis Gyll.*). He constructed phenocurves and superimposed them onto the heat resources nets of different parts of the Ukraine.

In the Kursk district, the new method was used by Yu. S. Vovchenko and A. S. Podolsky against the sugar-beet root aphid; the heat resources net was based on soil temperatures (see Fig. 1.95).

We end this section about forecasts with prolonged foresight (pheno-climatic forecasts) with a discussion of calculations for the developmental dates of the very serious citrus pest California red scale (*Aonidiella aurantii*). This work was carried out in Israel; apparently, it is also of direct interest for the United States.

---

*This expression may be correct if we take into consideration the fact that, on the basis of this general principle, many analogous methods can be derived—maybe as many methods as there are organisms to which the principle is applied. But the more correct expression "Methodology" is given by the reviewer of this book.

TABLE 4.5. Forecasts made on PTNs, and facts relating to the Colorado potato beetle (*Leptinotarsa decemlineata*).

				Date of Occurrence of the Stage				
					According to the Forecast		Errors in the Forecasts	
Generations	Stage of Development	Average Temperature of the Air (in °C)	According to the Observations	Without Amendment for Temperature	With Amendment for Temperature	Without Amendment	With Amendment
1	2	3	4	5	6	7	8
				Moldavia, 1962			
1.	Egg laying	—	19 V	—	—	—	—
	Larvae	19.1	28 V	28 VI	—	0	—
	Prepupa	16.0	17 VI	12 VI	—	+5	—
	Imago	22.6	27 VI	28 VI	—	−1	—
2.	Egg laying	—	22 VII	—	—	—	—
	Larvae	21.5	25 VII	28 VII	—	−3	—
	Prepupa	21.2	9 VIII	11 VIII	—	−2	—
	Imago	24.5	19 VIII	24 VIII	19 VIII	−5	0
				Moldavia, 1963			
1.	Egg laying	—	13 V	—	—	—	—
	Larvae	18.1	22 V	23 V	—	−1	—
	Prepupa	16.7	7 VI	8 VI	—	−1	—
	Imago	18.6	26 VI	26 VI	—	0	—
2.	Egg laying	—	4 VII	—	—	—	—
	Larvae	21.5	10 VII	10 VII	—	0	—
	Prepupa	21.5	24 VII	24 VII	—	0	—
	Imago	23.3	3 VIII	4 VIII	—	−1	—

3.	Egg laying	—	21 VIII	—	—	—	—
	Larvae	18.7	29 VIII	29 VIII	—	0	—
	Prepupa	20.7	14 IX	13 IX	—	+1	—
	Imago	14.5	17 X	19 X	21 VIII	-2	—

Moldavia, 1964

1.	Egg laying	—	26 V	—	—	—	—
	Larvae	16.1	1 VI	3 VI	—	-2	—
	Prepupa	20.8	15 VI	17 VI	15 VI	-2	0
	Imago	21.9	27 VI	4 VII	27 VI	-7	0
2.	Egg laying	—	12 VII	—	—	—	—
	Larvae	20.6	19 VII	18 VII	—	+1	0
	Prepupa	21.9	2 VIII	1 VIII	—	+1	0
	Imago	18.2	21 VIII	13 VIII	21 VIII	+8	0

Ukraine, Lvov (Obroshino), 1961

1.	Egg laying	17.7	18 VI	—	—	—	—
	Larvae, 1 age	18.4	25 VI	28 VI	26 VI	-3	-1
	Larvae, 2 age	19.5	28 VI	1 VII	28 VI	-3	0
	Larvae, 3 age	19.1	2 VII	6 VII	2 VII	-4	0
	Larvae, 4 age	19.1	6 VII	10 VII	6 VII	-4	0
	Prepupa	18.4	11 VII	14 VII	10 VII	-3	+1
	Imago	17.9	29 VII	3 VIII	2 VIII	-5	+1
2.	Egg laying	21.5	8 VIII	—	—	—	—
	Larvae, 1 age	23.0	12 VIII	17 VIII	15 VIII	-5	-3
	Larvae, 2 age	19.6	16 VIII	21 VIII	20 VIII	-5	-4
	Larvae, 3 age	17.5	21 VIII	25 VIII	23 VIII	-4	-2
	Larvae, 4 age	16.7	25 VIII	30 VIII	27 VIII	-5	-2
	Prepupa	16.7	3 IX	7 IX	2 IX	-4	+1
	Imago	15.3	27 IX	—	1 X	—	-4

TABLE 4.5. (*Continued*)

Generations	Stage of Development	Average Temperature of the Air (in °C)	Date of Occurrence of the Stage			Errors in the Forecasts	
			According to the Observations	According to the Forecast		Without Amendment	With Amendment
				Without Amendment for Temperature	With Amendment for Temperature		
1	2	3	4	5	6	7	8
			Byelorussia, Pruzhany, 1966				
1.	Egg laying	—	20 V	—	—	—	—
	Larvae, 1 age	—	5 VI	1 VI	—	+4	—
	Larvae, 2 age	—	8 VI	6 VI	—	+2	—
	Larvae, 3 age	—	12 VI	11 VI	—	+1	—
	Larvae, 4 age	—	15 VI	16 VI	—	−1	—
	Prepupa	—	22 VI	21 VI	—	+1	—
	Imago	—	10 VII	11 VII	—	−1	—
2.	Egg laying	—	19 VII	—	—	—	—
	Larvae, 1 age	—	25 VII	30 VII	—	−5	—
	Larvae, 2 age	—	30 VII	3 VIII	—	−4	—
	Larvae, 3 age	—	4 VIII	6 VIII	—	−2	—
	Larvae, 4 age	—	9 VIII	11 VIII	—	−2	—
	Prepupa	—	13 VIII	15 VIII	—	−2	—
	Imago	—	10 IX	4 IX	—	+6	—
			Buelorussia, Minsk, 1969				
1.	Egg laying	—	12 VI	—	—	—	—
	Larvae, 1 age	—	22 VI	22 VI	—	0	—

First group (continuation):

Larvae, 2 age	—	26 VI	26 VI	—	0	—
Larvae, 3 age	—	30 VI	1 VII	—	−1	—
Larvae, 4 age	—	5 VII	5 VII	—	0	—
Prepupa	—	9 VII	10 VII	—	−1	—
Imago	—	4 VIII	30 VII	—	+5	—

1. Byelorussia, Minsk, 1970

Egg laying	—	5 VI	—	—	—	—
Larvae, 1 age	—	15 VI	15 VI	—	0	—
Larvae, 2 age	—	20 VI	19 VI	—	+1	—
Larvae, 3 age	—	25 VI	24 VI	—	+1	—
Larvae, 4 age	—	—	28 VI	—	—	—
Prepupa	—	—	4 VII	—	—	—
Imago	—	—	24 VII	—	—	—

1. Byelorussia, Minsk, 1971

Egg laying	—	27 V	—	—	—	—
Larvae, 1 age	—	3 VI	8 VI	4 VI	−5	−1
Larvae, 2 age	—	7 VI	13 VI	9 VI	−6	−2
Larvae, 3 age	—	10 VI	18 VI	12 VI	−8	−2
Larvae, 4 age	—	13 VI	22 VI	17 VI	−9	−4
Prepupa	—	—	28 VI	26 VI	—	—
Imago	—	17 VII	18 VII	19 VII	−1	−2

1. Latvia, Riga, 1966

Egg laying	19.9	9 VI	—	—	—	—
Larvae, 1 age	18.5	17 VI	21 VI	17 VI	−4	0
Larvae, 2 age	18.5	20 VI	26 VI	20 VI	−6	0
Larvae, 3 age	20.2	24 VI	1 VII	23 VI	−7	+1
Larvae, 4 age	18.6	27 VI	4 VII	27 VI	−7	0
Prepupa	14.1	30 VI	10 VII	1 VII	−10	−1
Imago	20.9	19 VII	31 VII	20 VII	−12	−1

The females of California red scale are viviparous: instead of laying eggs they give birth to larvae.

Its reaction to heat is expressed by the biological curves in Figure 1.45. The initial data for constructing these curves were collected with the help of D. Rozen and H. Podoler (Jerusalem University).

As can be seen from Figure 1.45, the data of various authors coincide to a high degree. These data reconfirm the thesis that temperature is the main factor in determining the developmental dates (but not the numbers) of cold-blooded organisms, since, for example, the humidity in controlled temperature chambers and insectaries in experiments by various authors and in different regions could be different.

These phenological curves are put onto the heat resources net of Tirat-Zvi (Fig. 1.120).

The phenological problems are solved on a PTN, for the most part, as was done before. For example, the actual beginning of the appearance of the first-age larvae of that generation of California red scale, which we assume to be the *first*, was observed near Tirat-Zvi on September 25, 1977 by H. Podoler. When will the first-age larvae of the *second* generation appear, if the expected temperatures are close to the norm?

The solution is as follows: since the biological process that interests us here started on September 25, it is necessary to take into account, first, the heat resources over the period from this starting date (September 25) up to the next generation, the date of which is not yet known. For this purpose, we have to find in Figure 1.120 the line of average temperatures from the starting date of September 25 (25 IX), and look for the point where this line intersects phenocurve 9, which represents the period between one generation and another.

This would give an accurate answer if the phenological curve were constructed on the basis of phenological observations in the natural biotope on the one hand, and of air temperature in the shade at a height of 2 m above the soil surface on the other hand (2 m is the generally accepted height at meteorological stations). Since the heat resources net is also constructed on the basis of air temperatures at a height of 2 m, then the error in the heat resources would be approximately equal to the error in the requirements (i.e., in the reaction of the organism), and thus these errors would not have affected the result of the solution (Sections 2.4 and 6.3). This can also be seen to be true from the results of the wide application of this new method by various authors.

However, in the present case we use the phenocurve based on data from controlled temperature chambers. The temperature in a controlled temperature chamber corresponds to the temperature of the insect's body, as if it were in the biotope. Hence the heat resources net, which is constructed on the basis of temperatures at a height of 2 m, must correspond to the temperature in the biotope.

The biotope of the California red scale is the penumbra formed by the crown of citrus trees. The temperature of the insect in this penumbra is about 2° C higher than in the shade at a height of 2 m (meteorological station). For example, according to the data of D. Rozen and H. Podoler, the leaves of plants in the sun are 4° warmer than in the shade; hence in the penumbra it is warmer than in the shade by $4° \div 2 = 2°$.

Therefore, the heat resources net, that is, the line of average temperatures marked on the PTN by 25 IX, must be moved 2° to the right according to the scale—see the coarse dashed line in Figure 1.120. Then the ordinate of the intersection of this dotted line with phenocurve 9 gives $n = 57$ days. Therefore, the first-age larvae of the *second* generation will begin to appear on September 25 + 57 = November 21.

According to observations in 1977 near Tirat-Zvi (Podoler and Rozen), the actual date was close to the date calculated.

In order to avoid drawing dotted lines and overloading the net with phenological curves, we can draw the latter on tracing paper. Tracing paper is easier to move to either side relative to the net. In the present instance, the phenological curve should be moved horizontally 2° to the left; the same effect is obtained by moving the net 2° to the right.

Let us remember that the intersection of the heat resources line (which is a graphic reflection of the corresponding "meteorological" equation) with the phenocurve (which reflects the corresponding "biological" equation) gives the graphical solution of the system composed of these two equations.

Let us continue with the solution of problems. We find the date when the first-age larvae of the *third* generation begin to appear, in the same way as we did for the second generation: in Figure 1.120 we find the line of average temperatures with starting date of November 21. We move this line 2° to the right on account of the biotope's microclimate. Then we look for the intersection of the line that has been moved with the same phenocurve 9. We read on the vertical axis (Fig. 1.120) the ordinate of intersection. It is $n = 150$ days. Therefore, the first-age larvae of the *third* generation will appear on November 21 + 150 = April 21 of the next year. Observations made in 1978 by Rozen and Podoler showed that this date is also close to the actual date.

However, it is well known that during observations in nature it is not easy to recognize individual generations of insects, if there are several generations that overlap one another in time. The PTN makes it possible to discern the developmental dates of individual generations. For example, let us assume that September 25, November 21, and April 21 (see above) are not only the dates when the first-age larvae of the first, second, and third generations respectively appeared in 1977-1978, but are also the average multiannual dates near Tirat-Zvi. Then, in the same way as above, we obtain the *fourth* (June 15) and the *fifth* (August

1) generations. The fifth generation gives the date of the generation that we assumed to be the first: August 1 + 47 days = September 17, which is close to September 25.

Thus experimental-mathematical calculation shows that there are five generations, although some investigators had presumed that California red scale had only three to four generations in Israeli conditions. Such as error was possible because of the confusion of generations.

As can be seen, the calculation was somewhat complicated because the phenocurve was obtained from a controlled temperature chamber rather than from field observations. If we had not moved the average temperature line 2° to take this circumstance into account, calculations on the PTN would have given not less than four generations (i.e., 20% less). The difference is not very large, but it could be smaller if it were not for the acute angles of intersection between the autumn average temperature line and phenocurve 9. The acute angles are caused by peculiarities of the climate and of the phenocurve, that is, by the sharp deceleration of California red scale's development at low temperatures. Such situations will be discussed in the next section.

Since it is possible to put phenocurves for various organisms—cold-blooded animals, plants, and viruses—onto the same heat resources net, the Podolsky method is an aid to the study, and even the management, of various biogeocenoses. For example, on the PTN for Tirat-Zvi (Fig. 1.120), as well as the phenocurves for California red scale, phenocurves are plotted for species that prey on some citrus pests. This makes it possible to calculate parallel-associated pheno-forecasts for the purpose of biological pest control. However, this will be discussed in Chapter 5.

4.1c. High-accuracy Forecasts of the Developmental Dates of Pests (Phenoclimatic Forecasts with Shortened Foresight)

In the preceding section an attempt was made to determine, by means of phenoclimatic forecasts with prolonged foresight, whether the beginning of the fourth mass egg laying of the cotton bollworm in Kurgan-Tyube would occur, and when, if the beginning of the third mass egg laying was observed on August 13, 1959. We obtained an answer which meant that even if a fourth egg laying did occur, it would occur too late. In fact, the fourth egg laying occurred in 1959 on September 23. Hence the phenoclimatic forecast with prolonged foresight proved to be on the borderline between accuracy and error.

However, in Section 4.1.b, we paid attention to the following circumstance: *when the angle of intersection of the average tempera-ture line with the biological curve is narrow,* then even the slightest shift in the average temperature lines entails a considerable change in the result of a calculation. The shifting of average temperatures lines can occur, in particular, because the temperatures of a given year are not equal to the multiannual average temperatures (on the basis of which the heat resources net is constructed). Therefore, it is necessary to use the method of high-accuracy phenoclimatic forecasts with shortened foresight, which was described in detail for plants at the beginning of Chapter 3. This method involves the use of actual temperatures over a part of the forecast period in a given year.

Let us calculate a phenological forecast with shortened foresight for the date when the fourth mass egg layings of the cotton bollworm would begin in Kurgan-Tyube in 1959, if we know the actual date of the third egg laying (August 13), and the actual air temperature trend from August 13 to 31 inclusive, that is, over a 19-day period.

We will not work out all the 24-hour average air temperatures in Kurgan-Tyube over this period, but we will give their total, which is necessary for calculating the actual average temperature over this 19-day period:

$$t_{av} = \frac{\Sigma t_{av.\,24\text{-hour}}}{19 \text{ days (from August 13 to 31)}} = \frac{508.6°\,C}{19} = 26.8°\,C$$

We plot point b with the calculated coordinates $t_{av} = 26.8°\,C$ and $n = 19$ days onto the PTN for Kurgan-Tyube (Fig. 1.55). From the theory about PTNs, we already know that point b lies on a certain line of average temperatures actual for 1959, dated August 13. From point b we trace a dashed line bc parallel to the multiannual line of average period temperatures dated August 13, up to the intersection with phenocurve 12 representing the development of the cotton bollworm between generations. A period $n = 41$ days between the third and fourth egg layings corresponds to this intersection. This period is more real for 1959 than that calculated entirely on the basis of multiannual temperatures. Hence the forecast of the date when the fourth egg laying should begin will be August 13 + 41 days = September 23.

The foresight period of this forecast, starting from September 1 (since the last actual temperature used in the forecast relates to August 31), is 23 days; there is no error.

Thus the phenoclimatic forecast with shortened foresight proved to be much more successful than the phenoclimatic forecast with prolonged foresight.

As can be seen, forecasts for autumn generations of poikilothermic parasites must be checked by a calculation with shortened foresight.

In Table 4.1 it was also noted that the third egg laying of the cotton bollworm did not occur in 1959, according to the phenoforecast for Dushanbe. However, this egg laying was actually observed in the field on August 27. Therefore, the phenoclimatic forecast with prolonged foresight proved to be incorrect. Let us try, therefore, to give a forecast with shortened foresight, using the actual air temperatures in Dushanbe over the 22-day period from July 27 (the actual date of the second egg laying) to August 17 inclusive:

$$t_{av} = \frac{\Sigma t_{av. \, 24\text{-hour}}}{22 \text{ days (from July 27 to August 17)}} = \frac{586.4° \text{ C}}{22} = 26.7° \text{ C}$$

The actual point with coordinates $t_{av} = 26.7°\,$C and $n = 22$ days is plotted onto the PTN for Dushanbe (Fig. 1.56). Then we draw from this point a dotted line parallel to the multiannual line of average temperature dated July 27, up to the point of intersection with the phenocurve representing the cotton bollworm's development between generations (this phenocurve does not appear on Fig. 1.56—it must be copied onto tracing paper from Fig. 1.1; the tracing, which is to be done on the scale of Fig. 1.56, is to be superimposed onto Fig. 1.56). Then it will be possible to read, at the intersection, $n = 35$ days. According to this forecast, the date of the third egg laying will be July 27 + 35 days = August 31*.

The forecast of August 31 differs from the actual date, August 27 (Table 4.3), by 4 days. The main fact, however, is that the forecast with shortened foresight (10 days), in contrast to the one with prolonged foresight, proves without a doubt that a third egg laying of the cotton bollworm in Dushanbe in 1959 should occur. And, in fact, it did occur, although a third egg laying in the cotton bollworm does not occur in Dushanbe every year.

Calculations like the forecasts with shortened foresight must be used not only for the autumn generation of an insect pest, but also in cases when the foresight period of a phenoclimatic forecast turns out to be too long. If a given period in the year is noted for a large temperature anomaly, then when the forecast is given far ahead of time, the error can increase to a high degree.

The investigation of extensive data about the cotton bollworm in different regions of Tadzhikistan over the period of 1948 to 1961 showed that the root-mean-square error of phenoclimatic forecasts with shortened foresight $\sigma = \pm 3.3$ days, and the "guarantee" of errors less than or equal to 3.3 days, is about 70% (for plants, these figures are ±3.4 days and 80%, respectively).

In the case of the piorcer, we have approximately the same characteristics, or somewhat better. They can probably also be extended to a number of other cold-blooded organisms.

As can be seen, phenological forecasts for cold-blooded organisms

---

*The point with coordinates 26.7° and 22 days should be plotted by the reader on Fig. 1.56.

are a little less successful than for plants. This is reconfirmed by the forecasts in Kazakhstan for *Apamea anceps Shiff.-H. sordida* which is a parasite on spring wheat. According to V. K. Adzhbenov (1971, 1972), the σ of phenoforecasts with shortened foresight for *Apamea anceps Shiff.-H. sordida* is ±2.5 days, as compared with ±2 days for wheat.

Hence the curve—similar to that in Figure 3.1—representing the distribution of error frequencies according to their values obviously does not have a great direct excess for the cotton bollworm as compared with cotton. For the cotton bollworm, the frequency spectrum is closer to the theoretical curve of normal distribution (by Gauss) which, in the interval ±σ, covers 68% of all cases.

In Table 4.6 the results of forecasting the developmental dates of the cotton bollworm on the basis of totals of squared active temperatures are compared with those made on the basis of the Podolsky method. Here there will be a discussion about phenological forecasts with shortened foresight (using the former and the latter methods), as well as about those years and those regions of Tadzhikistan which were not compared above.

Forecasts with shortened foresight were also applied by V. A. Chekhonadskikh (1973) to the development of the eggs of the hawthorn tortricid. Chekhonadskikh constructed a phenocurve for this pest on the basis of observations in gardens in the Tula and Voronezh districts

TABLE 4.6. Cotton bollworm's developmental dates, obtained by pheno-forecast with shortened foresight made on the basis of totals of squared active temperatures and on the basis of Podolsky's method, for Ordzhonikidzeabad (Tadzhikistan).

Method	1951 (the second egg laying)			1952 (the second egg laying)		
	Forecast	Actual Date	Error	Forecast	Actual Date	Error
1	2	3	4	5	6	7
Totals of squared active temperatures	2 VIII	26 VII	−7	15 VII	11 VII	−4
Podolsky's method	31 VII	26 VII	−5	11 VII	11 VII	0

	1952 (the third egg laying)			1956 (the second egg laying)			Average Error (in days)
	Forecast	Actual Date	Error	Forecast	Actual Date	Error	
	8	9	10	11	12	13	14
	16 VIII	6 VIII	−10	18 VII	17 VII	−1	−5.5
	11 VIII	6 VIII	−5	17 VII	17 VII	0	−2.5

(see Fig. 1.47). The phenocurve was superimposed onto the heat resources net for the Tula district (Moscow suburbs). The errors in the forecasts based on Podolsky's method turned out to be +3 days in 1965, +5 days in 1966, +3 days in 1967; an average of +3.7 days. However, the errors in forecasts based on the method of effective temperatures (the lower threshold of +5°) were respectively 1, 15, and 8 days; an average of 8 days (the forecasts based on totals also had shortened foresight, and they used the actual temperatures over the same part of the forecast period as in the Podolsky method).

It is interesting, but difficult to explain, that a simple totaling of temperatures, reckoned from 0° C, proved to be more successful than the totaling of effective temperatures, but less successful than the Podolsky method (the latter fact is, however, explicable): the errors in the forecasts based on temperature totals from 0° C were 1, 8, and 4 days—an average of 4.3 days.

To conclude this section, we wish to point out that for forecasts with shortened foresight, it is not necessary to calculate the starting point on the PTN using actual temperatures (as was calculated for point b on the PTN for Kurgan-Tyube in Fig. 1.55). It is sufficient to ask the local meteorological station to provide the temperature anomaly over the recent period of time. The particular average temperature line necessary for the calculation must be moved to the extent equal to the value of this anomaly.

Finally, it is possible to make the forecast somewhat more exact without shortening the period of foresight, by using the temperature anomaly over the preforecast period. For example, if the starting date of the forecast is August 13 (as it was in the preceding example with the cotton bollworm in Kurgan-Tyube), then we can obtain from the meteorological station the temperature anomaly over not the period after August 13, but over a short period preceding August 13.

Last, *it is always advisable to give two forecasts: one forecast is preliminary—with prolonged foresight; the other is more exact—with shortened foresight.*

4.1d. Phenosynoptical Forecasts Based on the Podolsky Method

These are calculations using the long-range weather forecasts, which were described for plants in Section 3.1f. The method remains unchanged for cold-blooded organisms.

4.1e. Retrospective Calculations

These were also described for plants (Section 3.1h), and the method remains unchanged for cold-blooded organisms. We shall only point out that retrospective calculations will be of importance, in particular to biological pest control, in cultivating entomophages for a given date in a natural environment.

4.1f. Forecasting Developmental Stage/Temperature Combinations for Cold-Blooded Organisms

This problem is analogous to the problem that was discussed for plants (and which cannot be solved by means of the temperature totals method).

Let us assume that we need to determine when, in multiannual terms, an egg of the cotton bollworm must be put into a field insectary or a climatic chamber imitating the natural climatic conditions of, for example, Kurgan-Tyube, so that the developmental process from egg to egg will proceed at an average temperature of 23.1° C (given in advance) over this period. We need to know whether such a developmental stage/temperature combination is possible at all in the climatic conditions of Kurgan-Tyube in normal years.

From Figure 1.55 we can see that the vertical line drawn from $t_{av} = 23.1°$ C up to phenocurve 12, representing the cotton bollworm's development from egg to egg, intersects this curve within the bounds of the heat resources net, in contrast, for example, to the vertical line from $t = 32°$ C. The temperature of 23.1° C is higher than the temperature at which the cotton bollworm begins to develop in spring or goes into diapause in autumn (the diapause will be discussed below). Therefore, the given developmental stage/temperature combination will be possible in Kurgan-Tyube.

Through the point where the vertical line (starting from 23.1° C) intersects curve 12 run two lines of average temperatures, dated April 25 and August 7. This means that the necessary developmental stage/temperature combination can be obtained only twice a year, by putting an egg into the insectary on April 25 and August 7. Then the egg of the next generation will appear on April 25 + 50 days = June 14, or August 7 + 50 days = September 26, where $n = 50$ days is the ordinate of intersection of the average temperature lines from April 25 and from August 7 with developmental curve 12.

A situation similar to the latter example (August 7—September 26) was observed in Kurgan-Tyube in 1959. The early example (April 25—June 14), as is absolutely clear, is only possible provided that the initial egg is already available on April 25. In natural conditions, no generation of the cotton bollworm begins egg laying so early. However, in artificial conditions such an initial egg can probably be obtained if necessary.

Such calculations of developmental stage/temperature combinations can be important for the breeding of entomophages. Moreover, they are important for managing the vital activity and ratio of the sexes in, for example, the bombyx, according to B. L. Astaurov (1958).

4.1g. Daily Estimation of Developmental Rates of Pests

In 1959 the beginning of the first mass egg laying of the cotton bollworm was observed in the field in Kurgan-Tyube on June 13, and the second egg laying was observed on July 13. The question is does the cotton bollworm's development in 1959 lag behind the multiannual norms or is it ahead of them? If either one or the other, then by how many days, during which stages, and for what causes (temperature or nontemperature factors)?

As usual, we take the date when the 24-hour average air temperatures regularly exceed 15° C to be the date of spring awakening in the cotton bollworm. We see that, on the multiannual average, this date in Kurgan-Tyube is April 6: from the mark of 15° C on the horizontal axis of the multiannual temperature net runs the average temperature line corresponding to April 6 (Fig. 1.55). The line of April 6 intersects phenological developmental curve 11 for the cotton bollworm (the period from awakening to the first egg laying) at $n = 64$ days; hence the average multiannual date of the first egg laying will be April 6 + 64 days = June 9.

Since the actual date of the first egg laying in 1959 was June 13 we have a four-day lag in the cotton bollworm's development in 1959, in comparison with the multiannual norm, during the stage between awakening and the first egg laying.

We go on with the transitive phenoforecasting. The intersection of the average temperature line marked June 9 (the calculated multiannual date of the first egg-laying) with phenocurve 12, for development from egg to egg, gives $n_1 = 31$ days. Therefore, the second mass egg laying of the cotton bollworm, on the multiannual average, should begin on June 9 + 31 days = July 10.

Since the beginning of the second egg laying was observed on July 13, 1959, we see that there was still a three-day lag in the cotton bollworm's development in 1959 in comparison with the multiannual norms.

It now remains to answer the question about the causes of this slower developmental rate in the first half of 1959. As we already know from the solution of the analogous problem for plants (Section 3.1i), to answer this question we have to use the actual lines of average temperatures for 1959, and not the multiannual lines. In 1959 in Kurgan-Tyube the actual date when the 24-hour average air temperatures exceeded 15° C was April 6 (in the present instance, this coincided with the multiannual date). We construct, on the basis of actual temperatures, a small segment of the average temperature line from April 6, 1959. This line segment must intersect biological curve 11, and therefore, it is sufficient to construct it on the basis of only two points— one point higher than line 11, and the other point lower (see Section 3.1e). This phenosynoptical reliable forecast gives a first egg-laying date close to that obtained above by means of the phenoclimatic forecast, that is, close to June 9. So, the developmental lag of cotton bollworm in Kurgan-Tyube in 1959, in comparison to the multiannual norms, was not caused by insufficient heat during the first inter-developmental stage. The temperature in 1959 during this period were close to the norm, and so the first egg-laying date could have been the same as the multiannual. Therefore, the causes here are of a non-temperature nature—for example, undernourishment of the cotton bollworm.

Now we construct, on the basis of the actual temperatures in 1959, a segment of the average temperature line corresponding to the actual date of the first egg laying, June 13. The intersection of this segment with biological curve 12 gives a second egg-laying date, which is very close to the actual date—July 13. Therefore, during the period between the first and second egg layings, the cotton bollworm developed in accordance with actual temperatures, and without any distortions caused by nontemperature factors. This is why the previous developmental lag, as compared with the multiannual norm (July 10—13), remains.

Calculations for further generations show that the sharp increase in air temperatures, which began at the end of the first half of 1959, caused such a rapid increase in the developmental rates of the cotton bollworm that the actual development not only made up the lag, but even outstripped the multiannual norms, especially at the time of the fourth egg laying (when a warm, dry autumn favored this out-stripping).

4.1h. The Number of Generations of Pests, Numbers of Individual Pests, and Dates of Diapause

The way to calculate the number of a pest's generations over a season was already discussed in Section 4.1b. It consists of calculating the date of appearance of every generation. This method will also be used in the next section, and in others. Here we wish to point out again the following: to calculate the possible number of generations on the multiannual average, or in a certain year, is of great importance. It is especially important because in a number of cases it is impossible to calculate the number of generations with sufficient accuracy in the field, since generations overlap one another, e.g., in the malaria mosquito. It is even more difficult to discriminate in the field between the generations of the spider mite, which has about 20 generations. Moreover, even for the cotton bollworm, which has three to four generations per season, or for California red scale, which has four to five generations, a PTN helps to clarify the confusion of generations.

It is also important to note that calculations based on the PTN can sometimes discover generations that were not yet known from observations for a particular pest.

Calculating the number of generations on the basis of the effective temperatures totals is less reliable, since this "constant" (total) varies not only from generation to generation, but also within one generation, in relation to the temperatures making up the total.

Calculating the developmental dates of individual generations of a pest is useful for calculating the numbers of individuals of that pest. The fact is that the methods presently available for calculating the numbers of a pest do not take into account one very important fact: the numbers increase sharply as a result of the overlapping of different generations. By forecasting the developmental dates of individual generations, and the duration of the egg laying by each generation, it is possible to calculate the connections and the overlapping between generations, which are different in different years. Therefore, it is possible to calculate the numbers, which are related to these factors. If to this we add the possibility of making parallel forecasts on the same PTN for the developmental dates of plants (nutrient basis), then there is a good prospect of working out a new method for forecasting numbers of pests. In addition, we have to take into consideration the fact that a PTN makes it possible to calculate the limiting effect of photoperiodicity on the pubescence (for egg laying) of, for example, the Colorado potato beetle. This was already discussed in Section 2.3, and it will be discussed again later on.

Forecasting the numbers of agricultural pests on the basis of a PTN has already been begun by the Byelorussian Plant Protection Institute.

An example of calculating numbers of insects by means of totaling the generations will be shown for the malaria mosquito in Section 6.6c.

The Podolsky method also makes it possible to calculate the dates of insects' mass entry into the diapause, which marks the end of reproduction. As is known, the mass facultative diapause (i.e., the appearance of females that do not lay eggs) coincides with a sharp decrease in egg laying in nature. Therefore, the problem is to calculate the date of the mass cessation of egg laying. This date gives us the period that is necessary for the appearance of a developmental stage adjusted to wintering (for more details, see Section 6.6b).

4.2. BIOCLIMATIC FORECASTING, ZONING, AND MAPPING

4.2a. Behavior of Cotton Pests in New Agricultural Regions

We need to determine the dates of appearance and the number of generations of the cotton bollworm in the northwestern part of the

Yavan Valley in Tadzhikistan after the area was supplied with water and cotton was planted.

Before the Yavan Valley was supplied with water, the multiannual average heat resources of the northwestern section were represented by the net of average temperatures in Figure 1.50. After water was supplied in the northwestern area, the temperature should have decreased, beginning from June, by 1° to 1.5° C. Until June (in the rainy period) the temperature remained as it was before the supply of water. Hence the lines of average temperatures necessary for the calculation should be moved to the left by 1° to 1.5° C from June.

Thus from the point marked 15° C on the horizontal axis the line of average temperatures from April 11 rises (until June the net is unchanged). That means that, on the multiannual average, the spring awakening of the cotton bollworm in the northwest part of the Yavan Valley is to be expected on April 11.

The line of average temperatures from this date intersects pheno-curve 18, representing the cotton bollworm's development from awakening to the first egg laying, at $n = 65$ days. Hence the average multiannual date of the beginning of the first mass egg laying will be April 11 + 65 days = June 15.

At the point where the line of average temperatures from June 15 intersects phenocurve 19, representing the cotton bollworm's development between the egg layings of different generations, we would ordinarily have found that the second egg laying would be on June 15 + 31 days = July 16.

However, starting from June, all the lines of average temperatures, in particular the line dated June 15, are to be moved to the left by 1° C. Thus the date of the beginning of the second mass egg laying will be, more exactly, June 15 + 30 days = July 15.

We move the line of average temperatures, dated by the last result July 15, 1° C to the left and in the same manner we find the date of the beginning of the third egg laying: July 15 + 29 days = August 13; the date of the fourth egg laying is August 13 + 37 days = September 19.

The line of average temperatures from September 19 does not intersect phenocurve 19. Hence a fifth egg laying will not occur, in the multiannual terms, because there will not be enough heat. Moreover, even if the intersection were to occur in the autumn-frosts zone, the organism would perish from the frost.

Therefore, in the warmest, northwestern part of the Yavan Valley after it was supplied with water, the multiannual average number of generations of cotton bollworm is limited to four egg layings.

When this is calculated for the colder, northeastern part of the Yavan Valley (after it was supplied with water), the lines of average temperatures on the initial nomogram move to the left by 1° C, up to May inclusive (so that the awakening of the cotton bollworm here will be not on April 11, but on April 17), and by 2° C starting from June.* For the coldest, middle section of the Yavan Valley, the PTN for Kurgan-Tyube can be used (see Fig. 1.55), but the lines of average temperatures must be moved to the left by 1°C in all calculations. Finally, for the southern part of the Yavan Valley we use the PTN for Kurgan-Tyube without any changes. As can be seen, there is no need for separate nets for each mesoclimatic region.

In the above manner, and using phenocurve 22 (Fig. 1.50), it is also easy to calculate the dates and the number of generations of the spider mite, an important step for the estimating of its numbers.

Thus if cotton pests are not entirely annihilated or are not controlled on the necessary dates, their development in the Yavan Valley will occur according to the dates designated in Table 4.7.

These half-theoretical results confirm the actual data for regions adjoining the Yavan Valley. For example, E. P. Luppova states that in the south of Tadzhikistan there are, in fact, 16 to 18 generations of spider mite.

In spring, one spider mite's generation develops during 18 days or more (Table 4.7); in summer, when the temperature is higher, it develops in 8 to 9 days; in autumn, as the temperatures decrease again, it takes 14 to 41 days (depending on the region). Therefore, if we put onto the graph the multiannual average dates for the egg layings of the spider mite's different generations (using part of Table 4.7), then we obtain a very smooth *curve of the cubic parabola type* (Fig. 4.1). In different regions, the spider mite's generations of the same number appear on different dates. It is interesting that these differences in dates are accumulated from generation to generation, and that at the end of the season the differences between regions are as great as 20 to 47 days.

On the graph in Figure 4.1 we can easily determine how many generations of spider mite have already appeared on each given date, for example, on July 19. This can be done both in multiannual terms and for a particular year. In the latter case, on the basis of the actual temperatures in a particular year, a fragment of the heat resources net must be constructed; more exactly, small segments of average tempera-ture lines must be constructed, so as to intersect the phenological curve of the spider mite's development. Using this concrete net, rather than a multiannual one, we can draw an illustration like Figure 4.1 for the current year.

Here we shall solve the problem on a multiannual basis. Let us find, on the vertical axis of Figure 4.1, the date of July 19. From this point we

*Details of forecasts for mesoclimates in the Yavan Valley before and after the supply of water were discussed by Podolsky in 1963 (see the list of references at the end of this book).

TABLE 4.7. Average dates of development and number of generations of cotton pests in the Yavan Valley after it was supplied with water; all data were calculated on the PTN.

Part of the Valley	Date of Spring Awakening	Beginning of Mass Egg Laying				
		1st	2nd	3rd	4th	5th
		Spider Mite				
Northwest	26 II	16 IV	5 V	20 V	1 VI	12 VI
Northeast	7 III	24 IV	13 V	27 V	8 VI	19 VI
Middle	3 III	19 IV	7 V	21 V	2 VI	13 VI
South	23 II	13 IV	1 V	15 V	27 V	6 VI
		Cotton Bollworm				
Northwest	11 IV	15 VI	15 VII	13 VIII	19 IX	—
Northeast	17 IV	20 VI	20 VII	19 VIII	—	—
Middle	11 IV	15 VI	17 VII	20 VIII	—	—
South	6 IV	8 VI	9 VII	8 VIII	28 IX	—

Part of the Valley	Date of Spring Awakening	Beginning of Mass Egg Laying				
		6th	7th	8th	9th	10th
		Spider Mite				
Northwest	26 II	22 VI	1 VII	10 VII	19 VII	28 VII
Northeast	7 III	28 VI	7 VII	16 VII	25 VII	3 VIII
Middle	3 III	23 VI	2 VII	11 VII	20 VII	29 VII
South	23 II	16 VI	25 VI	3 VII	12 VII	21 VII
		Cotton Bollworm				
Northwest	11 IV	—	—	—	—	—
Northeast	17 IV	—	—	—	—	—
Middle	11 IV	—	—	—	—	—
South	6 IV	—	—	—	—	—

Part of the Valley	Date of Spring Awakening	Beginning of Mass Egg Laying			
		11th	12th	13th	14th
		Spider Mite			
Northwest	26 II	6 VIII	15 VIII	24 VIII	3 IX
Northeast	7 III	12 VIII	21 VIII	31 VIII	12 IX
Middle	3 III	7 VIII	17 VIII	28 VIII	10 IX
South	23 II	30 VII	7 VIII	16 VIII	26 VIII

TABLE 4.7. (*Continued*)

Part of the Valley	Date of Spring Awakening	Beginning of Mass Egg Laying			
		11th	12th	13th	14th
Cotton Bollworm					
Northwest	11 IV	—	—	—	—
Northeast	17 IV	—	—	—	—
Middle	11 IV	—	—	—	—
South	6 IV	—	—	—	—

Part of the Valley	Date of Spring Awakening	Beginning of Mass Egg Laying				Total Number of Generations
		15th	16th	17th	18th	
Spider Mite						
Northwest	26 II	15 IX	29 IX	24 X	—	17
Northeast	7 III	26 IX	19 X	—	—	16
Middle	3 III	27 IX	7 XI	—	—	16
South	23 II	7 IX	21 IX	12 X	—	17
Cotton Bollworm						
Northwest	11 IV	—	—	—	—	4
Northeast	17 IV	—	—	—	—	3-4
Middle	11 IV	—	—	—	—	3
South	6 IV	—	—	—	—	4

Notes. 1. Spring awakening of spider mite occurs at +8° C; spring awakening of cotton bollworm occurs at +15° C. 2. The first egg laying of the spider mite we consider to be the egg laying by pubescent individuals of the first generation. The first egg laying of the cotton bollworm we consider to be the egg laying by individuals of the wintering generation.

trace a horizontal line to the intersection with the curves on the graph, and from these points of intersection we trace vertical lines up to the axis of abscissae. On the latter we read that, for example, in the northeastern part of the Yavan Valley (curve IV), by July 19 the eighth generation of imagoes had already laid eggs which began to develop to produce a ninth generation of imagoes (and from these eggs, in all probability, larvae were already hatched).

Of course, on the given date of July 19 not only did pubescent spider mites of the eighth generation and larvae of the ninth generation coexist in the field, but also other developmental stages and generations: they overlap one another because of the duration of egg laying in each generation. For example, in Section 6.6c, we will see that, even in the malaria mosquito *Anopheles superpictus*, which in Tadzhikistan

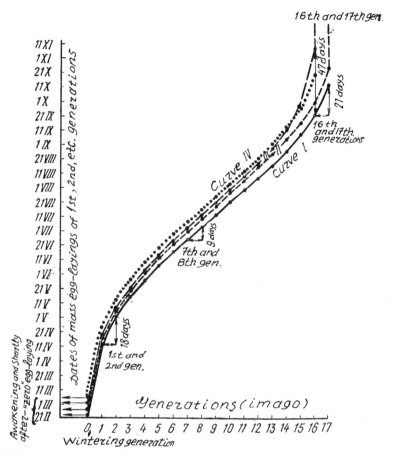

FIGURE 4.1. Average multiannual developmental dates and the number of generations of the spider mite (*Tetranichus urticae koch*) which appeared on a particular date in the Yavan Valley after the valley was supplied with water (bioclimatical forecast based on the PTN). Curve I: for the southern area of the valley; curve II: for the northwestern area; curve III: for the middle area; curve IV: for the northeastern area.

has 10 to 11 generations, by the end of a season there are 4 generations coexisting in the same developmental stages, at exactly the same time.

Hence it is clear how difficult it is in nature, and even in field insectaries, to distinguish the generations of some cold-blooded organisms. A mathematical method of approach, as can be seen, enables us to do this.

It was already mentioned that the development of ticks is influenced not only by temperature, but also by relative atmospheric humidity. The way to take this factor into account is described in Section 2.2.

We conclude this section with the words of the prominent physicists Heinrich Hertz:

> We cannot help feeling that mathematical formulas live their own, independent lives, that they are cleverer than their inventors, and that we obtain from them more than was initially put into them.

How right this is: you may not be an entomologist in the full meaning of the word, but the nomogram will make statements that you cannot. The author felt the same way when first introduced to the spider mite by severe testers of the Podolsky method.

Finally, it should be pointed out that the calculations described above (both here, and previously, in the section about agroclimatic division into regions for cotton) can be named a *multiannual* forecast as opposed to a short-term forecast, a seasonal forecast, or a forecast for the next year; the mentioned multiannual forecasts give average multiannual developmental dates, and for a biogeocenosis, such as a plant and the accompanying pests.

4.2b.   Bioclimatic Zoning and Mapping. Calculating the Height Limits of the Distribution of Poikilotherms.

In Section 3.2b some ideas were expressed about experimental-theoretical calculations for bioclimatic zones. These ideas could be realized on the basis of a PTN for the "base" points, on the one hand, and on the basis of climatic isotherm maps on the other hand.

However, in highlands, which are remarkable not for the horizontal, but for the vertical zoning of their climates, isotherms would help very little with our problem. This is because any horizontal temperature interpolation is risky in these conditions, if it is made without taking account of the heights. Therefore, it seems that in mountainous regions it would be possible to use base PTNs in only on way: the factual temperature differences must be calculated between each outlying station and the base station over the period of the year which interests us; then we calculate, using the base PTN, the dates for each outlying station; finally, we plot the isophenes between the stations on a geographical map.

However, it is possible to simplify this greatly, even in highland areas. It is known that in the warm half of a year the air temperature normally decreases with height, especially if we are considering average (over a certain period) multiannual temperatures. It is also known that the vertical gradient—the extent of the temperature decrease in the air per each 100 m of increased altitude—changes quite systematically from spring to summer, and from summer to autumn, according to average data. For example, the author's observations in Kara-Mazar at a height of 1600 m above sea level (the southern spurs of the Kuramin mountain range, in the Leninabad district in Tadzhikistan), when compared with simultaneous observations at a height of

414 m in the neighborhood of Leninabad, showed that in February the average vertical temperature gradient was $\gamma = -0.3°C/100$ m, in March it was $\gamma = -0.5°C/100$ m, in April it was $\gamma = -0.6°C/100$ m, in May it was $\gamma = -0.7°C/100$ m, in June it was $\gamma = -0.8°C/100$ m, and in July and August it changes little and then begins to decrease systematically (in absolute value). Approximately the same values can be obtained from the *Agroclimatic Reference Book* for Tadzhikistan (1959), with respect to other heights and other regions of Tadzhikistan.

If we know the air temperatures at the base station, then we can, with the help of the vertical temperature gradients, calculate the temperatures at different levels (lower or higher) in the vicinity of the base station. For our purposes it is enough to know only the temperature difference between the given level and the base station. To the extent of this difference we move those average temperature lines necessary for our calculation, on the nomogram of the base station, and thus obtain the dates of phenological phenomena at the level that interests us.

As can be seen here, two forecasts are given: a spatial climatic one (temperature at a given level), and a temporal bioclimatic one (date of a phenological phenomenon at this level). Such complex bioclimatic calculations do not require a wide network of meteorological stations surrounding the base station; in some cases it is possible to manage with only one base station. Nevertheless, these calculations can give a map that will serve as a guide to the bioclimate of the region.

Thus taking as an example the piorcer *Carpocapsa pomonella* (the distribution of which is almost universal), we choose three base stations in Tadzhikistan: Dushanbe as a center of southwestern Tadzhikistan, Kulyab as a center of the region transitional to the Western Pamirs, which is very different from Western Tadzhikistan, and Khorog as a center of the Western Pamirs (Table 4.8 and Fig. 4.2). The north of Tadzhikistan does not interest us for the present; therefore, among the stations mentioned there is no base station for this region.

We take standard levels above and below each base point (Table 4.8, column 2). For these levels we must find out the dates of the piorcer's spring awakening, which begins when the 24-hour average air temperatures exceed $10°C$. This can be done quite simply on the nomogram of the base station. For example, on the PTN of Dushanbe (Fig. 1.56), from the point marked $10°C$ on the horizontal axis runs the line dated March 25—this will be the date of awakening (Table 4.8, column 5). For other levels it is necessary to calculate, first of all, the temperature differences $(\Delta°C)$ from the base station, using the vertical temperature gradients for March and April, because we know that the awakening, at least in Dushanbe, occurs in March.

March and April temperature gradients (γ) are indicated in column 3

TABLE 4.8. Vertical zoning of the piorcer's behavior in Tadzhikistan in

Region and Base Meteorological Station	Height above Sea Level (m)	Vertical Temperature Gradient per 100 m, γ[a]	Temperature Difference from the Base Station[b]	Date of Awakening[c]		Vertical Temperature Gradient per 100 m, γ	Temperature Difference from the Base Station (Δ°C)	Length of the Period Between Awakening and Mass Flight[d]	Mass (50%) Flight of Butterflies of Wintering Generation (on PTN)
				Observed	Obtained Taking Δ°C into Account				
1	2	3	4	5	6	7	8	9	10
South-western Tadzhiki-stan. Dushanbe....	300	-0.5, Mar.	+2.5	—	7 III	-0.5, Mar.	+2.5	43	19 IV
	500	-0.5, Mar.	+1.5	—	16 III	-0.6, Apr.	+1.8	37	22 IV
	803	—	0.0	25 III	—	—	0.0	37	1 V
	1000	-0.5, Mar.	-1.0	—	30 III	-0.6, Apr.	-1.2	38	7 V
	1500	-0.6, Apr.	-4.2	—	14 IV	-0.7, May	-4.9	44	28 V
	2000	-0.6, Apr.	-7.2	—	27 IV	-0.7, May-June	-8.4	55	21 VI
Kulyab.... Transition to the Pamirs.	586	—	0.0	15 III	—	—	0.0	38	22 IV
	1000	-0.5, Mar.	-2.0	—	26 III	-0.6, Apr.	-2.5	40	5 V
	1500	-0.6, Apr.	-5.5	—	11 IV	-0.65, Apr.-May	-5.9	43	24 V
Western Pamirs. Khorog....	1500	-0.6, Apr.	+3.5	—	4 IV	-0.6, Apr.	+3.5	41	15 V
	2076	—	0.0	23 IV	—	—	0.0	41	3 VI
	2500	-0.7, May	-2.9	—	7 V	-0.7, May	-2.9	46	22 VI
	3000	-0.7, May	-6.4	—	25 V	-0.8, June-July	-7.2	63	27 VII
	3500	—	—	15 VI[i]	—	-0.8, June-July	-11.2	∞	No flight

of Table 4.8; they are taken from the observations mentioned above (in Kara-Mazar and Leninabad). We then multiply the difference between the heights (in hundreds of meters) of the base station and of the given level by γ, and obtain the temperature difference at these levels. For example, the level of 300 m (column 2) differs from Dushanbe (800 m) by 500 m. Since for each 100 m of altitude the temperature decreases by 0.5° C (column 3), or increases by the same value as altitude decreases, then for 500 meters it rises by Δ°C = 5 × 0.5°C = 2.5°C (column 4).

Then, on the PTN for Dushanbe, we count from the point 10° C on the axis of abscissae 2.5° C to the left; we see that from this latter points runs the line of average temperatures dated March 7 (column 6); this is exactly the date when the average 24-hour air temperatures exceed 10° C, that is, the date when the piorcer awakens at a height of 300 m in western Tadzhikistan. It should be noted that in Shaartuz, which is also located in western Tadzhikistan but at a height of 363 m, and where, therefore, the temperature must exceed 10° C somewhat later than at the height of 300 m, the date when the temperature exceeds multiannual terms (experimental-theoretical investigations).

Height above Sea Level (m)	Vertical Temperature Gradient per 100 m, γ	Temperature Difference from the Base Station (Δ°C)[b]	Length of the Period Between Flights[e]	Dates of Mass Flight[f]	Vertical Temperature Gradient per 100 m, γ	Temperature Difference from the Base Station (Δ°C)	Length of the Period Between Flights[g]	Date of Mass Flight[h]	Mass Flight of the Third Generation (on PTN)
2	11	12	13	14	15	16	17	18	19
300	−0.7, May	+3.5	50	8 VI	−0.8, June-July	+4.0	48	26 VII	Possible on quince
500	−0.7, May-June	+2.1	55	16 VI	−0.8, June-July	+2.4	47	2 VIII	
803	—	0.0	58	28 IV	—	0.0	46	13 VIII	
1000	−0.7, May-June	−1.4	61	7 VII	−0.7, July-Aug.	−1.4	47	23 VIII	
1500	−0.7, June-Sept.	−4.9	110	15 IX	—	—	∞	No flight	Impossible
2000	−0.6, July-Oct.	−7.2	∞	No flight	—	—	—	No flight	
586	—	0.0	55	16 VI	—	0.0	47	2 VIII	
1000	−0.7, May-June	−2.9	59	3 VII	−0.7, July-Aug.	−2.9	46	18 VIII	Impossible
1500	−0.8, June-July	−7.3	87	19 VIII	−0.6, Aug.-Sept.	−5.5	∞	No flight	
1500	−0.8, June	+4.7	54	8 VII	−0.7, July-Aug.	+4.2	46	23 VIII	
2076	—	0.0	73	15 VIII	—	0.0	∞	No flight	
2500	−0.7, July-Aug.	−2.9	∞	No flight	—	—	—	No flight	Impossible
3000	−0.6, Aug.-Sept.	−5.4	∞	No flight	—	—	—	No flight	
3500	—	—	—	No flight	—	—	—	No flight	

[a] Minus designates decrease of temperature with height.
[b] +Δ°C if t° is higher than the base, −Δ° if t° is lower.
[c] i.e., when average 24-hour temp. exceeds 10°C (on PTN).
[d] Of wintering generation on PTN (n days).
[e] Of wintering generation and first generation (n).
[f] Of the first generation (on PTN).
[g] Of the first and second generations.
[h] Of the second generation (on PTN).
[i] Actual average date, when the temperatures exceed 10°C, for a number of years' observation in Murgab (3577m).

10°C was actually observed on March 9 according to the multiannual data.

Now, when we know the awakening dates (column 5 and 6), we can calculate the dates of the mass flight of butterflies from the wintering generation. On the PTN, as we already know, this can be done very simply. For Dushanbe, as an example, we trace the line of average temperatures dated March 25 (column 5) up to the intersection with phenological curve 17 (Fig. 1.56), where we find $n = 37$ days (column 9), giving us a result of May 1.

In order to use the Dushanbe PTN (800 m) for the level of 300 m, we again have to know the temperature difference between these levels. Since at the height of 300 m awakening occurred on March 7 (see above), we may suppose that the process from awakening to the flight of the wintering generation will take place mainly in March. Therefore, γ remains as before—0.5°C/100 m (column 7). Hence the temperature difference between the 800-m Dushanbe base station and the 300-m level will be Δ°C = + 2.5°C (column 8). We move the legs of a drawing gauge apart by a distance that corresponds to 2.5° on the scale of the Dushanbe PTN; then we put the left leg of the drawing gauge onto the line of average temperatures for March 7 (in Fig. 1.56 this line is not plotted, but it is clear that it will be traced parallel to the line of March 5 and a little to the right of the latter); we move the left leg of the drawing gauge along the line of March 7, maintaining the horizontal position of the legs. The gauge is moved to the right and upward, up to the moment when the right leg touches phenocurve 17. By this means the line of average temperatures from March 7 for Dushanbe is transferred to the right, that is, to the side of the higher temperatures, such as characterize the heat resources net for the 300-m level; but at the same time, this net must have the same shape as the net for Dushanbe (as the annual trend at the height of 300 m is similar in form to the annual trend at the height of 800 m in Dushanbe, but differs from it by 2.5°C). The right-hand leg of the drawing gauge touches phenocurve 17 at $n = 43$ days (the first line of column 9); hence the date of the mass flight of the butterflies from the wintering generation at the height of 300 m will be on March 7 + 43 days = April 19 (column 10).

It is clear that for levels higher than Dushanbe the temperature difference from Dushanbe ($\Delta°C$) will be negative, and therefore, we can expect the phenological curve to be intersected not by the right leg of the drawing gauge, but by the left leg. For example, at the height of 1000 m (column 2), the piorcer's awakening occurs on March 30 (column 6), and the April temperature difference between this level and Dushanbe is 1.2°C (column 8). Therefore, we move the legs of a drawing gauge apart by 1.2°C; then we put the right leg onto the average temperature line of March 30, and we move the gauge along this line up to the point where the left leg touches phenocurve 17. This occurs at $n = 38$ days (column 9). Hence the flight of the butterflies of the wintering generation at the level of 1000 m will be on March 30 + 38 days = May 7 (column 10), that is, somewhat later than at the level of Dushanbe.

In a similar manner we calculate, using phenocurve 18, the dates of the flights of butterflies from the first generation (columns 11-14) and from the second generation (columns 15-18). We find that (because the average temperature line corresponding to the date of the flight of the first generation does not intersect phenocurve 18) the flight of the second generation is, on the multiannual average, impossible at the heights of 1500 and 2000 m in districts near Dushanbe (column 18). However, before this (in column 14) it can be seen that the flight of the first generation also does not occur at the height of 2000 m, except in years that are remarkable for a considerable temperature anomaly (in comparison with multiannual average data).

Before we begin to analyze the results obtained in Table 4.8, we have to make sure that these results agree with the facts. For this purpose, in

Table 4.9, we compare pairs of dates as follows: on the one hand, the dates when 24-hour average temperatures exceed 10°C—these dates can be obtained from the data in Table 4.8 for different places in Tadzhikistan by means of interpolation; on the other hand, the actual dates observed at the corresponding meteorological stations, on the multiannual average.

The actual dates in Table 4.9 were taken from the *Agroclimatic Reference Book* for Tadzhikistan, and the dates interpolated with the help of Table 4.8, or, more exactly, extrapolated from the base stations, were obtained in the following way. Let us assume that it is necessary to determine the date when the average 24-hour air temperature exceed 10°C for Kirovabad, which is located at a height of 363 m above sea level. In Table 4.8 we see that, in the regions served by the base station of Dushanbe at the 300-m level, the temperature exceeds 10°C on March 7 (column 6), on the basis of extrapolations from observations in Dushanbe; at the 500-m level the temperature exceeds 10°C on March 16. Therefore, 200 m will give a difference of nine days, which gives three days for 63 m. We add these three days to the date of March 7 (which corresponds to 300 m) and obtain March 10 for the 363-m level. We write the date of March 10 in column 3 of Table 4.9 opposite Kirovabad, as well as opposite Shaartuz, since the latter lies at the same height.

As can be seen from the data in Table 4.9, the extrapolated dates and the actual dates almost exactly coincide, with the possible exception only of the Western Pamirs. Therefore, the vertical temperature gradients that were used for extrapolations give good results. Hence it also follows that the dates of the piorcer's awakening as taken in Table 4.8 for further calculations are close to the actual dates.

To what degree can these further calculations in Table 4.8 be considered real?

If the calculations had been made using only concrete PTNs for each level, there would be no cause for doubt. However, we used only PTNs related to the base stations, and extrapolated these PTNs for the different levels; thus extra proof of the reality of the results obtained is required.

Therefore, let us compare the results of calculations for some levels on the basis of extrapolated PTNs with the results obtained on the basis of real PTNs relating to these levels. For example, for the 427-m level with the base station of Dushanbe, the interpolation in Table 4.8 gives the awakening on March 12, the flight of the wintering generation on April 20, the flight of the first generation on June 12, and the flight of the second generation on July 29-30. The 427-m level is exactly the height of the meteorological station of Kurgan-Tyube, for which we have a special PTN (Fig. 1.55). Using this PTN we can obtain the dates of the awakening on March 10, the flight of the wintering generation on

TABLE 4.9. Dates extrapolated from base meteorological stations, and the actual average mutliannual dates, on which average 24-hour air temperatures exceed 10° C in different points of Tadzhikistan.

Meteorological Station	Height above Sea Level (m)	Dates Extrapolated on PTN	Actual Dates (*Agroclimatic Reference Book*)	Comment
1	2	3	4	5
Southwestern Tadzhikistan				
Dushanbe, base station	803	—	20 III	The PTNs net for Dushanbe is based on observations made in different years from those in the *Agroclimatic Reference Book.* Therefore, on the PTN we have not March 20, but March 25; this somewhat decreases the similarity of the successive extrapolations to the data in the *Agroclimatic Reference Book.*
Kirovabad	363	10 III	12 III	
Shaartuz	363	10 III	9 III	
Dzhilikul	340	9 III	13 III	
Kurgan-Tyube	427	12 III	12 III	
Shakhrinau	853	26 III	23 III	
Khodzha-Obigarm	1807	22 IV	25 IV	
Garm	1319	8 IV	10 IV	
Obigarm	1387	11 IV	14 IV	
Kulyab Group of Districts (transition to the Pamirs)				
Kulyab, base station	586	—	15 III	According to the Kulyab PTN it is also March 15.
Dangara	660	17 III	20 III	
Kangurt	908	23 III	23 III	
The Western Pamirs				
Khorog, base station	2076	—	16 IV	The PTNs net for Khorog was constructed (at the Institute of Medical Parasitology) on the basis of a small number of years in comparison with the data in the *Agroclimatic Reference Book.* Therefore, on the nomogram for Khorog we have April 23 instead of April 16; this decreases the similarity of the extrapolations for Irkht to the data in the *Agroclimatic Reference Book.*
Irkht	3290	6 VI	1 VI	

April 19, the flight of the first generation on June 11, and the flight of the second generation on July 28.

As can be seen, the results of extrapolation from the Dushanbe PTN for the level of Kurgan-Tyube differ from the calculations made directly on the Kurgan-Tyube PTN by no more than 1 to 2 days. Nevertheless, we have to take into consideration that individual, large errors are possible during calculations of the flight dates of the last generations, not because of extrapolating the net of the base station for other levels, but because of interpolation in Table 4.8 between levels too remote from each other. For example, it turns out that interpolations over differences of 500 m (such as are met in Table 4.8) are risky with respect to the last generations. Hence it is possible to avoid such errors, especially in flat country, when working with isotherms that are not distant from the base stations.

Now we can start directly with the analysis of the results in Table 4.8. From this table it follows, first of all, that the dates when phenological phenomena in the life of the piorcer occur are systematically later with increasing height above sea level. This is natural since the higher the altitude, the colder it is during the warm half of the year.

The difference beween the flight dates of the wintering generation is more than one month on the levels of 300 and 1500 m (southwest), and more than two months on the levels of 300 and 2000 m (southwest), or 1500 and 3000 m (the Pamirs) (column 10).

Apparently, the vertical boundary of the piorcer's distribution in the Western Pamirs is 3000 to 3200 m (column 10); this approximately coincides with the vertical boundary of the distribution of apple trees and pear trees in the Pamirs. Cold-blooded organisms are not always able to penetrate so high; for example, in Section 6.5 we will see that the vertical boundary for the development of the tick *Hyalomma detritum* in Tadzhikistan is about 1500 m. It is no mere chance that the piorcer has the widest distribution area in the world.

At the same heights phenological phenomena occur first in the Western Pamirs, then in the transitional zone of Kulyab, and finally in the southwest of Tadzhikistan. For example, at a height of 1500 m, the piorcer's awakening occurs in these regions on April 4, 11, and 14 (column 6), and the flight of the wintering generation occurs on May 15, 24, 28, respectively (column 10).

Apparently, this result is not a chance one: the dry mountains and uplands in the Pamirs have relatively high temperatures during the warm half of the year in comparison with locations at the same heights in the mountain systems more to the west, including, for example, the Caucasus Mountains (this is why the level of mountain glaciers in the USSR decreases from east to west).

The difference in phenological dates between locations at similar heights in the southwest of Tadzhikistan and in the Pamirs is equivalent to the difference in phenological dates produced by a difference in heights of 400 m (the dates in the southwest at a height of 1100 m are approximately equal to the dates in the Pamirs at a height of 1500 m).

Since phenological dates are progressively later with increasing height, it is natural that the higher the altitude, the fewer the number of the piorcer's generations. As can be seen from the dates in Table 4.8, up to a height of 1000 m or a little more, the piorcer develops in three generations, including the wintering one, both in the west and in the Kulyab group of districts (column 18); at a height of about 1500 m, the piorcer develops in two generations in these same regions (column 14); at a height of 2000 m in Western Tadzhikistan and 2500 to 3000 m in the Pamirs, it has only one generation (column 10). Since the heights of 1000 m correspond to valleys (Hissar, Vakhsh, Fergana), the heights of about 1500 m correspond to mountainous regions (Garm, the mountainous part of the Leninabad districts), and the heights of 2500 to 3000 m and more correspond to the high-mountain region of Tadzhikistan, then the results obtained in Table 4.8 coincide with the actual data:

"In Tadzhikistan the piorcer has from one to three generations, in relation to the climatic conditions. In the valley regions (Hissar, Vakhsh, and the valley part of the Leninabad zone) it has three generations, in the mountainous regions (Garm and the mountainous part of the Leninabad zone) two generations, and in the high-mountain regions (Mountainous Badakhshan autonomous district)—one generation." (Ryabtzeva et al., 1964). This reconfirms the accuracy of the phenological values calculated in Table 4.8.

It has already been shown above that with the help of Table 4.8 it is easy to obtain, by means of interpolation, phenological data for concrete levels, that is, for certain points in Tadzhikistan. For this purpose it is necessary only to know the height of the point above sea level, as well as to what group of regions it is related (southwest, Kulyab Group, the Pamirs). The heights of a number of points and their division according to regions are given in Table 4.9. As well as this, we shall need for further calculations Aivadj—319 m, Kalininsky—391 m, Pakhtaabad—641 m, Gushary—1859 m, Faizabad—1216 m, Komosomolabad—1259 m, and Yavan—664 m (all these points are in the southwest of Tadzhikistan); Kolkhozabad—472 m, Iol—1308 m, and Khovaling—1437 m (Kulyab Group); Rokharv—1751 m, Rushan—1979 m, Ishkashim—2564 m, Irkht—3290 m, Dzhaushangoz—3410 m, Murgab—3577 m, and Kara-Kul—3930 m (the Pamirs). For all these points (including those in Table 4.9) we interpolate the flight dates of the wintering generations using the data from Table 4.8.

The dates obtained for the flight are then put onto a blank geographical map (preferably a physical map). We draw lines to connect the identical dates (conforming to the relief of the region), and we thus obtain the *map of isophenes* for the mass flight of piorcer butterflies of wintering generations in Tadzhikistan (Fig. 4.2).

Thus we have achieved our objective—bioclimatic mapping on the basis of the Podolsky method—and it has been achieved with the help of only three base stations for the very complicated, although not large, mountain area of Tadzhikistan.

In Figure 4.2 we see, first of all, that a small, northern part of Tadzhikistan is not covered by isolines. Since the objective was, mainly, the vertical zoning of the bioclimate, this absence of isolines need not hinder us—quite the contrary: without the northern part, the area covered by isolines has a range of latitude of no more than 2°; therefore, we can disregard latitudinal changes in the bioclimate, an aspect that does not interest us at this stage.

Therefore, in the given case the isophenes mainly "trace" the relief of the region. In fact, the early dates of the piorcer's flight, April 19 to 30, cover the low-lying part of Tadzhikistan. As we move to the north and east, that is, toward the mountain and high-mountain areas, the butterflies' flight is 1 to 2 months later: on the spurs of the Hissar mountain range (to the north of Dushanbe), butterflies of the wintering generation fly at the end of May, and even in the first half of June (Gushari, Khodzha-Obigarm), and in the Western Pamirs they fly up to the end of June (Ishkashim) and later. To the east of 72° E the Pamirs become so high that the piorcer probably loses its ability to develop at all.

Of course, the map could be made more exact. For example, for a number of places there was no need to calculate the dates when the average 24-hour temperatures exceed 10° C—they appear in the *Agroclimatic Reference Book*, which is compiled on the basis of concrete observations; but the author wished to show how this calculation is made when such actual data are not available.

There was no need in all cases to extrapolate the phenotemperature nomograms for different points in Tadzhikistan from the base stations, since we have special PTNs for Kurgan-Tyube, Kirovabad, Yavan, Dangara, and Garm; but it was necessary to show how this is done when we have only a few base stations (for want of a chain of meteorological stations).

In many cases there was also no need to calculate the temperature differences between the base point and the points in outlying districts, since the actual differences could be obtained directly from observations at the base meteorological station and at those in the outlying districts. However, then we would not have obtained Table 4.8, which represents the vertical zoning of the bioclimate at equal intervals of height; this table has particular importance.

The experimental-theoretical method for bioclimatic division into

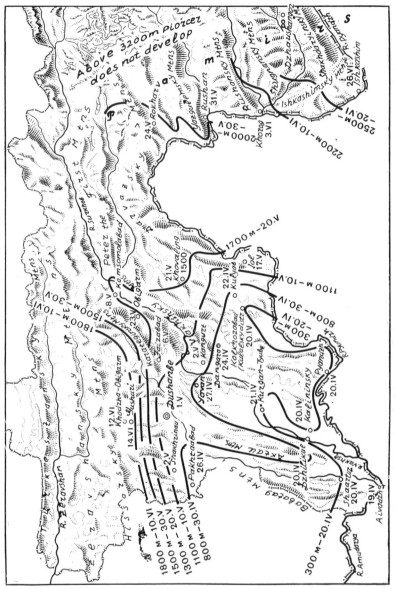

FIGURE 4.2. Map of isophenes representing the average multiannual dates of mass (50%) flight of piocer butterflies (*Carpocapsa pomonella L.*) of wintering generation in Tadzhikistan (according to the results of experimental-theoretical analyses carried out using PTNs). The interpolation between the isophenes can be made only by taking into account the height of the point to which the interpolation refers. For this purpose, near to each isophene is designated the approximate height above sea level to which this isophene refers (one and the same isophene in the west and the east of Tadzhikistan may relate to somewhat different heights—see the corresponding text).

regions, and mapping, requires relatively little labor. For instance, calculating and compiling the map takes one person 1 or 2 days. To this we have to add, of course, years of meteorological observations at three base stations (Dushanbe, Kulyab, and Khorog), which, however, were already made and used for other purposes, and several years of entomological observations (at two to three plant protection stations), which were used for constructing phenological curves of the piorcer's development. However, all of this cannot be compared with the enormous amount of work that would have had to be invested all over Tadzhikistan in order to construct multiannual isophenes on the sole basis of empirical observations at many stations.

In conclusion, we must point out that agroclimatic mapping for agricultural crops does not differ in principle from the bioclimatic mapping described above (see also Section 3.2b).

In addition, it should be said that maps of isophenes (both for pests and for crops) are useful, apart from other purposes, for planning work in agricultural aviation, as well as for deciphering data from "satellite phenology."

4.2c. Calculating Latitudinal Borders for the Possible Distribution of Cold-Blooded Organisms and Plants (especially quarantine objects)

In the previous section we explained how to obtain, in mountainous conditions, the height limit of the piorcer's distribution on the pheno-temperature nomogram (PTN). This was done by moving the average temperature line according to the temperatures at different levels, and thus finding, little by little, the height where development is impossible because of lack of sufficient heat.

As a matter of fact, in the same way, using a number of PTNs and a map of isotherms (or by constructing a series of PTNs in order of longitude), we could calculate a latitudinal limit of distribution (except in the case of mountainous areas). However, we tried to find a shorter way to achieve our aim.

The example used will be calculating the western section of the northern limit of distribution of a quarantine object—the fall web-worm. Of course, the principle of such a calculation can be used for plants too.

The phenoforecasting method has already been used for the fall webworm in Section 4.1b. There we saw that the errors in operational forecasts are within the range of 1 to 3 days. Such high accuracy gives us reason to hope that, in cases where we have nothing with which to compare the results of calculations (in the north of the USSR there is no fall webworm for the present), these calculations will, nevertheless, reflect the real potential state of affairs.

There are indirect ways for determining the northern border of the fall webworm's distribution in the USSR. For this purpose the general climatic conditions in the regions of Canada, where the pest lives, are compared with various places in the USSR. The totals of active temperatures are also compared, and in addition, the northern limit of the fall webworm's distribution is associated with the southern limit of coniferous forests, since such a correlation was noted in Canada.

On the basis of these data, the northern limit of the possible distribution of the fall webworm in the USSR is plotted along the line Leningrad-Vologda-Perm-Kurgan-Karkaralinsk-Zaisan (Churayev, 1962).

In order to determine the northern limit by means of the Podolsky method, and with the minimum number of PTNs, we choose a starting point that lies approximately in the region of the conjectural border (this limit is assumed on the grounds of the indirect indications mentioned above). For example, let us take Volkhov, which is near Leningrad, but is less subject to the micro- and mesoclimatic influence of the water surface. We construct a PTN for Volkhov (Fig. 1.89).

We take the lower temperature threshold for the beginning of the development of *Hyphantria cunea* to be 9°C (I. A. Churayev suggests 9°-10°C, and Sevesku suggests 11°C; reckoning on phenotypical acclimatization of the fall webworm in the north, we take the lower temperature threshold—the more so, since the PTN confirms low thresholds). Then the multiannual average date of the spring regeneration of the first pupae will be May 13 (see, in Fig. 1.89, the heat resources line starting from the mark 9°C on the horizontal axis; this line is also an averge multiannual line).

We begin the calculation from the pupa, because only this developmental stage can be viable in spring. In all other developmental stages the fall webworm would not be capable of surviving the winter; such is the biology of this pest, as is known.

The line of May 13 intersects phenocurve 3 representing the period from pupa to imago, at $n = 49$ days. Hence the beginning of the butterflies' flight will be on May 13 + 49 = July 1. This is again a multiannual date, since the PTN was based on multiannual data, and we did not take the starting dates observed in any concrete year.

We assume that the butterflies lay eggs, on the average, two days later, that is, July 1 + 2 days = July 3. Therefore, the first caterpillars will appear on July 3 + 18 days = July 21, where $n = 18$ days is the ordinate of intersection of the meteorological line from July 3 with biological curve 1.

In the same way we find, using curve 2, the beginning of pupation: July 21 + 80 days = October 9.

The average temperature line of October 9 does not intersect phenocurve 3 (or any other curve for the fall webworm). Therefore, this time there will be no flight of butterflies, and the pupae will begin to winter, especially since, at the beginning of October in the Volkhov region, the autumn frosts begin at just this time (an important coincidence).

Let us follow the development of a caterpillar that hatched from the egg on July 21 (for this date, see above). As can be seen on the axis of abscissae (Fig. 1.89), the multiannual average 24-hour temperature of the air is close to $+17°C$, which is a temperature very favorable to the development of the caterpillars from one age to another. The intersection of the meteorological line of July 21 with biological curve 2 occurs at $t_{av} = 12°C$ (see the abscissa of the intersection). This is an average temperature over the period from caterpillar to pupa. This temperature is also comparatively favorable for the development. Pupation occurs on October 9 at $t_{av} \approx 5°C$ (see the point on the axis of abscissae where the line of October 9 begins). Before October 9 the caterpillar can still develop slowly in the direction of the pupa stage, thanks to the daytime temperatures. But on the whole, the temperature, as can be seen here, approaches a level that can be disastrous for younger caterpillars. However, this temperature occurs at the time of pupation, and for pupae, $5°$ are not dangerous, and even lower temperatures can be tolerated (again an interesting coincidence).

Thus in Volkhov only one generation of fall webworm is really possible.

However, is there any point further north than this where one generation of fall webworm is possible? To answer this question we have to note that caterpillars which hatched in Volkhov district on July 21 (the date obtained above) or even later were able to pupate, but those which hatched later than indicated by the average-temperature line tangential to phenocurve 2 were not (e.g., caterpillars which appear on July 26 do not pupate, since the line of July 26 does not intersect, in fact, it does not even touch phenocurve 2). This tangent is represented in Figure 1.89 by a dashed line, and it refers approximately to July 24. The dashed line is $0.5°C$ (horizontally) from the line of July 21 (when the first caterpillar in Volkhov should usually appear), in the direction of decreasing temperatures. Therefore, pupation would be impossible not only for caterpillars that appeared on July 24, but also for caterpillars that appeared, as above, on July 21, if the air temperature in Volkhov were just $0.5°C$ lower (since in this case, the line of July 21 should be moved to the place of the tangent).

Hence we draw the conclusion that, in another place, where the temperature is more than $0.5°C$ lower than in Volkhov, pupation of the caterpillars will be impossible, and therefore, the development of the fall webworm will be impossible even in a single generation. We look in the *Climatic Reference Book* and find such a place using the summer isotherms: Olonetz in Karelia, located at a latitude of $61°$, a little to the north of Volkhov. This is the most northern point, in the west of the

USSR, up to which the distribution of the fall webworm can extend in normal years.

Now we can finish the calculation, but let us check it with the data for another place, Petrozavodsk, which is located to the north of Olonetz. We construct a PTN for Petrozavodsk (Fig. 1.90), and make the same calculations as on the PTN for Volkhov (Table 4.10). We arrive at the conclusion that in Petrozavodsk a full generation of all webworm is impossible, and that a place where it is 2° C warmer than Petrozavodsk in August through September (or approximately 1°C warmer on average during the whole spring/summer/autumn period) will be the most northerly point to which the distribution of the fall webworm can extend (Fig. 1.90). Such a point, according to climatic maps, will be Olonetz in Karelia. Therefore, the last calculation brought us back to our initial result.

The correctness of this result is reconfirmed by the fact that Olonetz is located at the southern limit of the distribution of dark-coniferous taiga woods. In all probability, as we move further to the east, the northern border of the fall webworm's distribution will be further south, the causes of this being the Atlantic Ocean and the Gulf Stream: the further east, the less effective they are in making temperatures milder. The same is true of the permafrost in the USSR: its limit moves south as we go further east. This general theoretical statement ("spatial forecast") was later reconfirmed by the concrete investigations of E. A. Sadomov (1975), to which we will return later.

We can, therefore, assume that the point calculated (Olonetz) is the most northerly point in the fall webworm's distribution in the USSR (especially in warm years).

Finally, a theoretical conclusion seems to be possible, as follows. In the calculations, the factor of nutrition was not taken into account directly. Nevertheless, we obtained a point on the southern limit of the dark-coniferous taiga woods. Thus the relationship between the northern limit of the fall webworm's distribution and the southern limit of the dark-coniferous taiga woods is determined, not so much by nutritional relationships, as by the climatic relationships between these two biological organisms (the fall webworm's development stops at the point where the development of these woods begins).

Among the advantages of determining such limits using PTNs, as opposed to determining them according to indirect indications, are the following.

1. Determining these limits on the basis of vegetable or other indirect indicators is, obviously, less exact. Moreover, if there are no such indicators, and if special experiments at different latitudes are forbidden (in the case of quarantine objects), or difficult, then the only method remaining is that of calculation.

2. Indirect calculations of limits using active or effective temperatures, and in particular a simple comparison of the main climatic

TABLE 4.10. Average dates of the development of the fall webworm within the limits of the USSR (according to calculations on PTNs).

Date of Spring Regeneration of Pupae (at $t_{av} = 9°C$)	Date of Appearance			
	Butterflies	Eggs	Caterpillars (of the first age)	Pupae
1	2	3	4	5

Petrozavodsk, φ (latutide) = 62°

24 V	24 V	11 VII	13 VII	Do
	+ 48 days	+ 2 days	+ 19 days	not
	= 11 VII	= 13 VII	= 1 VIII	occur[a]

Volkhov, φ = 60°

13 V	13 V	1 VII	3 VII	21 VII
	+ 49 days	+ 2 days	+ 18 days	+ 80 days
	= 1 VII	= 3 VII	= 21 VII	= 9 X[b]

Grodno, Byelorussia, φ = 54°

25 IV	25 IV	12 VI	14 VI	2 VII
	+ 48 days	+ 2 days	+ 18 days	+ 47 days
	= 12 VI	= 14 VI	= 2 VII	= 18 VIII[c]

Kursk, φ = 52°

25 IV	25 IV	9 VI	11 VI	28 VI
	+ 45 days	+ 2 days	+ 17 days	+ 42 days
	= 9 VI	= 11 VI	= 28 VI	= 9 VIII[d]

Eisk, Krasnodar district, φ = 47°

Wintering generation		First generation		
13 IV	13 IV	26 V	28 V	10 VI
	+ 43 days	+ 2 days	+ 13 days	+ 35 days
	= 26 V	= 28 V	= 10 VI	= 15 VII

First Generation		Second generation		
—	15 VII	2 VIII	4 VIII	13 VIII
	+ 18 days	+ 2 days	+ 9 days	+ 43 days
	= 2 VIII	= 4 VIII	= 13 VIII	= 25 IX[e]

[a] Caterpillars are doomed to death since they do not reach (because of insufficient heat) the stage of pupae which are able to survive hibernation.

[b] Autumn butterflies will not fly. Pupa goes into hibernation.

[c] Autumn butterflies can fly on 18 VIII + 49 days = 6 X; they perish simultaneously with egg laying because there is no succeeding pupa stage.

[d] Autumn butterflies can fly on 9 VIII + 35 days = 13 IX; they perish simultaneously with the egg laying because there is no succeeding pupa stage.

[e] Butterflies of the second generation do not fly, pupa goes into hibernation (butterflies can fly only with a positive anomaly of 2°-3°C, but they are still doomed to death).

characteristics in the organism's home area and in the new geographical region, are less concrete, less exact, and less strict, scientifically speaking, than the direct solution of systems of ecological equations on a PTN, which gives the dates of occurrence of individual developmental stages, average temperatures over periods between developmental stages, and so on. In addition, let us remember that PTNs are based exclusively on empirical data.

3. Apparently, calculations of southern limits cannot be obtained using the temperature totals method, as opposed to the Podolsky method, since temperature totals are not applicable to above-optimal branches of phenological curves.

As for the given concrete example of calculation (for the fall webworm), it should be said that direct solutions on the PTN here cannot also claim absolute accuracy. There are not enough actual observations in the low-temperature section of the phenological curve to enable 100% accuracy: since the intersections of meteorological lines with biological curves in the second half of the summer at relatively low temperatures occur at narrow angles, even a small shift of the biological curve influences the result of the calculations to a considerable degree.

Besides this, phenotypical or ecotypical changes in a phenological curve are possible in the north.

Therefore, the immediate problem (which E. A. Sadomov solved in 1975) is to collect the maximal quantity of phenological observations in various regions and countries, and at different latitudes, for the fall webworm.

This author wished primarily to show, in principle, how to use a PTN for determining the distribution limits of organisms.

In Table 4.10, we can see together the calculations not only for Petrozavodsk and Volkhov, but also for a series of other places, from the north to the south (all the PTNs are given in Chapter 1). As can be seen, in Petrozavodsk there cannot be even a single generation to fall webworm; from Volkhov (more exactly, from Olonetz) up to Kursk, or somewhat to the north of Kursk, a single generation is possible (and in Kursk the development somewhat outstrips the dates for Grodno, since Kursk is located further south than Grodno); finally, in Eisk two generations are possible, as actual observations confirm (A. E. Stadnikov).

4.2d. Potential Zoning and Mapping for Harmful (quarantine) Organisms, as well as for Useful (introduced) Organisms

The Colorado potato beetle was a quarantine object until recently. Despite its mass penetration further and further east in the USSR, this pest is still not found in the entire USSR. We wish to determine its potential "possibilities" not only in the regions where it is long established, but also on its new regions. This problem was solved by L. I. Arapova (1972, 1974) on the basis of Podolsky's methodology.

In the same way as was shown in Table 4.10, that is, by means of consecutive calculations on PTN for various geographical ponits, Arapova obtained the developmental dates and the number of generations of the Colorado potato beetle for the whole European section of the USSR. This was done using a combination of her own observations, this author's work, and the well-known investigations made by A. S. Danilevsky on the photoperiodicity of insects (1961) (see Section 2.3).

As a result, Arapova published (1974) the following bioclimatic map (Fig. 4.3). Onto this map this author has also put the isolines relating to another quarantine object—the fall webworm—using both the author's works and those of E. A. Sadomov (1974, 1975, 1976, 1977).

Looking at the map we see, first of all, that the number of generations of Colorado potato beetle increases, naturally, from north to south, but that it has no more than three generations in the USSR. The zones of the beetle's distribution lie mainly in a latitudinal direction, like the climatic zones.

The isophenes calculated for the beginning of the appearance of the second-age larvae of the Colorado potato beetle are also plotted on the map. The control of the larvae of the second age is the most effective measure for protecting the potato yield.

Although these isophenes relate to the average multiannual dates, they guide the plant protectors as to the most effective dates. However, the forecast isophenes can be calculated for each year, according to the concrete dates of potato planting and the concrete temperatures.

As can be seen, developmental dates of Colorado potato beetle larvae in the north and in the south of the USSR differ by one month or more.

More detailed calculations were made by Arapova for Byelorussia (1972, 1974). Here she plotted on the geographical maps not only egg-laying dates, and the developmental dates of larvae, pupae, and imagoes, but also isophenes for the potato; these latter isophenes were also calculated on PTNs on the basis of the corresponding phenocurves (for the PTNs see Chapter 1). The maps are divided into the early, mass, and late sproutings of the potatoes. In this combined bioclimatic forecasting the Podolsky method demonstrated its possibilities.

In Figure 4.3 the division into regions is also given according to the number of generations of the fall webworm in the European section of the USSR. As can be seen, in ordinary years no more than two to three generations of this pest are possible. The northern limit of its distribution stretches along the line between Olonetz, Vologda, and Perm. In fact, it coincides with the southern limit of the dark-coniferous taiga woods. The "spatial forecast" was also accurate (Section 4.2c): the

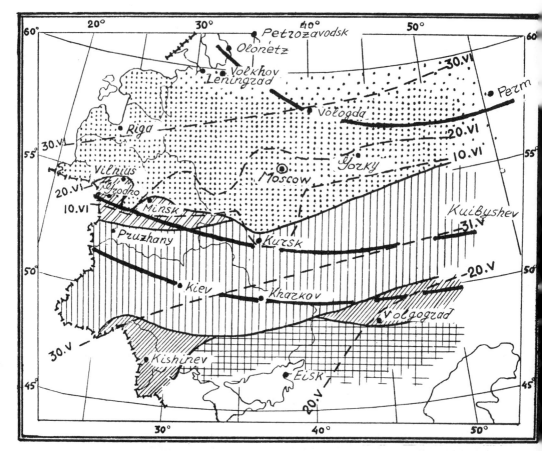

FIGURE 4.3. Division of the European section of the USSR into regions. I: according to the number of generations of the Colorado potato beetle and according to the dates when larvae of the second age appear: ▦ —one generation, ▨ —one generation and, a second, incomplete, generation, ▥ —two generations, ▨ —two, plus a third, incomplete, ▦ —three generations. Dashed lines with dates designate isophenes of the dates when the second-age larvae (of the first generation) begin to appear. II: according to the number of generations of the fall webworm. Northern limits of zones are designated by bold, solid lines: at the foot (in the south)—the zone with two complete generations and a third incomplete generation of this pest; in the middle—with one complete generation and a second incomplete generation; at the top—the zone with one generation.

northern limit of the fall webworm's distribution lies further to the south as we move further east.*

It is obvious that, in the same way, bioclimatic maps can be

*It should be noted that only a small fraction of the calculations for the fall webworm is presented in Table 4.10; the map is somewhat more exact than this table, because the map is based on more reliable phenocurves.

calculated to define the potential of useful, introduced organisms: for example, entomophagous species brought from other countries.

4.3. SHORT-TERM FORECASTING FOR DEVELOPMENTAL DATES OF CROP DISEASES

The developmental rates of pathogens causing fungal diseases in crops are dependent, as is that of the "host" crop, on the ambient temperatures. For example, in Figure 1.36 we see curves 3 and 3a; based on observed data they show in a very clear manner the existence of relationship between the length of the incubation period for vine mildew and the temperature. We also determine this factor for apple scab (Fig. 1.93, line 9), for the phytophthora pathogen in tomatoes (Fig. 1.73, curves 12 and 13), for the duration of the development of the uredostage of brown rust in winter wheat (Fig. 1.82, curve 7), as well as for the incubation period of stem rust pathogenes (Fig. 1.99, curve 5).

Humidity is a more important factor at the time when the plant is infected by the fungus than it is during the period of the disease's development, although during this period it is still important. At the same time, however, the temperature also has importance during the period of infection (Stepanov and Chumakov, 1967; Stakman and Harrar, 1957; and others).

Apparently, a close relationship between development and temperature can be found not only for fungi, but also for many viruses. At any rate, viruses that are parasitic on animal organisms develop and exist in close relation to the temperature (Fig. 1.38, curve 3; Fig. 3.12, curve 1).

Therefore, if onto the heat resources net we superimpose temperature curves for the development of pathogens of fungal and viral diseases (these curves characterize the heat requirements of these organisms), we can forecast the developmental dates of the corresponding plant diseases.

For example, onto the PTN for Dushanbe (Fig. 1.56) temperature curve 19 is plotted, representing the length of the period between two outbreaks of vine mildew, the agent of which is the parasitic fungus *Plasmopara viticola*. Let us say that an outbreak of vine mildew in one of the vineyards in Dushanbe occurred on April 25. When could the next outbreak be expected, provided that there is sufficient moisture, and that the temperatures are close to the multiannual ones?

As before, we find on Figure 1.56 the line of average temperatures dated April 25, and we trace it up to the intersection with biological curve 19.

The intersection has an ordinate of $n = 7$ days; hence the next outbreak of vine mildew should be on April 25 + 7 days = May 2. Therefore, spraying in the vineyard should begin on May 1.

If we had temperature phenocurves for various gradations of moisture, then the latter could be taken into account as described in Section 2.2.

Of course, Moldavia (where biological curve 19 was obtained) and Tadzhikistan are rather distant from each other, but, provided the distances are not too great, phenological curves can be transferred, owing to the stable relationship between the temperature and the organisms' development (Section 2.2). The fungus *P. viticola* itself shows remarkable stability in its phenological-temperature "passport" (compare, in Fig. 1.36, curve 3 for Moldavia with curve 3a for Czechoslovakia).*

The forecast of mildew development given above is based on the multiannual temperature trend over the period between the initial date and the forecast date. However, it is clear that for plant diseases, as well as for the plants themselves and for their pests, a high-accuracy forecast or a phenosynoptical forecast may be needed.

The two latter kinds of forecasts are calculated in a way similar to that used for plants and insects. It should only be pointed out that it is advisable here to obtain the high-accuracy forecasts, not by taking into account the actual temperatures over part of the forecast period (because in this case the forecast's foresight period will be short), but by taking into account the actual temperatures over the *period preceding the forecast period*.

Clearly, for phytopathological forecasting, phenosynoptical forecasts can also be important: weather forecasts for a week in advance (and for the incubation of diseases such a period is often enough) are much more successful than long-term weather forecasts.

Finally, it should be noted that, over a considerable range of temperatures, the phenocurves of a fungus's development display a rising above-optimal branch (e.g., curve 3a in Fig. 1.36). The method of temperature totals is not applicalbe to such curves, in contrast to the Podolsky method.

The new phenoforecasting method was used not only in Tadzhikistan, but also in central regions of the USSR. For example, the author, together with V. I. Bichkova (1974), applied it for forecasting the developmental dates of apple scab. In Figure 1.93 the PTN for Kursk is given, with line 9 representing the duration of the scab's incubation. For example, at the training farm of the Kursk Agricultural Institute, the beginning of the scab's incubation was recorded on May 11, 1969. A

---

*It should be noted that cases of vine mildew in Tadzhikistan are not very frequent: this region is too dry for the disease to thrive (even though during the winter-spring period alone in Dushanbe there were more than 600 mm of precipitation, that is, almost as much as in the humid northwest of the USSR during an entire year; the total precipitation in April in Dushanbe is 114 mm, with rain falling, on the average, almost every second day). Nevertheless, it is important to demonstrate the principle of how to use the Podolsky method for forecasting the developmental dates of crop diseases.

forecast was needed of the date when the incubation would end. At the intersection of the average temperature line of May 11 with biological curve 1 we obtain May 11 + 13 days = May 24. The actual date when the incubation ended was recorded in 1969 on May 25. The error of the forecast is thus + 1 day.

For forecasting the ripening dates of ascospores of the apple scab's agent in Byelorussia, the author's method was applied by Academy Member N. A. Dorozhkin and his colleagues L. V. Bondar and N. A. Shtirenok (1977). It is very interesting that the phenological curve from the beginning of the ripening of the perithecia to the ripening of the ascospores is constructed on the basis not of growth chamber experiments, but of *observations in the field*. Observations from a period of 16 years were used (1961-1976). The investigators assumed that the perithecia begin to ripen on the date when the air temperature exceeds 0°. The equation of the phenocurve turned was as follows:

$$n = \frac{226.66}{t_{av} + 9.33} \tag{16}$$

This curve was then superimposed onto the heat resources net for the Minsk region (Byelorussia).

According to the opinion of the authors of the article, the reliability of the forecasts based on PTNs proved to be high: errors were no greater than 0 to 3 days.

These authors point out (in agreement with L. V. Kuntcevitch, 1963) that the first early-spring spraying with 4% Bordeaux at the time when the ascospores ripen (but before the first dispersion of ascospores) is more effective than spraying even four times with 1% Bordeaux during the subsequent development of the apple tree. It is thus necessary to have a forecast for the ripening dates of the ascospores.

In the control of phytophthorosa in tomatoes (the agent is *Phytophthora infestans Mont.*) Podolsky's forecasting method was used both at the All-Union Plant Protection Institute (in Russian—VIZR) and at the Byelorussian Plant Protection Institute (V. V. Psareva and L. S. Kononuchenko, 1978). The difficulty here was to define the initial date of the forecast—the date of infection. This must be defined according to the "critical" conditions. For Byelorussia these critical conditions are as follows: if during two consecutive days the minimum air temperature at a height of 2 m does not fall below 9° C, the maximum air temperature does not exceed 22° C, the 24-hour average relative atmospheric humidity is not less than 84%, and the minimum humidity is not less than 60%, then the second of these two days will be the date when the tomatoes are infected.

Having determined the date of infection, we find the average temperature line that corresponds to this date on the PTN, and we obtain the ordinate of the intersection of this meteorological line with the phenocurves of the incubation period. There are two phenocurves—for leaves and for fruit (Fig. 1.73). The ordinate of *n* days is added to the date of infection, and we thus obtain the date when the disease of the tomatoes becomes apparent.

The error of the phenoforecasts in Byelorussia is ± 0 to 1 day. The method also showed high efficiency in Kabardino-Balkaria (Caucasus), where it was approved for use.

Podolsky's method was also applied for short-term forecasting of cereal crop diseases.

For example, V. V. Porodenko (1972) forecasted in the Moscow suburbs the length of the uredostage of stem rust (*Puccinia graminis Pers.*). A PTN was used similar to that which was given in Figure 1.99. The results are given in Table 4.11.

As can be seen from Table 4.11, the errors of the forecasts are minimal, and those forecasts based on more precise data are almost error free (of 10 forecasts, 7 are error free, and 3 have an error of ± 1 day). The more precise base data consisted of the actual deviation of the air temperatures from the norm during the first days of the forecast period or even during the preceding period, as described above both for plants and for cold-blooded organisms.

V. V. Porodenko was the first phytopathologist to obtain pheno-forecasting calendars with temperature anomalies on the basis of PTN. The procedure for the calculation is explained in Section 3.1i. In Table 4.12 a variant of the calendar for stem rust is given, greatly abridged.

The calendar was abridged by the author for the sake of economy in space. In Porodenko's calendar, a calculation is given for every day of the period during which the disease may possibly continue.

Finally, in accordance with K. P. Shashkova (1973), the author gives some results of phenological forecasts for brown rust on winter wheat (the agent is *Puccinia triticina Erikss.*), made using the PTN (Fig. 1.82).

Here the Podolsky method was approved for organizing the chemical plant protection of wheat variety Mironovskaya 808. For the experiments they used 80% Cineb (derivative of carbamic acid) in a 0.5% concentration at the rate of 600 litre per hectare (10,000 m^2). Two methods of chemical protection were compared: treatment of crops according to a "strict" timetable, with intervals of 10 days, and treatment according to the phenological method using the PTN.

The results were as follows. After spraying three times on the basis of the phenological method, the development of rust was reduced from 54.0 to 4.3% with a technical efficiency of 93%. Spraying on the basis of the "strict" timetable reduced the disease only to 28.9%, with a technical efficiency of 51.6% (these are all average values).

A number of authors have pointed out the common principle of the new methodology as used for various species of plants and pests. Now

TABLE 4.11. The duration of the uredostage of stem rust (causal agent in wheat), both actual and calculated using Podolsky's method, in the Golitzino-Moscow suburb, the All-Union Phytopathological Research Institute (according to V. V. Porodenko, 1972).

Years	Actual Data			Forecasts Based on Multiannual Temperatures		Forecasts Based on more Precise Data		Temperature Deviation from the Norm (in °C)
	Date of Inoculation	Date of Appearance of Illness	Duration of Uredostage (in days)	Duration of Uredostage (in days)	Errors (in days)	Uredostage	Errors	
1960	20 VI	28 VI	8	9	−1	9	−1	+0.5
	27 VI	5 VII	8	8	0	8	0	0.0
	18 VII	24 VII	6	8	−2	6	0	+5.7
1961	23 VI	1 VII	8	9	−1	8	0	+1.5
1962	19 VI	29 VI	10	9.3	+0.7	10	0	−0.8
1963	23 VI	3 VII	10	9	+1	9	+1	−0.5
1964	11 VI	20 VI	9	9.7	−0.7	8	+1	+3.0
1965	25 V	5 VI	11	11	0	11	0	0.0
	8 VI	18 VI	10	10	0	10	0	+0.5
	18 VI	27 VI	9	9	0	9	0	+1.0

Average errors, without distinction of positive and negative · · · /0.64/ · · · /0.30/

TABLE 4.12. Phenoforecasting calendar for the end of the incubation period of stem rust (causal agent in wheat) in *Zaporozhye* (Ukraine), in relation to the date of infection.

Date of Infection	Temperature Anomalies (in °C)						
	0	−1	−2	−3	+1	+2	+3
1 IV	23 IV	26 IV	29 IV	3 V	22 IV	19 IV	17 IV
10 IV	28 IV	1 V	2 V	6 V	26 IV	24 IV	23 IV
20 IV	3 V	4 V	6 V	9 V	2 V	1 V	1 V
30 IV	10 V	11 V	12 V	13 V	10 V	9 V	9 V
10 V	19 V	20 V	20 V	21 V	19 V	18 V	18 V
20 V	29 V	29 V	30 V	30 V	28 V	28 V	27 V
31 V	8 VI	8 VI	9 VI	9 VI	7 VI	7 VI	6 VI
10 VI	18 VI	18 VI	19 VI	19 VI	17 VI	17 VI	16 VI
20 VI	27 VI	27 VI	28 VI	28 VI	26 VI	26 VI	26 VI
30 VI	7 VII	7 VII	7 VII	8 VII	6 VII	6 VII	6 VII

we see that the same principle can be used in forecasting developmental dates, not only for plants and their pests, but also for plant diseases. This circumstance becomes very important when we are dealing with viral diseases in plants, which are transferred by cold-blooded organisms, especially if the plants are affected by the disease only during a particular developmental stage. For example, in western Siberia, in some years the entire crop of oats dies because of a virus that causes the "pupation" disease. This virus, it is known, develops first in the body of the leafhopper (*Calligypona striatella Fall.*), during the period when the leafhopper develops from larva to imago. If the temperature conditions are favorable and the virus has time to develop in this period of the leafhopper's development from larva to imago, up to the infection stage, then the leafhopper's imago transfers the infectious pupation virus onto the oats. If they are at a certain developmental stage, the oats become diseased and die. Thus with the help of the PTN we can calculate, on the basis of multiannual average data or on the basis of the data for an individual year, the correlation of the developmental dates of the leafhopper, of the pupation virus, and of the oats. With these results, we can determine which period is the most dangerous for cereals, and which dates of sowing and of further development could "lead" the crops away from the pupation virus, and we can forecast the possibility of the crop's dying in a given year.

To conclude the phytopathological section of this book, we should point out that many phenological problems involving short interdevelopmental stages (the incubation periods of plant diseases are very often short) can be solved directly using the annual temperature trend, without the need for a net. However, this is an entirely new application of the method, which will be described in Chapter 10.

CHAPTER FIVE

Combined Phenological Forecasting as a Method for the Study and Management of Agroecosystems (Agrobiogeocenoses)

In 1960, on a particular farm in the Moscow suburbs the infection of spring barley with stem rust was observed on June 27. When should we expect to see the first signs of the illness? Can we apply chemical poisons against the disease on crops which began to head on June 25?

The solution is as follows: the ordinate of the intersection of phenocurve 5* with the line of average temperatures from 27 VI is $n = 8$ days (Fig. 1.99). Hence the first signs of the illness are to be expected on June $27 + 8 =$ July 5 (according to the observations of V. V. Porodenko in the Moscow suburbs, the inoculation of stem rust carried out on June 27, 1960 caused the appearance of the illness exactly 8 days later—see Table 4.11). On this date, or close to it, fungicides are to be used.

On the other hand: the intersection of the meteorological line of 25 VI with phenocurve 6 gives the ripening of spring barley on June $25 + 31$ days = July 26.

It is known that the treatment of cereals by pesticides is forbidden later than 20 days before harvesting. If pesticides are used on the date calculated above, then from this date to the date of harvesting there are July $26 -$ July $5 = 21$ days, or somewhat more. It is clear that we must lose no time in treating the crops with fungicides, so that the chemical poisons will not penetrate into the grain and the straw.

Here we see a calculation for the purposes of managing an agrobiogeocenosis, where the dominant factors are a parasite fungus, a higher plant, and an agricultural animal or human users of barley.

*We assume that the development of stem rust on wheat and barley differs little.

Now we go to another agrobiocenotic problem. According to the multiannual average data, the spring awakening of the piorcer in the Moscow suburbs occurs on May 6 (in actual fact, the awakening occurs at +10°C; in Fig. 1.99 we see that from the mark 10° on the axis of abscissae runs the multiannual line of average temperatures from May 6). What is the average date for the beginning of the flight of the piorcer's wintering generation? When is it necessary to release *Trichogramma* (entomophagous species) in the orchards? How many generations of *Trichogramma* can be born in the orchards, if we take the generation that was let out as the initial generation?

The solution is as follows: the ordinate of the intersection of the meteorological line dated 6 V with phenocurve 7 (Fig. 1.99) is $n = 36$ days, hence the wintering generation begins to fly out on May $6 + 36 =$ June 11. It is known that the piorcer lays eggs on the 8th day after flying out, that is, June $11 + 8 =$ June 19. For this date we have to prepare *Trichogramma* for release in the orchards. The succeeding processes are known: *Trichogramm* lays its eggs into relatively large piorcer eggs, and from *Trichogramma's* eggs hatch larvae that eat away the piorcer's eggs from the inside.

By this first generation of the entomophagous species having laid eggs on the date of the flight, the flight of the next generation follows on June $19 + 22$ days $=$ July 11, where $n = 22$ is obtained at the point of the intersection of the meteorological line for 19 VI with phenocurve 8 for *Trichogramma*. With the date 11 VII, and with the same phenocurve 8, we obtain the flight of the next generation: July $11 + 20$ days $=$ July 31; and so on, as long as the meteorological lines still intersect phenocurve 8. In total we obtain five generations. It should be noted that the natural reproduction of *Trichogramma* adds to the effect of the *Trichogramma* bred artificially.

Here we see another original biocenosis: apple-tree piorcer, *Trichogramma*, and man, who, by virtue of *Trichogramma*, can eat apples not treated with chemical poisons, thus presenting no risk to his health.

O. S. Yermakov and the author of this book (1973) compiled, on the basis of PTNs, phenoforecasting calendars for the development of the piorcer and of the yellow *Trichogramma* (*Trichogramma cacoecia Pallida*) in the Kursk district (450 km to the south of Moscow). These calendars were compiled in the same way as for plants. A fragment of these calendars is given here (Tables 5.1-5.3), greatly abridged (in the 1973 work, calendars are given for each day in the initial column).

Identical phenoforecasting calendars were compiled by Yermakov and this author (1973) for the turnip moth (*Agrotis segetum Schiff*) and for the common *Trichogramma* (*Trichogramma evanescens Westw.*), which is parasitic on the eggs of the turnip moth.

Let us consider another problem. On a particular farm in the Moscow suburbs, the biggest harvest of honey is obtained from the flowering of

TABLE 5.1. Phenoforecasting calendar for the beginning of the flight of the piorcer's wintering generation for Kursk district, in relation to the date of the beginning of pupation and to temperature anomalies.

The Actual Date of the Beginning of Pupation	Temperature Anomalies (in °C)						
	0	+1	+2	+3	−1	−2	−3
20 IV	14 V	12 V	11 V	10 V	15 V	—	—
30 IV	21 V	19 V	18 V	17 V	22 V	22 V	24 V
10 V	29 V	27 V	26 V	25 V	30 V	31 V	1 VI
15 V	2 VI	1 VI	30 V	29 V	3 VI	4 VI	6 VI

buckwheat. For which date is it necessary to prepare the strongest bee families, if buckwheat was sown on May 22, 1970 and temperatures are close to the norm? With phenocurve 9, (Fig. 1.99) and the meteorological line dated 22 V, we find the beginning of buckwheat flowering to be May 22 + 43 days = July 4. We must prepare the bee families for this date.

At the end of Section 4.3 we discussed the agrobiocenotic problem of the system: oat pupation virus/leafhopper insect which transfers the virus/oats. This agrobiocenosis can be managed by calculating the sowing dates that will "lead" the oats away from the virus and from the leafhopper.

As can be seen, biogeocenoses are considered here in a broad sense. We relate to biogeocenoces, for example, the coexistence, in certain geographical conditions, of the malaria plasmodium, the malaria mosquito, and man. The links of this biogeocenosis *interact closely within a single trophic chain.* This system was discussed superficially in Section 2.1, but it will be discussed much more fully in a speical section about malaria in Chapter 6.

Furthermore, sometimes biogeocenoses are bound up with agro-biocenoses within a single system. For example, the development of the Yavan and Dangara Valleys in Tadzhikistan for cotton cultivation involves the transition of the cotton bollworm and the spider mite from

TABLE 5.2. Phenoforecasting calendar for the beginning of the flight of the piorcer's first generation.

Actual Beginning of the Flight of the Wintering Generation	Temperature Anomalies (in °C)					
	0	+1	+2	+3	−1	−2
15 V	25 VII	19 VII	14 VII	9 VII	1 VIII	9 VIII
25 V	30 VII	24 VII	19 VII	15 VII	6 VIII	14 VIII
5 VI	7 VIII	2 VIII	27 VII	23 VII	15 VIII	25 VIII

TABLE 5.3. Phenoforecasting calendar for the flight of *Trichogramma cacoecia Pallida* for Kursk district, in relation to the date of its egg laying and to temperature anomalies.

Actual Egg Laying	Temperature Anomalies (in °C)						
	0	+1	+2	+3	−1	−2	−3
11 V	5 VI	3 VI	2 VI	1 VI	6 VI	—	—
20 V	12 VI	11 VI	10 VI	8 VI	13 VI	15 VI	—
31 V	22 VI	21 VI	19 VI	18 VI	23 VI	25 VI	26 VI
10 VI	1 VII	29 VI	28 VI	27 VI	2 VII	3 VII	5 VII
20 VI	10 VII	9 VII	7 VII	6 VII	11 VII	13 VII	14 VII
30 VI	19 VII	17 VII	16 VII	15 VII	20 VII	22 VII	23 VII
10 VII	28 VII	27 VII	25 VII	24 VII	30 VII	31 VII	2 VIII
20 VII	7 VIII	6 VIII	4 VIII	3 VIII	8 VIII	10 VIII	11 VIII
31 VII	19 VIII	17 VIII	16 VIII	14 VIII	21 VIII	22 VIII	24 VIII
10 VIII	31 VIII	29 VIII	28 VIII	26 VIII	2 IX	4 IX	5 IX

wild producers to cultured cotton. On cotton these pests can propagate in great numbers unless we know in advance the possible dates of their development, the number of generations, and so on, and unless we prepare in advance for control measures against these pests on those dates. Multiannual forecasting for the system of cotton and its pests in the Yavan Valley was discussed in full detail in Chapters 3 and 4 (in particular in Table 4.7).

On the other hand, Yavan and Dangara when supplied and irrigated with water may provide reservoirs where malaria mosquitoes will develop. The inhabitants of these areas have no natural immunity—neither those who were born there (malaria was never here before) nor those who have recently arrived from central areas of the USSR (in Yavan large industrial complexes have been built). Hence malaria can prove to be an undesirable by-product of civilization. Therefore, we needed to study in advance the problem of malaria, especially for Yavan; this is described in Section 6.6.

At the same time, we have to take into consideration the fact that, after a certain period of development, malaria will subside again: chemical poisons liquidate not only pests, but also mosquitoes, as we know from experience in other regions of Middle Asia.

Thus a complex problem for Yavan was discussed, which concerns a whole system of food sources and consumers: cotton/its pests/the malaria mosquito/the malaria plasmodium/man.

This chain may also be made longer. In Yavan livestock farming has always been practiced. As in other regions of Middle Asia and Kazakhstan, livestock suffers from piroplasmoses, and in particular from theileriasis. Theilerias are carried by *Hyalomma detritum* ticks.

The mutual relationships between ticks and livestock are calculated in Table 6.5, which makes it possible to forecast dates for the treatment of cattle against ticks, and thus to control the numbers of the carrier as well as the pathogen of the disease.

Beef tapeworm, other pathogens of helminthoses, leischmaniasis, fleas as carriers of the plague, and fasciola/mollusc systems will be discussed in Chapter 6; these are all components of biogeocenoses.

Finally, can we consider biogeocenosis not only as a spatial phenomenon but also as a temporal phenomenon? By this we mean a concentrated (in time) succession of crops on the same field in one year as follows: cotton after cereals (Section 3.2b) and soya after potato or barley (Section 3.2a). Refer to the crop successions in "green conveyors" in the *canning industry* (for the uninterrupted provision of raw materials) and in bee keeping (Section 3.2a).

A very important problem was solved on the PTNs relating to the consortium consisting of the pest *Hadena sordida Bkh* and spring wheat (V. K. Adzhbenov—1973, 1979; A. S. Podolsky—1974). This pest is found in the European section of the USSR, the Caucasus and the Crimea, in the Southern Urals, and in Siberia. The zone where it is found most consistently, and causes the most damage, is in Northern Kazakhstan.

It is known that the butterflies of *Hadena sordida Bkh* lay their eggs in the heads of spring wheat. In the years when the butterflies' flight and egg laying coincide with heading, there are great harvest losses from this pest. For example, in Northern Kazakhstan in 1957 harvest losses from *H. sordida* over an area of 7.3 million hectares were 2.4 million tons of grain (Grigoryeva, 1965).

Thus we need to determine to what degree the mass egg layings of *H. sordida's* butterflies coincide with the heading of spring wheat variety Saratovskaya 29, sown on May 10, 20, and 30, 1966 in Yesil in Kazakhstan (Fig. 1.66).

First, let us determine the heading dates of the wheat. For this purpose we use the actual (for 1966) temperatures over part of the forecast period (high-accuracy phenoforecast).

For sowing on May 10: We can already estimate on May 10 approximately when heading would occur if the temperatures in 1966 were similar to the multiannual temperatures. The multiannual line of average temperatures from May 10 intersects phenocurve 5 (sowing-full heading) at $n \approx 56$ days. Thus heading will occur on May 10 + 56 = July 5. Therefore, if the actual temperatures are taken into account, for example, 41 days after sowing, that is, on July 20, then we shall be in agreement with the forecast. Thus on June 21, 1966 it was known that the actual average temperature over the period from May 10 to June 20 inclusive, that is, over 41 days, was 16.2°C. We put the corresponding point (*a*) onto Figure 1.66. In fact, we are not late with this forecast—point a lies comparatively far from phenoline 5. Then, probably, we can venture to calculate on the morning of July 1, point b for the 51-day period. The actual temperature over this period is 16.0°C. We put onto Figure 1.66 point b with coordinates $t_{av} = 16.0°C$ and $n = 51$ days. From point b we plot a dotted line parallel to the multiannual line of average temperatures dated May 10. The dotted line intersects phenoline 5 at $n = 57$ days. Thus heading will occur on May 10 + 57 = July 6. This date coincided with the actual date.

For sowing on May 20: A fragment of the actual line of average temperatures from the initial date of May 20, 1966 is shown by line segment cd. From point d we trace a dotted line parallel to the multiannual line of May 20. The dotted line intersects phenoline 5 at $n = 52$ days. Thus heading should occur on May 20 + 52 days = July 11. It actually occurred on July 10.

Sowing on May 30: We construct a fragment of the actual line of average temperatures from the starting date of May 30 (ef). The dotted line parallel to the multiannual line of May 30 gives a heading date of July 17, which coincided with the actual date.

Now we forecast the date of the mass flight of $H.$ $sordida$'s butterflies, on the basis of the actual date of mass pupation—May 29, 1966. Similar to the above, the fragment of the actual line of average temperatures from the initial date of May 29, 1966 is shown by segment qh. From point h we trace a dotted line parallel to the line of average temperatures from May 29. The dotted line intersects phenocurve 8 (mass pupation-mass flight of the butterflies) at $n = 30$ days. Hence the mass flight will be on May 29 + 30 = June 28. This date is close to that observed in 1966.

Using the date of June 28 and phenocurve 9 (mass flight-mass egg-laying), we obtain the date of the mass egg laying: June 28 + 6 days = July 4 (here there is no need to take into account the actual temperatures).

By comparing the calculations for wheat with the calculations for its pests, we see the following. Crops sown on May 20 and 30, which give, according to the calculations, full heading on July 10 and 17, respectively, "avoid" the sphere of influence of $H.$ $sordida$ (July 4). However, crops sown on May 10, which give full heading on July 6, run great danger, according to the calculation. In fact, in 1966 maximum egg laying was observed on the spring wheat sown on May 10 to 20. Crops sown on May 20 and later turned out to be only partly covered with eggs.

All of this became evident from calculations we made at the end of June (see the beginning of the calculations), but, if we had made calculations with a longer foresight period, we could have known earlier. Then we would have had sufficient time to rationalize the dates for examining the fields for the presence of $H.$ $sordida$ caterpillars, and to prepare for the pest's presence on the crops sown on May 10.

If such calculations are made in multiannual terms, it is possible to know which sowing dates for wheat will "lead" this crop away from H. sordida (using the same kind of forecasts it is possible to "lead" crops away from autumn frosts).

V. K. Adzhbenov showed (1979) that the expense saved by introducing Podolsky's method in the Turgai region of northern Kazakhstan was 2,330,000 roubles in 1974 to 1976, that is, according to the official exchange rate at that time, more than $2,500,000—and this was only a superficial saving. The average profitability is 350%, since expenses associated with this forecasting method are very low; the main expense is training people to use the method. Adzhbenov intends to apply Podolsky's method in Kazakhstan on an area of 10 million hectares, and he also recommends the method for western Siberia.

The next important problem is determining permissible dates for the chemical treatment of spring wheat. This treatment must be stopped 15 to 20 days before harvesting. It is clear that forecasting the dates when wax ripeness occurs will give the necessary orientation. During one conference Adzhbenov said that after using the new method, we no longer had to eat bread "with a percentage of poison."

Let us continue this chapter concerning bio- and agrogeocenoses with an experimental-mathematical investigation of the consortium of Colorado potato beetle (*Leptinotarsa decemlineata*) and potato.

In Figure 1.73 we see a PTN for Pruzhany (Byelorussia). All the curves are empirical. Using this PTN, we need to calculate the dates of the first egg layings of the Colorado potato beetle on the sprouts of potatoes sown on May 5, and the dates of the beetle's subsequent development (of course, most harmful is the beetle's development directly on the potato, and not in the environment). All the calculations are to be made using phenoclimatic forecasts with prolonged foresight, that is, on the basis of multiannual temperatures.

The solution is as follows: sprouting will occur 30 days after planting on sandy soil, and 32 days after planting on loamy soil, where $n = 30$ days and $n = 32$ days are the coordinates of the intersection of the meteorological line starting May 5 with phenocurves 7 and 8, respectively. Thus sprouting will occur on sandy soil on May 5 + 30 days = June 4, and on loamy soil on May 5 + 32 days = June 6.

As is known, wintering beetles from surrounding areas come immediately to the potato sprouts, among them beetles ready for egg laying. Thus the first egg layings on the potato sprouts (this is an average sprouting date for Byelorussia) will be on June 4 to 6. We take June 5.

The subsequent stages will occur as follows: the second-age larvae will occur on June 5 + 14 days = June 19, where $n = 14$ days is the ordinate of the intersection of the meteorological line from 5 VI with phenocurve 2 (from egg to second-age larva). The intersection occurred after dotted line a (but before dotted line b). According to the significance of the dotted lines (see the end of Section 2.3), this means that second-age larvae will appear on the potato sprouts. The pupation of larvae will occur on June 5 + 28 days = July 3 (according to phenocurve 5), and this will be at the beginning of budding, since the intersection is located after dotted line *b*. Young beetles will begin to come out from the soil on June 5 + 48 days = July 23. This will occur at the beginning of flowering, since the intersection of the meteorological line of 5 VI with phenocurve 6 is located after dotted line *c*.

All of these young beetles will be unable to lay eggs, since the intersection mentioned occurs in the shaded area I where the day length is critical (see the explanations in Section 2.3).

Of course, the same calculations could be made for a number of other geographical regions and biological organisms. For example, with regard to *Hadena sordida*, it should be noted that the areas of spring wheat heading for early, middle, and late sowings could be constructed on a PTN in the same way as was done for the potato.

Observations in themselves, that is, without the mathematical theory, would hardly make possible such a detailed and many-sided study of the ecology of given organisms on a large geographical scale.

In Tables 5.4 and 5.5 we give an abbreviated variant of pheno-forecasting calendars for the Colorado potato beetle in the Kursk region (A. S. Podolsky, 1975).

TABLE 5.4.   Phenoforecasting calendar for the beginning of the appearance of the second-age larvae of the Colorado potato beetle in Kursk (Russia), in relation to the actual date of the beginning of the egg laying and to temperature anomalies (for any generation).

Actual Date of the Beginning of Egg Laying	Temperature Anomalies (°C)						
	0	+1	+2	+3	−1	−2	−3
10 V	28 V	27 V	26 V	25 V	30 V	31 V	2 VI
15 V	1 VI	31 V	30 V	29 V	3 VI	4 VI	6 VI
20 V	6 VI	4 VI	3 VI	2 VI	7 VI	8 VI	10 VI
24 V	9 VI	8 VI	7 VI	6 VI	10 VI	12 VI	13 VI
25 V	10 VI	9 VI	8 VI	7 VI	11 VI	13 VI	14 VI
31 V	15 VI	14 VI	13 VI	12 VI	17 VI	18 VI	19 VI
10 VI	25 VI	23 VI	22 VI	21 VI	26 VI	27 VI	28 VI
20 VI	4 VII	3 VII	1 VII	30 VI	5 VII	6 VII	7 VII
30 VI	13 VII	12 VII	10 VII	9 VII	14 VII	15 VII	17 VII
10 VII	22 VII	21 VII	20 VII	19 VII	23 VII	25 VII	26 VII
20 VII	1 VIII	30 VII	29 VII	28 VII	2 VIII	3 VIII	5 VIII
23 VII	4 VIII	2 VIII	1 VIII	1 VIII	5 VIII	6 VIII	8 VIII
31 VII	12 VIII	11 VIII	9 VIII	8 VIII	14 VIII	15 VIII	16 VIII
5 VIII	18 VIII	17 VIII	15 VIII	13 VIII	19 VIII	20 VIII	22 VIII

Phenoforecasting calendars for the Colorado potato beetle were compiled in the same manner as was done for plants (Section 3.1i), or, for example, for the turnip moth. We put three phenocurves for the Colorado potato beetle onto the heat resources net of Kursk (Fig. 1.93). The comparison of phenocurves for Byelorussia and for Kursk showed their closeness, that is, the stability of the genotype (Fig. 1.41). Therefore, the author of this book used phenocurves for Byelorussia that are based on more complete data. On the PTN the initial dates were given, taken from the first column of the calendar, and the derived dates were obtained (from the PTN) for the text columns. Originally, calendars were compiled for each initial day.

The calendars for the Colorado potato beetle are to be used in the following way: let us assume that the beginning of egg laying was observed in the Kursk district in a certain concrete year on May 25. We wish to determine when the second-age larvae will come out, if we assume that the temperature trend from May 25 to the larvae's appearance will be similar to the multiannual average. In the first column of Table 5.4 we find May 25, and in the next column, which corresponds to a 0° anomaly, we read June 10; this will be the date that interests us. (The same result can be obtained at the intersection of the average temperature line from 25 V with phenocurve 10 on the PTN for Kursk—Fig. 1.93.)

We consider the procedure described above as a preliminary forecast made on May 25. To make it more accurate we can, for example, in 7 days' time, take the actual average air temperature over the period from May 25 to 31 inclusive, and compare it with the average multiannual temperature over the same period. In this way we can obtain the temperature anomaly for the current year. Information about this anomaly can also be obtained from the nearest meteorological station. Let us assume that the anomaly proved to be +1° (the temperatures in the given year are higher than the multiannual average temperatures). Then, opposite the date of May 25, we read (Table 5.4) the phenoforecast in the column headed +1°—June 9. If the anomaly proved to be −1°, then the second-age larvae will begin to come out on June 11, and so on. It is clear that, the more days taken for calculating the actual average temperature, the more exact the phenoforecast will be, but the shorter the foresight period of this forecast will be. For interdevelopmental stages of 30 to 40 days, actual temperatures over 15-to-20-day periods can be taken into account. A subsequent and even sharp change in temperature cannot significantly change the developmental trend that has begun to show because of "biological inertia." Of course, if +1° or −1° are designated in a long-term weather forecast (for the end of May and the beginning of June), then this forecast must be taken into account in the same way described above.

These calendars can be used not only in Kursk, but also in other

TABLE 5.5. Phenoforecasting calendar giving the beginning of the appearance of the Colorado Potato Beetle's Imago in Kursk (Russia), in relation to the actual date of the beginning of the appearance of the Second-age Larvae, and in relation to temperature anomalies (for Any Generation).

Actual Date of the Beginning of the Appearance of the Second-Age Larvae	Temperature Anomalies (°C)						
	0	+1	+2	+3	−1	−2	−3
1	2	3	4	5	6	7	8
31 V	6 VII	3 VII	1 VII	28 VI	9 VII	13 VII	17 VII
2 VI	8 VII	5 VII	3 VII	30 VI	11 VII	14 VII	18 VII
4 VI	9 VII	6 VII	4 VII	1 VII	12 VII	16 VII	20 VII
5 VI	10 VII	7 VII	5 VII	2 VII	13 VII	16 VII	20 VII
7 VI	12 VII	9 VII	7 VII	4 VII	15 VII	18 VII	22 VII
9 VI	13 VII	10 VII	8 VII	5 VII	16 VII	19 VII	_23 VII_
10 VI	14 VII	11 VII	9 VII	6 VII	17 VII	20 VII	24 VII
12 VI	16 VII	13 VII	11 VII	8 VII	19 VII	22 VII	26 VII
13 VI	17 VII	14 VII	11 VII	9 VII	19 VII	_23 VII_	27 VII
15 VI	18 VII	16 VII	13 VII	11 VII	21 VII	24 VII	28 VII
17 VI	20 VII	17 VII	15 VII	12 VII	_23 VII_	26 VII	30 VII
19 VI	22 VII	19 VII	16 VII	14 VII	24 VII	27 VII	31 VII
20 VI	_23 VII_	20 VII	17 VII	15 VII	25 VII	28 VII	1 VIII
22 VI	24 VII	22 VII	19 VII	17 VII	27 VII	30 VII	2 VIII
23 VI	25 VII	_23 VII_	20 VII	18 VII	28 VII	31 VII	3 VIII
25 VI	27 VII	24 VII	22 VII	20 VII	30 VII	1 VIII	4 VIII
26 VI	28 VII	25 VII	_23 VII_	21 VII	31 VII	2 VIII	5 VIII
28 VI	30 VII	27 VII	24 VII	_23 VII_	1 VIII	4 VIII	7 VIII
30 VI	31 VII	28 VII	26 VII	25 VII	3 VIII	6 VIII	9 VIII
10 VII	9 VIII	6 VIII	4 VIII	3 VIII	13 VIII	16 VIII	20 VIII
20 VII	21 VIII	17 VIII	15 VIII	14 VIII	25 VIII	30 VIII	5 IX
27 VII	30 VIII	26 VIII	22 VIII	21 VIII	3 IX	9 IX	17 IX
28 VII	1 IX	27 VIII	24 VIII	23 VIII	5 IX	11 IX	_20 IX_
30 VII	4 IX	30 VIII	27 VIII	25 VIII	9 IX	15 IX	27 IX
31 VII	6 IX	1 IX	28 VIII	26 VIII	11 IX	18 IX	30 IX
2 VIII	9 IX	4 IX	31 VIII	28 VIII	15 IX	23 IX	4 X
4 VIII	12 IX	7 IX	3 IX	31 VIII	19 IX	_28 IX_	11 X
5 VIII	14 IX	9 IX	5 IX	1 IX	21 IX	2 X	∞
7 VIII	18 IX	13 IX	8 IX	4 IX	27 IX	14 X	∞
8 VIII	20 IX	15 IX	10 IX	6 IX	_30 IX_	22 X	∞
10 VIII	25 IX	19 IX	14 IX	10 IX	6 X	∞	∞
12 VIII	1 X	23 IX	18 IX	13 IX	14 X	∞	∞
14 VIII	_8 X_	28 IX	22 IX	17 IX	25 X	∞	∞
15 VIII	12 X	1 X	24 IX	19 IX	∞	∞	∞

TABLE 5.5. *(Continued)*

Actual Date of the Beginning of the Appearance of the Second-Age Larvae	Temperature Anomalies (°C)						
	0	+1	+2	+3	−1	−2	−3
1	2	3	4	5	6	7	8
17 VIII	21 X	7 X	28 IX	22 IX	∞	∞	∞
18 VIII	25 X	10 X	30 IX	24 IX	∞	∞	∞
19 VIII	30 X	13 X	2 X	26 IX	∞	∞	∞
20 VIII	∞	16 X	5 X	28 IX	∞	∞	∞
22 VIII	∞	24 X	11 X	2 X	∞	∞	∞

Notes. (1) Below the solid lines dates are that related to beetles that do not lay eggs (beetles go into diapause); (2) below the dashed lines are dates that relate to beetles that totally unable to lay eggs, which could appear after the beginning of the first autumn frosts if the latter did not disturb the beetles; (3) the symbol for infinity (∞) designates the impossibility of the appearance of beetles under any circumstances, because of the lack of sufficient heat.

districts of the Kursk region, and even in neighboring regions, if their temperature differences in comparison with Kursk are known. Let us suppose that the previous problem is to be solved for a region that is 1° warmer than Kursk during the growing period. Then, even with a zero temporal anomaly in the temperature, we use the +1° column. If the temporal anomaly is +1°, we use the +2° column (Table 5.4) and read opposite May 25 the date of June 8; that is the date when the second-age larvae begin to appear. Here +2° consists of a +1° anomaly in space and a +1° anomaly in time. On the other hand, if in a given region the temporal anomaly is −1°, we use the zero anomaly column, since +1° + (−1°) = 0°.

Calendars can be compiled for each initial day. In the calendars given here, in the majority of cases, the initial (for the phenoforecast) dates are not given for each day. Here an elementary interpolation is required.

For forecasting the larvae's appearance, it is better to use accurate actual observations of the first egg layings on wild solanaceae and on potato sprouts (in this case we can give two separate phenoforecasts). However, if we have no alternative, then we can assume that the egg layings of the Colorado potato beetle in the field begin on the day when the 24-hour average air temperatures regularly exceed +15°.

The dates when the imagoes will appear can be forecasted using Table 5.5. Let us assume that the beginning of the appearance of the second-age larvae was actually observed on June 12. Then, young beetles of the first generation begin to appear, with a zero temperature anomaly, on July 16 (the same result can be obtained on the PTN with curve 11 in Fig. 1.93); with an anomaly of $-2°$ this will occur on July 22; and so on. Since for Kursk the critical day length occurs on July 23 (after this date the days become shorter than required for the beetles' pubescence for egg laying, see Section 2.3), then we come to the conclusion that both in the first case (July 16) and in the second (July 22) there will be beetles that are able to lay eggs. These beetles will constitute 20 to 25% of this generation, as is usually the case with the first summer generation at the latitude of Kursk (Fig. 2.19). In fact, the meteorological line of 12 VI intersects phenocurve 11 beyond zone I, the zone of the critical day length (Fig. 1.93), although it does lie close to it. With an anomaly of $-2°$ the intersection occurs even closer to zone I (it should be not forgotten that zone I moves, during this process, to the left to the same extent that the lines of average temperatures move).

If the second-age larvae begin to appear, for example, on June 22, then the young beetles will appear (with a zero anomaly) on July 24; with an anomaly of $+1°$ they will appear on July 22; with an anomaly of $-2°$ they will appear on July 30. Therefore, in the first and last cases (July 24 and 30), 100% of the beetles will be unable to lay eggs: the dates for their egg layings are located below the uninterrupted line (Table 5.5) which is drawn under the date 23 VII. The corresponding intersections on the PTN occur in zone I. With a positive temperature anomaly, 20 to 25% of the beetles are able to lay eggs. (In order to avoid misunderstandings and to understand completely the "mechanism" of the PTN, it should be noted that although the lines of average temperatures, for example, of 15 VIII, 26 VIII, and so on, lie entirely beyond zone I, they will be related entirely to the critical zone, since these lines, even the initial segments, relate to the after-critical date of July 23.)

Calculations of average multiannual developmental dates, and of the number of generations, are made in the following way: the average multiannual date when the 24-hour average air temperatures regularly exceed $+15°$ in Kursk is May 24 (we obtain this from the *Agroclimatic Reference Book* or from the point on the Kursk PTN where the line of average temperatures begins at the 15° mark). We take this date to be the date of the first egg laying. Opposite the date of May 24 we find, in the column with a zero anomaly (Table 5.4), June 9. This will be the average multiannual date of the appearance of the second-age larvae. Opposite the date of June 9 we find, in the column with a zero anomaly (Table 5.5), July 13, which will be the multiannual date when young beetles of the first summer generation begin to appear. This date is not critical (see above). Therefore, not all the beetles will go into diapause: 20 to 25% of the beetles of the first summer generation begin to lay eggs, but 10 days later than this, that is, on July 13 + 10 days = July 23. These eggs will give the second summer generation.

In the same manner as that described above, opposite the date of July 23 (Table 5.4) we find August 4; this will be the multiannual date when the second-age larvae of the second summer generation begin to emerge. With the date of August 4 we find, in Table 5.5, September 12 as the multiannual date when young beetles of the second summer generation begin to appear. The latter date is located far below the uninterrupted line, which means that all the beetles will go into diapause. However, it lies above the dotted line; therefore, the beetles will appear almost a month earlier than the first autumn frosts, which occur in Kursk, on the average, on October 9. This will make it possible for the beetles to find food and to prepare for hibernation.

Thus in Kursk, in most years two summer generations are possible for the Colorado potato beetle. However, in cold years only one generation develops. We find here complete agreement between the results of our calculations and the bioclimatic map of the USSR (Fig. 4.3).

Besides the problems described above, other problems can be solved using the calendars. For example, we can give combined forecasts for the system of beetle and potato or beetle and its entomophagous species (provided that calendars are compiled both for the potato and for the entomophagous species); we can also compare the daily estimation of the pest's development with the multiannual norms; we can make retrospective calculations in which, on the basis of the subsequent developmental stage, we find the date of the previous stage. The list of problems that can be solved using the calendars could be made even longer.

Phenological forecasts based on the above calendars for the Colorado potato beetle in the Kursk region were made by the author of this book over a period of several years. They proved to have good accuracy. Particular interest was aroused in everyone by the fact that forecasts for the *non*appearance of the beetle's second generation in cold years, and in relation to the critical day length, proved to be correct.

The Podolsky method was recommended by the All-Union Institute of Experimental Meteorology for the Hydrometeorological Network of the USSR, for forecasting the Colorado potato beetle's development (V. V. Volvach, 1975).

Finally, it should be noted that the Podolsky method has still far from exhausted the possibilities for investigating the system of "harmful poikilotherms/useful fauna and flora" from the point of view of integrated pest control. This, in particular, chemical control measures, should be organized in such a way that the useful organisms do not suffer along with the harmful organisms. For this purpose we have to forecast parallel development of both harmful and useful organisms.

Thus a great majority of living organisms on the Earth are poikilotherms and plants, the developmental dates of which are determined primarily by solar heat and a number of other environmental conditions. Therefore, the experimental-mathematical method of phenological forecasts, and bioclimatic estimations as described here, makes it possible to study many biogeocenoses: *a PTN filled with 24 phenocurves, for example, contains within the area of a sheet of writing paper information consisting of approximately 11,000 daily forecasts*, with a considerable foresight period, for 24 organisms and developmental stages, provided that for each organism calculations are limited to a two-month range of initial dates and to six temperature anomalies.

CHAPTER SIX

The New Experimental-Mathematical Method for Phenological Forecasting and Bioclimatic Estimations in Medical and Veterinary Parasitology and Virology

This discussion covers pathogens of human diseases and of farm animal diseases: pathogens of *invasive* diseases (helminths and, among the protozoa, the malarial plasmodium), as well as pathogens of some *infectious* illnesses (the viruses of foot-and-mouth disease and of yellow fever).

We will also discuss arthropod carriers of *vector-born* diseases (ticks as carriers of theileriasis in cattle, malaria mosquitoes as carriers of malaria, mosquitoes as carriers of leishmaniasis, fleas as carriers of the plague, and *Aëdes aegypti* mosquitoes as carriers of yellow fever).

Invasions are represented, first of all, by roundworms (class *Nematoda*), tapeworms (*Cestoidea*), and flatworms (*Trematoda*). Among them are *geohelminths*, which do not need intermediate hosts and, furthermore, do not need to have any biological connection with the hosts (*Haemonchus contortus, Bunostomum trigonocephalum, Chabertia ovina, Ascaris lumbricoides*, and *Trichuris trichiurus*), and *biohelminths*, which have intermediate hosts (beef tapeworm; fasciola).

Among the biohelminths, the author would give particular emphasis to *biogeohelminths*, such as *Fasciola hepatica*—pathogens of fasciolasis. Fasciola has an intervening host—the mollusc—but this host is a cold-blooded organism in which fasciola continues to develop as it does in the environment. In addition, fasciola's eggs also develop in the environment itself (in ponds, bogs, and pools).

6.1. PHENOLOGICAL FORECASTING OF DEVELOPMENTAL DATES OF ROUND GEOHELMINTHS IN THE ENVIRONMENT: *Haemonchus contortus, Bunostomum trigonocephalum*, and *Chabertia ovina* (intestinal strongyls in sheep)

It is known that the nematode *Haemonchus contortus* in the invasive phase of larvae (of the third age) causes haemonchosis in sheep. During this process the development from eggs (which are excreted by the sheep together with the feces) to larvae of the third age proceeds in the environment (on pasture) at a rate that is dependent on temperatures. This dependence is represented on the PTN for Dushanbe (Fig. 1.56) by phenological curve 10, which is transferred here from Figure 1.34; Figure 1.34, in turn, was constructed by the author of this book on the basis of field experiments carried out by I. F. Pustovoi (1969). Thus the temperature developmental curve is constructed on the basis of field phenological observations (rather than observations in a controlled temperature chamber) and on the basis of temperatures that are taken from the local meteorological station at a height of 2 m. The heat resources net is also constructed using temperatures at a height of 2 m. Therefore, we do not have to take special account of the biotopes' microclimates (see Sections 2.4 and 6.3; for a discussion of taking into account the 24-hour fluctuations in temperature and in this connection, the minimum threshold for the *H. contortus* development, see section 2.4). Phenological curve 10, mentioned above, has also been transferred to other nomograms, relating to various regions in Tadzhikistan.

When, in various regions of Tadzhikistan, will the third-age larvae of the wireworm appear, if the first pasturing of sheep after the summer sterilization (purification) of the pasture (sterilization by high temperatures) occurred on January 15, on February 25, on March 15, and so on? Pasturing dates are the egg-laying dates. The problem is to be solved in multiannual terms.

This solution can be obtained, as before, in quite a simple way. We find on the PTNs for the region that interests us the lines of average temperatures from January 15, February 25, March 15, and so on corresponding to the pasturing dates. Next, we find the ordinates of the intersection of these lines with the phenological curve of the wireworm's development from egg to larva. We add these ordinates to the dates of pasturing, and thus obtain the dates when larvae of the third age appear. For example, for Dushanbe we have the following results, obtained with the help of Figure 1.56: January 15 + 32 days = February 16; February 25 + 13 days = March 10; March 15 + 12 days = March 27; and so on.

At the same time we must pay attention to the phenological law: the difference between egg layings on January 15 and February 25 is 41 days; whereas the difference between the appearance of larvae corresponding to these egg layings is only 22 days (February 16 and March 10). Therefore, in March we can expect an accumulation of the wireworm's larvae from the February and March emergences on the pasture. It is no mere chance that the actual curves representing the prevalence of hemonchosis in sheep in Tadzhikistan invariably show a sharp maximum in March and April.

As can be seen in Figure 1.56, any line of average temperatures intersects phenocurve 10 for the wireworm's development. The same can be seen on the PTNs for Yavan, Kurgan-Tyube, and other regions. This means that in the valleys of Tadzhikistan (not in the high mountains), wireworm eggs can develop annually, with the exception, of course, of the hot part of the summer, when the pastures are sterilized. This is confirmed by Pustovoi's field observations.

Eggs and larvae of the wireworm perish at approximately the same temperature as the eggs of ascarides, that is, at $+60°C$. Such a temperature on the sunlit soil surface (this temperature is, at the same time, an indicator of strong ultraviolet radiation) exists in the conditions of Dushanbe approximately from June 25 to August 25. Since the development to the third-age larvae from eggs requires, in summer, about 6 days, then the pastures are sterilized in Dushanbe approximately from the middle of June to the beginning of September. The above can be considered as the simplest way of taking into account the critical temperatures for the wireworm's existence (for a more accurate method of taking this into account, together with nontemperature components, see Section 2.3).

Actual temperatures over a part of the forecast period, or the long-term weather forecasts, are taken into account during calculations for the wireworm in the same manner as that described for other organisms.

The results of the calculations for the developmental dates of wireworm larvae, in multiannual terms, for several natural climatic zones of Tadzhikistan, are given in Table 6.1.

The data in Table 6.1 make it possible to draw conclusions as follows:

1. If on a particular pasture the first (after the summer sterilization) pasturing was, for example, on November 15, then on this pasture, in the majority of years, the period when the sheep cannot become infected with hemonchosis lasts only until the end of November in Kurgan-Tyube, Yavan, and Dushanbe. In Garm, however, this pasture would remain safe until the beginning of spring pasturing in April of the next year. The last eggs from which larvae of the third age can appear in the conditions of Garm must be laid, in multiannual terms, not later than November 10, because the line of average temperatures from November 10 (Fig. 1.57) is the last line that touches phenocurve 5,

TABLE 6.1. Development of geohelminth *Haemonchus contortus* from egg to larva of the third invasive stage, in relation to the dates of pasturing or driving of sheep in different climatic zones of Tadzhikistan (average data; results of calculations on PTNs).

Zone	Dates of Pasturing (dates of egg layings)													Notes
	15 I	25 II	15 III	15 IV	15 V	15 VI	15 VII	15 VIII	15 IX	15 X	10 XI	15 XI	15 XII	
						Dates of Larvae Appearance								
Vakhsh Valley, Kurgan-Tyube area	13 II	9 III	26 III	23 IV	22 V	21 VI	Sterilization of pastures				23 XI	29 XI	25 I	Eggs and larvae can develop during entire autumn/winter/spring period
Yavan	7 II	10 III	27 III	23 IV	22 V	Sterilization of pastures					23 XI	28 XI	10 I	
Hissar Valley, Dushanbe area (for comparison)	16 II	10 III	27 III	23 IV	22 V	21 VI	Sterilization of pastures				24 XI	30 XI	No pasturizing	
Garm, area of meteorological station (track of sheep driving)	No driving of flocks			26 IV	23 V	22 VI	21 VII	22 VIII	22 IX	26 X	18 XII	Larvae do not appear because of heat deficiency	No driving of flocks	Appearance of larvae is interrupted from the second half of December to the end of April

354

which represents the wireworm's development (at $n = 33$ days, this gives the appearance of larvae on December 18).

For other pasturing dates, there are other date limits for the safety of the pastures. As can be seen in Table 6.1, these limits are close to those mentioned by Pustovoi (1963) on the basis of two years of observations.

2.   The calculations given above, therefore, make it possible to work out a well-founded plan, and a schedule for rotating pasturing between the different quadrants, of the pasture for the purpose of protecting the flocks against hemonchosis and its pathogens. For example, the pasture where the first autumn pasturing took place on November 15 should be replaced by another pasture in Kurgan-Tyube, Yavan, and Dushanbe, at the end of November. Pastures in Garm, where first pasturings took place after November 10, do not need any change.

3.   The dates given in Table 6.1 can be a guide for protection from *Haemonchus contortus* larvae.

4.   Because of the phenological principle, on pastures in the Hissar, Vakhsh and Yavan Valleys, larvae of February and March origin accumulate in March. Therefore, maximal measures against larvae (prophylactic and medicinal measures) must be taken in early spring.

5.   Phenotemperature nomograms make it possible, on the basis of relatively few experiments in one particular place, to make calculations relating to what will occur in other climatic regions without the need to resort to numerous experiments.

The same calculations can be made for other intestinal strongyls in sheep—for the geohelminths *Bunostomum trigonocephalum* and *Chabertia ovina* (Figs. 1.35 and 1.57).

Of course, calculations may be made not only in multiannual terms, but also for each concrete year, taking into account the weather forecast or the actual temperatures over a part of the forecast period or over the preceding period. This process has been described above several times.

6.2.   PHENOLOGICAL FORECASTING OF THE DEVELOPMENT OF ROUND GEOHELMINTHS OF HUMANS [*Ascaris lumbricoides* and *Trichocephalus (Trichuris) trichiurus*] IN THE ENVIRONMENT

The wide distribution of the complaints caused by these geohelminths is well known. Adult *A. lumbricoides* live in the small intestine of man, and *T. trichiurus* live in the large intestine. Their eggs enter the environment along with human feces, and there they develop until the larva stage.

In Figure 1.67 a PTN is given for Azerbaidzhan with phenocurves for the development of the eggs of these geohelminths in the environment. The PTN was constructed by R. E. Chobanov (1978) (after A. S. Podolsky).

The phenocurve for ascarides, drawn here on the basis of data obtained earlier by Chobanov and N. G. Dashkova for Azerbaidzhan, is very close to the phenocurve obtained by I. M. Suslov (1973) for the central regions of the USSR, using data from D. G. Timoshin.

In Table 6.2 we give a number of results. As can be seen, Chobanov carried out special field experiments in Azerbaidzhan for the purpose of testing the new phenoforecasting method.

The experimental site was located in penumbra. It was systematically irrigated. Eggs of *A. lumbricoides* and *T. trichiurus* were put onto disks of plankton filter. These disks were covered by other disks, and were sewn in pairs onto fiberglass. The fiberglass was put into soil at a depth of 5 cm on various dates which are indicated in the first column of Table 6.2. From February to September 1973 a total of 15 layings of the invasive material were made. Once or twice a week, and daily during the period of the maturation of invasive larvae, two filters were taken out of the soil and the eggs were washed off carefully. The eggs were examined under a microscope.

It should be remembered that phenoforecasts with prolonged foresight (phenoclimatic) are given using the intersections of multiannual average temperature lines (relating to the dates of experiments) with the phenocurves (see the dotted line from April 20 in Fig. 1.67). More accurate forecasts, with shortened foresight, are made using actual temperatures over a part of the forecast period (point *a*, Fig. 1.67).

In Table 6.2 it can be seen that the errors in the forecasts are very small, with infrequent exceptions. This brought Chobanov to the following conclusion: *The study of geohelminths in different climatic conditions using PTNs can replace the use of special, laborious, and long-term experiments, which are not always feasible.*

Note that field experiments of the same type, but for different organisms, carried out by Chobanov, Pustovoi, Babayeva, Karelin, and Slepov (see below for the latter three), in turn, make it possible to construct valuable phenocurves, based on the field data.

Such experiments can be carried out during one season in a number of locations (sometimes even in one place), and they can be used, with the help of a PTN, for many regions and over many years.

Suslov (1973) investigated the epidemiology of ascaridiasis in the central regions of the USSR (Kursk), using Podolsky's method. He also took into account social factors.

Like Chobanov, Suslov calculated and constructed the heat resources net of the soil at a depth of 5 cm. In doing this, he excluded from the calculations the winter period: when calculating the average temperature lines from autumn initial dates, he totaled the autumn and spring temperatures, disregarding the winter.

TABLE 6.2. Duration of the development of eggs of *A. lumbricoides* and *T. trichiurus* in soil of the Apsheron peninsula in Azerbaidzhan at a depth of 5 cm, according to the 1973 field experiments and to the PTN (after R. E. Chobanov).

	Ascaris lumbricoides					*Trichurus trichiurus*				
	Duration (in days)					Duration (in days)				
		After Phenoforecasts on PTN					After Phenoforecasts on PTN			
Experimental Dates when Eggs Were Put into Soil	In Field Experiment	With Shortened Foresight (15–60 days)	Errors of Forecasts (in days)	With Prolonged Foresight (18–101 days)	Errors (in days)	In Field Experiment	With Shortened Foresight (15–60 days)	Errors (in days)	With Prolonged Foresight (19–103 days)	Errors (in days)
1	2	3	4	5	6	7	8	9	10	11
15. II	101	102	−1	112	−11	103	105	−2	114	−11
25. II	90	92	−2	101	−11	93	94	−1	104	−11
10. III	77	76	+1	85	−8	80	78	+2	87	−7
26. III	62	63	−1	66	−4	64	65	−1	68	−4
5. IV	52	53	−1	54	−2	54	55	−1	56	−2
20. IV	38	39	−1	41	−3	40	41	−1	43	−3
5. V	29	29	0	30	−1	31	31	0	32	−1
26. V	23	23	0	24	−1	24	24	0	25	−1
15. VI	18	18	0	19	−1	19	19	0	20	−1
25. VI	18	18	0	18	0	19	19	0	19	0
5. VII	18	18	0	18	0	20	19	+1	19	+1
26. VII	18	18	0	17	+1	20	19	+1	19	+1
5. VIII	19	18	+1	17	+2	21	20	+1	19	+2
15. VIII	20	19	+1	19	+1	22	21	+1	20	+2
5. IX	26	26	0	30	−4	29	29	0	31	−2
Average errors without distinction of positive and negative			/0.6/		/3.3/			/0.8/		/3.3/

357

Next, Suslov put onto such a net a phenocurve based on the data of Timoshin (1957, 1967). To the results of calculating the developmental periods of ascarides, which enter the environment in late summer and in autumn, Suslov added the winter period, which was previously omitted, and thus obtained the spring dates for the maturation of the eggs.

Then he took, for the dates when the eggs enter the environment, dates at 10-day intervals during the entire year. Using a PTN, he obtained the corresponding dates of the maturation of eggs over the entire year, as well as the indicators of infectibility and of the curing (from this invasion) of a certain proportion of an affected population over each 10-day period.

The final curves, representing the annual epidemiological process, calculated by Suslov, are close to the results of the coprological examination of the same population given by A. A. Filipchenko and V. N. Dansker (1935).

6.3. FORECASTING THE SURVIVAL DATES OF THE BEEF TAPEWORM'S ONCOSPHERES IN THE ENVIRONMENT (TAPE BIOHELMINTH *Taeniarhynchus saginatus*)

It is known that the beef tapeworm causes teniasis in man and cysticercosis (measles) in cattle. In this process the definitive (final) host is man and the intermediate host is cattle. Thus cattle get this infection from man and noninfected humans get the infection from cattle (apart from exceptional cases).

The eggs and oncospheres of cestodes enter the environment from man. The oncospheres spread in the environment, and can survive for a very long time. The oncospheres enter the bodies of livestock (as well as the bodies of buffalo, zebu, and yaks) along with food and water, and then develop in the animals' muscles up to the stage of larvae-measles. Humans become infected when eating meat that has not been adequately treated.

In the human body, the beef tapeworm reaches the stage of pubescence, at a length of 10 m.

The forecasting method described below for *Taeniarhynchus saginatus* was worked out by the author together with R. I. Babayeva (Uzbekistan Institute for Scientific Research in Medical Parasitology, USSR).

Babayeva, in her *field* experiments in Samarkand, put oncospheres into different biotopes in various months, and systematically observed their viability.

The results of the observations, that is, the length of the survival

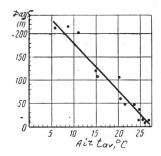

FIGURE 6.1. Biological line giving the survival dates of the oncospheres of beef tapeworm (*Taeniarhynchus saginatus*) on soil with grass in shade, in relation to air temperatures at the height of 2 m.

time, were combined with the average (for these periods) air temperatures taken from the Samarkand meteorological station in exactly the same way as was described at the beginning of the book, in the section concerning the construction of phenological curves. Next, the combined values (survival period and temperatures) were put onto graphs. Two of the lines are given separately in Figures 6.1 and 6.2, and all four together are given in Figures 1.59, 1.60, and 1.61.

In these figures, the straight lines expressing the relationship between the oncospheres' survival in days (n) and the average air temperature over this period (t_{av}) are obtained as a result of a simple linear regression. The straight line equations, the correlation coefficients (r), and the coefficients of determination (r^2) are as follows:

For the first experimental version, with oncospheres in shaded (by trees or mountains) grass:

$$n_1 = 280 - 10.2 \, t_{av}; \quad r = -0.96 \pm 0.08; \quad r^2 = 0.92 \qquad (17)$$

For the second version, with oncospheres in sunlit grass:

$$n_2 = 243 - 9.0 \, t_{av}; \quad r = -0.96 \pm 0.07; \quad r^2 = 0.92 \qquad (18)$$

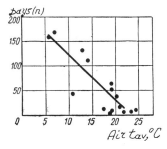

FIGURE 6.2. Biological line giving the survival dates of the oncospheres of beef tapeworm on a sunlit, bare soil surface, in relation to air temperatures at the height of 2 m.

For the third version, with oncospheres put onto an overshaded surface of bare soil:

$$n_3 = 210 - 8.3\ t_{av}; \qquad r = -0.84 \pm 0.15; \qquad r^2 = 0.71 \qquad (19)$$

For the fourth version, with oncospheres on sunlit bare soil:

$$n_4 = 217 - 9.4\ t_{av}; \qquad r = -0.79 \pm 0.17; \qquad r^2 = 0.62 \qquad (20)$$

As can be seen, the coefficients of determination show a rather high percentage of variations in the oncospheres' survival periods that can be explained by the variations of the temperature factor: in the first two experiments, 92%; in the third, 71%; in the fourth, 62%. The latter percentage can be explained in part by the replacement of the curvilinear relation (which can be seen in Fig. 6.2) by the rectilinear relation.

The results of the first two experiments, which are close to natural pasture conditions (i.e., the experiments with grass in the shade and in the sun), were examined using not only simple correlation, but also multiple correlation.

If we designate soil moisture by q, relative air humidity by ρ, air temperature by t_{av}, and the duration of the oncospheres' survival (as before) by n_1 and n_2 for grass in the shade and for grass in the sun, respectively, then the multiple regression equations will be as follows:

$$n_1 = 244 - 9.70t_{av} + 0.8q + 0.2\rho \qquad (21)$$
$$n_2 = 143 - 7.50t_{av} + 0.004\ q + 1.2\rho \qquad (22)$$

The plus or minus signs of the coefficients show that the survival periods decrease as the air temperatures increase, and they increase as the soil moisture and atmospheric humidity increase. The correlation coefficient between the survival periods and the three mentioned arguments turned out to be $R = 0.976$, that is, even higher than with the analysis of the simple correlation. The values for the partial correlation confirmed the key importance of the heat factor.

Of course, for practical purposes we can avoid cumbersome calculations of equations, and use only the graphical method for finding the lines of relations. In order to do this, we count the number of points above and below the line, both along the separate segments and along the entire line, as was described for phenological curves. This kind of graphical method can sometimes give an even more real line of relation than the analytical method; this is true, for instance, in Figure 6.2, where the curvilinear relation is more appropriate.

Thus let us compare, in Figure 1.59 (or in Fig. 1.60 or 1.61), the four biological straight lines that represent the oncospheres' survival in different biotopes. First, we see that in all versions of the experiment, a temperature decrease is accompanied by an increase in the survival period of the oncospheres. Next, we see that at the same air temperature (at a generally accepted height of 2 m), oncospheres have longest life in shaded grass (line I), have a shorter life in sunlit grass (line II), have an even shorter life on overshaded bare soil (line III), and have the shortest life on bare sunlit soil (line IV). This is understandable: at the same air temperature, shaded grass will have the lowest temperatures (both as a result of the shade and as a result of heat losses through transpiration), and sunlit bare soil will have the highest temperatures, especially at the warmest time of a 24-hour period or of a year, as is proved by parallel thermometric observations, which are not given here.

The lines of the oncospheres' survival (they are a part of the oncospheres' biological "passport"), which were obtained in Samarkand with the help of a simple regression, are superimposed, as before, onto various heat resource nets. In the present case, eight nets were constructed for different climatic regions of Uzbekistan, three of which are given in Figures 1.59 to 1.61. Phenotemperature nomograms are also given (in this case, they are called ecological-temperature nomograms).

Forecasting calculations are made with these nomograms in the same way as before. Let us assume that beef tapeworm eggs were deposited on a shaded pasture in Samarkand on April 12, 1963. We need to forecast the date when the eggs will perish. We find, on the net of the ecological-temperature nomogram for Samarkand (Fig. 1.61), the line of average temperatures from the initial date of the forecast, April 12. We trace this line, representing the heat resources, up to the intersection with biological line I (representing the eggs' survival in the shaded pasture); from the point of the intersection we trace a horizontal line to the vertical axis of the graph, where we find $n = 75$ days. Therefore, the eggs will perish on April 12 + 75 days = June 26. Such a forecast, of course, is called a phenoclimatic forecast with prolonged foresight. In the experiment in which oncospheres were put into a shaded grassy surface on April 12, 1963, they perished on June 25, 1963, that is, the error of the forecast was -1 day (Table 6.3).

In the present instance, the intersection of the meteorological line representing the heat resources from April 12 (up to the moment when the eggs perish) and the biological line representing the eggs' survival gave the solution to this system of ecological equations, in graphical form.

To what degree were we right when superimposing the lines of the egg's survival onto the net of average temperatures? This question arises because the oncospheres are located on and perish on the surface of the soil (or of plants), whereas the net is based on air temperatures recorded by a meteorological station at a height of 2 m above the soil.

In order to reply to this question, we use Figure 2.5. As can be seen, the temperatures in vegetation are lower than the air temperatures, because of evaporation and transpiration, but the two change quite synchronously (this can be seen on a graph that relates directly to the experiments with oncospheres—a graph that is not given here).

Therefore, if biological line I were based not on air temperatures, but on the temperatures in the biotope, then this new line would lie somewhat lower than line I on the nomogram (Fig. 1.61). In that case, the heat resources line from April 12 would intersect biological line I, not at $n = 75$ days, as before, but at a lower value of n.

However, having constructed the biological line on the basis of the biotope, we should also construct the heat resources net on the basis of temperatures in the soil's plant cover. Since these latter temperatures are lower than the air temperatures, the line of average temperatures from April 12 (together with the net) would move to the left—in the direction of the lower temperatures. In the latter case, the line of April 12 would again intersect the new biological line I at a higher n, that is, at $n = 75$ days, as before.

Thus the real result of the forecast, as can be seen, does not change, regardless of whether we construct the biological and meteorological lines on the basis of air temperatures or on the basis of biotope temperatures.

Since lines I to IV reflect the behavior of the same organism in different biotopes, then the results of calculations made on the basis of these lines on the nomogram reflect the different behavior of the organism in different biotopes.

This is how the PTN, although constructed on the basis of air temperatures both in its biological and meteorological sections, takes into account the microclimate of biotopes.

For example, in Samarkand on April 12, the egg put not into the shaded grass, but onto a sunlit bare soil surface perished not in 75 days, as was obtained before, but in $n = 48$ days, where n is the ordinate of the intersection of the resources line dated April 12 with biological line IV for a sunlit bare surface (Fig. 1.61) on such a surface; the oncospheres perish 27 days earlier (75-48) than in grass, in accordance with the fact that the temperature on a sunlit bare surface in spring-summer time is 5° to 15°C higher than the temperature in grass.

Since the heat resources net is based on average multiannual temperatures, one might think that the foregoing calculations on the basis of nomograms bear no relation to the forecasts of survival for individual years. However, in each year and on each farm there are particular dates when the oncospheres of the beef tapeworm enter the environment. If it is possible to take into account these concrete dates (e.g., on the basis of the dates when water-carrying oncospheres enter the irrigation canals, or on the basis of other signs), then the calculations on PTNs (or ecological-temperature nomograms) become forecasts. These forecasts are concrete for a given place and time, because the initial forecasting date is concrete—the date when the eggs enter the environment.

Nevertheless, with the Podolsky method it is possible to take into account the concrete temperatures of a current year, if this is necessary. Let us assume that we have a long-term weather forecast for May and June, which predicts that the air temperature in this period will be $2°C$ lower than the norm. Then, in the above problems, we have to move the line of average temperatures dated April 12 to the left by $2°C$, in accordance with the scale of the graph. In that case we find, for shaded grass, $n = 88$ days. It is clear that no corrections are needed for converting the air temperatures to pasture temperatures, since the entire PTN is based on air temperatures.

We can also manage without any long-term weather forecast, provided that we take into account the actual air temperatures over part of the forecast period. Let us assume that the average temperature in Samarkand over the period from April 12 to May 20 in a given year turned out to be $19.8°C$. Then we construct on Figure 1.61 a point k with coordinates $t_{av} = 19.8°C$ and $n = 38$ days (between April 12 and May 20); we see that this point lies to the right of the multiannual line dated April 12, because the given year was, obviously, warmer than the "multiannual average year." From point k we trace a dotted line parallel to the multiannual line from April 12. The intersection of the dotted line with biological line I gives $n = 59$ days, that is, the oncospheres will perish on April 12 + 59 days = June 9. Although the foresight period of the forecast is reduced here to 20 days (June 9-May 20), such a forecast may prove to be more exact than the phenoclimatic forecast with prolonged foresight.

Forecasts with shortened foresight are most advisable in the second half of summer and in autumn, when the average temperature lines intersect the biological lines at such acute angles that even a small shift in the average temperature line can cause a considerable change in the value of n.

Finally, it should be noted that here, as before, nontemperature factors can be taken into account when forecasting on the nomogram. For example, in equations such as (21) and (22), it is possible to give several constant gradations of soil moisture and atmospheric humidity. Then, for each gradation of these nontemperature factors, it is necessary to obtain equations $n = f(t_{av})$. Finally, on the basis of the equations $n = f(t_{av})$, we have to draw lines representing the relationships between n and t_{av}. In this way, it is possible to obtain for a given biotope, for example, four phenolines from equations $n = f(t_{av})$: two

TABLE 6.3. Forecast dates (with prolonged foresight) and actual dates of death of the beef tapeworm's oncospheres in grass, in the conditions of Samarkand (1963).

Dates when Oncospheres Were put into Grass	In Shade				In Sun			
	Forecast of Dates of Death		Actual Date of Death	Error of the Forecast (in days)	Forecast of Dates of Death		Actual Date of Death	Error of the Forecast (in days)
	n Days	Date			n Days	Date		
14 I	157	20 VI	—	—	144	8 VI	4 VI	−4
14 II	131	25 VI	13 VI	−12	120	14 VI	4 VI	−10
2 III	114	24 VI	18 VI	−6	104	14 VI	7 VI	−7
12 IV	75	26 VI	25 VI	−1	66	17 VI	13 VI	−4
20 IV	66	25 VI	18 VI	−7	60	19 VI	28 VI	+9
7 V	50	26 VI	25 VI	−1	44	20 VI	3 VII	+13
20 V	41	30 VI	3 VII	+3	33	22 VI	19 VI	−3
1 VI	31	2 VII	6 VII	+4	23	24 VI	25 VI	+1
17 VI	18	5 VII	2 VII	−3	14	1 VII	2 VII	+1
20 VII	20	9 VIII	30 VII	−10	10	30 VII	29 VII	−1

lines for two constant gradations of atmospheric humidity, and two lines for two constant gradations of soil moisture. All four phenolines are superimposed onto the heat resources net. When calculating we use the line that is closest to the state of the nontemperature factors in a particular problem. This is one way of solving multiple regression equations in a plane, which were already mentioned in previous chapters.

Let us compare with the facts the phenoclimatic forecasts with prolonged foresight, which are the simplest, although the least precise (Table 6.3).

As can easily be calculated in Table 6.3, the average (without distinction of plus or minus) error in a forecast with a foresight period of 9 to 50 days,* that is, with an average foresight period of 30 days, is /3/ days. With a foresight period of 60 to 141 days, that is, an average of 100 days foresight, the average error is /7.3/ days. Since the errors are equal to or less than 10% of the whole foresight period, the forecasts can be classified as excellent (according to the scale of appraisal given in Table 3.1). We can assume that if a greater amount of experimental data were available, making it possible to establish a more real relationship between temperatures and survival periods on the biological lines, then the forecasts would be even more exact.

For spring-summer dates when the eggs enter the environment,

---

*The foresight period of the forecast is the difference between the date when the oncospheres were put into the biotope and the actual (but not the forecast) date when the oncospheres perished.

phenological curves (but not straight lines) would make the forecast more exact. On the other hand, for example, forecasts for cases when the eggs enter the environment in August would simply prove to be wrong if straight phenolines were constructed. For instance, on the nomogram for Samarkand (Fig. 1.61), the meteorological line from August 5, if continued to the left, intersects biological straight line I approximately at $n = 160$ days, which means that the eggs will perish only in the following year. A biological curve, on the other hand (as opposed to a straight line), would show that this upper intersection gives only one of the two possible solutions of the quadratic equation which represents a joint analytical solution of the equations of a meteorological line and 'of a biological curve. The second solution would be found at the lower intersection. It is clear that only the lower intersection has biological meaning, since temperature conditions that are enough to cause the death of the eggs already appear in the first days designated by the lower intersection (see Section 2.1).

In Figure 1.61 we can see, finally, that the average temperature line, for example, from September 5 does not seem to intersect biological curve I at all, or, in mathematical language, it "intersects" the biological curve at infinity. Hence we can draw the conclusion that the eggs that enter the environment on September 5 will never perish. The error in such an opinion lies in the fact that the line of September 5, if calculated and continued up to $n = 200\text{-}300$ days, would intersect the biological straight line not at infinity, but at an n equaling the number of days that would precede the death of the eggs in the following year.

On the basis of ecological-temperature nomograms referring to different physical-geographical regions of Uzbekistan, and on the basis of a number of dates for the eggs' entering the environment (with equal intervals), which are common to all these regions, we can construct a phenological calendar (Table 6.4).

It is clear that a calendar, like an ecological-temperature nomogram, relates to multiannual dates. Nevertheless, provided that we know the concrete date, for each year and each farm, when the eggs enter the environment, we can, on the basis of the calendar, obtain a forecast of the survival periods for a given year. For example, in an experiment in 1963 in Samarkand, oncospheres of beef tapeworm were put into shaded grass on June 1. How many days will the oncospheres live? In the calendar we see that oncospheres that entered the environment on May 15 will live for 44 days; but oncospheres that entered the environment on June 15 will live only 20 days. Interpolating, we find that the period for June 1 is 32 days. In the field experiments they lived for 35 days (Table 6.3). Thus calendars maximally simplfy the use of nomograms.

However, such calendars also provide data that can be used for the bioclimatic division of a territory into regions. It is clear that, in the

TABLE 6.4. Calendar of survival dates of the beef tapeworm's oncospheres in various regions of Uzbekistan (in days).

Physical-Geographical Region	Date when Oncospheres Enter the Biotope	In Grass	
		In Shade	In Sun
1	2	3	4
Nizhne-Surkhandaryinsky	15 I	138	126
	15 II	110	101
	15 III	83	75
	15 IV	52	46
	15 V	21	17
	15 VI	3	Sterilization
Kitabo-Shakhrizyabsky	15 I	150	138
	15 II	123	113
	15 III	96	87
	15 IV	66	59
	15 V	36	31
	15 VI	10	—
	15 VII	Sterilization	
Samarkandsky	15 I	157	144
	15 II	131	120
	15 III	102	93
	15 IV	72	65
	15 V	44	38
	15 VI	20	15
	15 VII	13	<13
Karadaryinsky	15 I	137	126
	15 II	105	96
	15 III	74	67
	15 IV	42	37
	15 V	12	10
Chirchiksky	15 I	143	132
	15 II	115	106
	15 III	87	78
	15 IV	56	49
	15 V	26	22
	15 VI	4	Sterilization
	15 VII	Sterilization	
	15 VIII	9	<9
Bukharo-Karakulsky	15 I	145	134
	15 II	116	107
	15 III	86	78
	15 IV	54	47

TABLE 6.4.   (*Continued*)

Physical-Geographical Region	Date when Oncospheres Enter the Biotope	In Grass	
		In Shade	In Sun
1	2	3	4
Bukharo-Karakulsky	15 V	21	18
	15 VI	3	1
	15 VII	Sterilization	
	15 VIII	17	<17
Khorezmsky	15 I	134	124
	15 II	104	96
	15 III	73	67
	15 IV	37	33
	15 V	4	2
Tamdinsky	15 I	134	124
	15 II	104	96
	15 III	73	67
	15 IV	37	33
	15 V	6	4
	15 VI	Sterilization	

conditions of Uzbekistan, not only the horizontal, but also the vertical (mountain) differentiations in climate are of great importance. No nomograms for mountainous regions have been constructed for the present, but those which have been constructed for the plain and foothill regions are also quite revealing. For example, in Table 6.4, in the physical-geographical foothill region of Samarkand, with a height above sea level of 500 to 1000 m, the eggs of the beef tapeworm that enter the enviornment on March 15 remain viable for 102 days in shaded grass and 93 days in sunlit grass. On the other hand, in the low-lying and plain region of Khorezm, with a height above sea level of 100 to 500 m, the eggs put into the environment on March 15 have survival periods of 73 and 67 days, respectively, that is the oncospheres live almost 30 days less. Therefore, for each additional 100 m above sea level, the duration of the oncospheres' survival in spring increases by 5 to 6 days. Since, at this time of the year, the temperature decreases by 0.4° to 0.5°C with a 100-m increase in altitude (as is well known in meteorology), then an 8 to 10 day increase in the egg's survival period corresponds to a temperature decrease of 1°C. This can also be seen clearly from the angular coefficients in the linear regression equations (17) and (18) (Section 6.3); these equations relate the length of the survival period to the temperature during the first and second versions of the experiment. However, it must be said that at great heights over sea level, and on a sunlit bare surface, the *sterilizing properties of ultraviolet radiation* and of frosts (which increase with height) may be the dominant factor.

In regions located at similar heights above sea level, the survival dates of oncospheres are closer than in regions of different heights, although there is some influence from the region's latitude. For example, among the foothill regions, the most southern is the Nizhne-Surkhandaryinsky region: to the north lies the Kitabo-Shakhrizyabsky region and further north the Samarkandsky region. The survival periods of the oncospheres correspond to the latitudinal location: when eggs are put into the environment on March 15, the survival period in the Nizhne-Surkhandaryinsky region is 83 days in shaded grass and 75 days in sunlit grass; in the Kitabo-Shakrizyabsky region, the survival periods are 96 and 87 days, respectively; in the Samarkandsky region, the survival periods are 102 and 93 days, respectively. There is, consequently, almost a 20-day increase of survivability in the north, apparently owing to the lower temperatures. It may be that here the geographical latitude of the place is not the sole influence.

As can be seen, however, in any nomogram the difference between the oncospheres' survival periods in shaded grass and on a sunlit bare soil surface is 25 to 30 days, if the eggs were put into the environment simultaneously. Such a difference can exceed the latitudinal and longitudinal (but not altitudinal) differences between survival periods within Uzbekistan.

From the calendars given above it can be seen that the length of the oncospheres' life in the environment decreases in all regions as summer approaches. This happens because of the increase both in temperature and in ultraviolet radiation. The temperature increase causes the "spring" and, in particular, the "summer" eggs to age quicker than the "winter" eggs; owing to this fact, the dates when the eggs die quickly become closer and closer to one another. For example, eggs that entered an environment of shaded grass in Samarkand on February 14, 1963 perished, in the field experiment, on June 13; but eggs that entered the environment on March 2, that is, 16 days later, perished on June 18, that is, only 5 days later (Table 6.3). This is a manifestation of the general phenological principle which will be discussed in Chapter 8.

In summer, the self-purification (self-sterilization) of soil from oncospheres takes place; these oncospheres arrived on the soil surface in autumn and winter in the preceding year and in the spring and summer of the year in question.

On the whole, all the eggs that enter the environment between January and the middle of May die off during one summer month—June (Table 6.3). The length of their lifetime, therefore, decreases from

4 to 5 months (for eggs that entered the environment in winter) to several days (for eggs that entered the grass in June).

Thus, on the basis of actual observations of the oncospheres' behavior in Samarkand, the experimental-mathematical "heat-phenology" method helped us to estimate the oncospheres' behavior in other climatic regions, using a considerably reduced number of purely empirical observations.

In conclusion, we should probably use this same technique in order to work out a method for forecasting survival dates in the environment for other disease pathogens. This will be demonstrated using the example of foot-and-mouth disease (Section 6.8).

6.4. DATA AND IDEAS FOR USE IN PHENOLOGICAL FORECASTING FOR THE FLAT BIOHELMINTH *Fasciola hepatica* AND ITS INTERMEDIATE HOST, THE MOLLUSC *Galba truncatula*

This work was carried out by S. T. Karelin (Kursk, Russia) with the guidance and participation of the author of this book in regard to phenological forecasting (1973-1978).

It is known that fasciolasis causes great damage in both small and large domestic livestock and in other animals (both domestic and wild). Humans also suffer from fasciolasis.

The eggs of *Fasciola hepatica* (the common fasciola, family of *Fasciolidae*, class of *Trematoda*) enter the environment along with animal feces. A humid biotope is required for the eggs' development, such as a freshwater pond, bog, or pool. Inside the egg forms a miracidium, which emerges at a temperature-related rate.

When it finds a small mollusc, the miracidium penetrates actively into the mollusc's body. In the mollusc's body the miracidium (larvae) develop up to the stage of cercariae at a rate dependent on the temperature of the environment (since the mollusc is a poikilothermic creature, and its temperature is the same as the temperature of the environment). From the mollusc the cercariae come out into the environment, and in several hours turn into adolescariae.

Animals become infected with fasciolasis when they swallow adolescariae along with their water and food. Then the cycle begins again.

Of course the development of the intervening host—the cold-blooded mollusc—is also dependent on the temperature of the environment.

Here we can see yet another interesting *biogeocenosis* (Chapter 5), the components of which are bound to each other through the rotation of substances on the basis of the food chain. The study and management of such biogeocenosis is a challenging task for a phenoforecaster!

As we already know well, phenocurves are the important part of a PTN. It is clear from the above that it is necessary to draw three phenocurves for the given biogeohelminth: for fasciola's development from egg to miracidium in the environment, for the development of the mollusc's eggs, and for the development of fasciola larvae in the mollusc.

In order to do this, in 1973 on the bank of the Kur River (near the town of Kursk), the following experiment was conducted. Artificial biotopes were made: wooden boxes, without lids, each with an area of 1.5 m² and a height of 0.25 to 0.30 m, were put into the soil at a depth of 5 to 7 cm and filled with earth free of fasciolae. The space inside each box was divided into several compartments, corresponding to the different dates when the biological material was introduced into the biotope. In this way several biotopes were made, both in the shade of trees and in the sun.

The fasciola eggs (50,000 to 60,000) were taken from a meat-packing plant, from the bile of cattle. These eggs were put, in different months (from June until August, and in following years, from May until October) and therefore at different temperatures, into the biotopes mentioned above on the surface of moist soil, as well as into a ditch filled with water and earth. Observations of the dates of the miracidia's emergence were made daily or every second day using a microscope or a magnifying glass.

In a parallel process, molluscs were bred in a natural, fasciola-free biotope, and their eggs were obtained. Batches of mollusc eggs laid during a single 24-hour period were also put *several times per season* into artifical biotopes. The development of the mollusc eggs was also examined daily or every second day with a microscope or a magnifying glass.

Young molluscs that hatched from the eggs were infected at the age of 3 to 4 weeks in the artificial biotopes by being placed alongside the fasciola eggs with the emerging miracidia. Forty to forty-five days after the infecting, the investigators began to open three to four molluscs every 1 to 2 days in order to define the dates when the cercaria emerged and turned in to adolescaria.

For the purpose of defining the date of the appearance of adolescaria, and in order to count them, the investigators put into the artificial biotope of the infected molluscs small pieces (3 × 7) of photographic film, cleaned of emulsion, as well as food in the form of lettuce leaves. The pieces of film and the leaves were examined before the adolescaria's appearance every 1 to 2 days, and afterward every 3 to 5 days, (S. T. Karelin noted that the film "worked" much more intensively than the lettuce leaves; on the film 95-97% of the adolescaria gathered, and on the leaves only 3-5% gathered).

The period of development for fasciola eggs was defined as the time necessary for the miracidia to emerge from 50% of the eggs. An analogous criterion was used for the emergence of young molluscs from the eggs. The period of development for the fasciola larvae in the molluscs was defined as the period up to the maximum emergence of cercaria and the formation of adolescaria.

We related all these phenological dates to the air temperatures at the meteorological station closest to the site of the experiments.

Unfortunately, the author has only one year's data available, but these alone are sufficiently revealing (Fig. 6.3 and 6.4).

In spite of the scarcity of data, in Figures 6.3 to 6.5 we can clearly see the relationship of the developmental periods to temperatures in the cases of both fasciolae and molluscs. The field experiments of the succeeding years reconfirmed this principle. The marked temperature dependence of fasciola's development in molluscs was also reconfirmed.

The similarity between the phenocurves for biotopes in the shade and in the sun is quite remarkable (Figs. 6.3 and 6.4). Apparently, in the middle zone of the USSR the temperatures of moist biotopes and especially the temperatures of water differ little in the shade and in the sun.

We transfer these phenocurves—average between the solid and the dotted lines—onto the PTN, for example, of Lgov, in the Kursk district (Fig. 1.94), although the phenocurves were obtained in Kursk. As the reader already knows, this is permissible, since a phenocurve—an organism's "passport"—remains stable in similar climatic zones.

Let us assume that in Lgov the first pasturing of sheep suffering from fasciolasis was on a sterile pasture on May 26. This date can be accepted as the beginning of the fasciola's egg laying on the pasture, when the eggs are excreted along with the sheep's feces.

Thus from these eggs the mass appearance of miracidia, in condi-

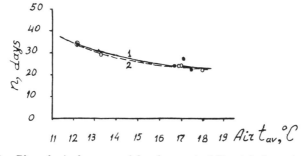

FIGURE 6.3. Phenological curves of development of *Fasciola hepatica* from egg to mass emergence of miracidia (larvae) on moist soil in artificial biotopes. Results are from field experiments in Kursk, Russia, 1973. Curve 1: for a biotope in the shade (the solid line through black points); curve 2: for a biotope in the sun (dotted line through white points) (from S. T. Karelin and A. S. Podolsky).

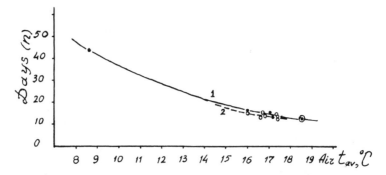

FIGURE 6.4. Phenological curves of development of a small mollusc *Galba truncatula* from egg to mass emergence of young molluscs in the artificial biotope. Results are from field experiments in Kursk, Russia, 1973. Curve 1: in the shade; curve 2: in the sun (from S. T. Karelin and A. S. Podolsky).

tions of sufficient moisture, must occur as follows:

$$\text{May } 26 + 23 = \text{June } 18$$

where $n = 23$ days is the ordinate of the intersection of the average temperature line from May 26 with phenocurve 4 in Figure 1.94.

For their further development, the miracidia must meet molluscs. However, there may be no mature molluscs at this time even if suitable biotopes (ponds, pools, or bogs) are available. In this case the epizootological chain may be interrupted.*

Let us see whether this is correct. For example, the first egg layings

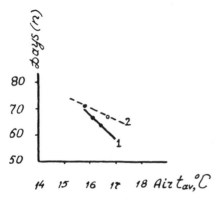

FIGURE 6.5. Phenological lines of development of fasciolae in molluscs from mass penetration of miracidia into molluscs to mass emergence of cercaria from molluscs and formation of adolescaria in the artificial biotope. Results are from field experiments in Kursk, Russia, 1973. Curve 1: in the shade; curve 2: in the sun (from S. T. Karelin and A. S. Podolsky).

*On the other hand, it may not be interrupted, since miracidia continue to emerge, with less intensity, after June 18, and those which emerged earlier can survive for a long time, especially if it rains.

of the mollusc were observed on April 25. The intersection of the meteorological line from April 25 with phenocurve 6 (Fig. 1.94) gives the date when the young molluscs emerge from these eggs:

$$April\ 25 + 27\ days = May\ 22$$

However, the molluscs must mature in order to be effective hosts for the miracidia. This takes about 20 days. Thus the miracidia can "effectively" (in regards to the continuation of their development) penetrate into the mollusc on June 11:

$$May\ 22 + 20 = June\ 11$$

Therefore, on the date of the mass emergence of miracidia—June 18 (see above)—there are already mature molluscs, and the miracidia penetrate actively into their bodies on June 18.

Cercaria will emerge from the molluscs on

$$June\ 18 + 50\ days = August\ 7$$

where $n = 50$ days is the ordinate of the intersection of the meteorological line from June 18 with phenocurve 5 (Fig. 1.94).

On August 7 the pasture will become a health risk, and will infect livestock. The facts corroborate this date: in the middle zone of Russia the maximum infection of molluscs is noted in July and August, and the *beginning* of the infection of animals is observed from the second half of July.

It is known that the incubation of fasciolasis in infected livestock develops over approximately a month and a half. Therefore, the first manifestations of the disease will occur on

$$August\ 7 + 45\ days = September\ 21$$

In reality, the first fasciolasis cases in animals are observed at the end of the summer and at the beginning of autumn. Here we also see agreement between the calculation and the facts.

After approximately two months, the acute fasciolasis turns into the chronic form which, therefore, begins on

$$September\ 21 + 60\ days = November\ 20$$

and the chronic form reaches mass prevalence in winter. This is also confirmed by the known facts: for example, in 1974 the forecast dates of the outbreak of fasciolasis in sheep coincided with the actual dates. The forecast made it possible to take measures in time to prevent the disease in a number of districts in the Kursk region.

The following situation still requires clarification: a PTN is based both in its meteorological part (net) and in its biological part (phenocurves) on the air temperatures at a meteorological station at the standard height of 2 m above the soil surface. These temperatures, of course, differ from the temperatures in the biotopes. However, the "error" in the construction of the net is approximately equal in value, and inverse in sign, to the "error" in the construction in the phenocurve. As a result, these errors cancel each other out, and the result of the calculations on the PTN proves to be correct. Section 2.4 and part of Section 6.3 (concerning the beef tapeworm) were devoted to this particular situation.

Finally, it should be noted that all of the above is not appropriate to arid regions, and especially in arid years. In steppe and semidesert zones there is no fasciolasis. However, even in the forest-steppe zone, which has a variable pattern of humidity, for example, in the Kursk region, the phenological forecast described above has already saved 30,000 roubles in 1976, 40,000 roubles in 1977, and similar sums in the preceding and the succeeding years (i.e., hundreds of thousands of dollars in only one small region) by preventing the loss of sheep and cattle.

Finally, in accordance with the above, would the reader agree with the distinction of the term "biogeohelminth," parallel to "geohelmith" and "biohelminth"?

6.5. TRANSMISSION OF INVASIONS BY ARTHROPODS. FORECASTING THE DEVELOPMENTAL DATES OF *HYALOMMA DETRITUM* TICKS AND THE DATES WHEN FARM ANIMALS WILL BE INFECTED BY THESE TICKS

The development of ticks, like that of other poikilothermic organisms, depends first on the temperature factor. Therefore, the dates of diseases, the pathogens of which are transferred by ticks, are connected with the annual temperature trend. For example, P. M. Mordasov (1957) wrote that the seasonal dynamics of hemosporidiosis in cattle are influenced basically by temperature and by the availability of pastures of different types. According to Mordasov's observations in the Moscow region, the precipitation, atmospheric humidity, and so on do not exert any appreciable influence on the dynamics of the disease.

If we know the temperatures and how to take into account some nontemperature factors, we can calculate the developmental dates for the carriers and pathogens of diseases.

It is known that the family of ticks known as *Ixodidae* are carriers of a number of very serious diseases of man and livestock. Taiga ticks

Ixodes persulcatus infect people with encephalitis, and the pasture ticks *Ixodes ricinus, Dermacentor pictus, Hyalomma detritum, Hyalomma anatolicum*, and *Boophilus calcaratus* transfer hemoparasitic diseases of cattle, horses, sheep, and deer, such as theileriasis and pyroplasmosis. Ticks also transfer relapsing ("Middle-Asian") tick typhus, lymphocytic choriomeningitis, tick rickettsiosis, and hemorragic fevers.

Let us focus on *Hyalomma detritum* ticks as the main carrier of cattle pyroplasmidoses in Tadzhikistan (although it should be said that the activity of *H. anatolicum* has also increased).

As is known, the bulk of *H. detritum* nymphs fall from the cattle's bodies during the spring pasturing, and develop in the environment up to the imago stage. At this stage the ticks attack the cattle and infect them with theileriasis.*

On the PTN for Dushanbe (Fig. 1.56), phenocurve 8 is given, which represents the length of the period from the time when the nymphs fall off the cattle until the imago stage, in relation to the temperature. This curve was constructed by the author in Figure 1.38 on the basis of observations published by I. G. Galuso (1947). Let us assume that in Dushanbe, the first pasturing of the year and, therefore, the time when the nymphs fall off the cattle occurred on February 25; when will imagoes emerge from the nymphs?

On the heat resources net in Figure 1.56, we find the line of average temperatures dated February 25 (this line must be interpolated between the lines from February 20 and March 5); we trace this line up to the intersection with phenocurve 8, and then from the point of this intersection we move the pencil strictly horizontally to the vertical axis, where we see that the period between the date when the nymphs fall off the cattle and the appearance of the imagoes is $n = 80$ days. Therefore, imagoes will emerge from these nymphs, the pubescent ticks will attack animals, and the animals can become infected with theileriasis. All of this will occur on February 25 + 80 days = May 16.

In other words, the pasture, when the first pasturing of the year (after the summer sterilization of the pasture) was on February 25, will become a health risk from May 16 onward. This is the situation with multiannual average temperatures, and with a concrete (for a given year and for a given pasture) date for the pasturing of the cattle. This forecast, as already mentioned, is called a pheno-climatical forecast with prolonged foresight.

If, from the long-term weather forecast, we know that in March and April the temperatures are expected to be 1°C higher than the norm, then the line of average temperatures from February 25 for the given

---

*Theileriasis is a severe cattle pyroplasmidosis that often results in death. The pathogen is *Theileria annulata*, a hemosporidium from the protozoa, like the malaria plasmodium.

year will be located 1°C to the right of the multiannual line, in accordance with the scale. The period is $n = 76$ days, and the pasture will be infected by ticks on February 25 + 76 days = May 12; that is, 4 days earlier than with multiannual temperatures. This will be a phenosynoptical forecast.

Finally, we can give a high-accuracy forecast with shortened foresight, using the air temperatures of a given year over a part of the forecast period. For example, in a given year in Dushanbe the meteorological station recorded in its basic table the following 24-hour average air temperatures: on February 25—9.1°C; on February 26—8.7°C; on February 27—9.5°C; on February 28—7.5°C; the total of the 24-hour average temperatures over March was 330.2°C, and the total over the first 10 days in April was 145.3°C. Thus the average temperature over the 45-day period from February 25 to April 10 inclusive will be

$$t_{av} = \frac{9.1°C + 8.7°C + 9.5°C + 7.5°C + 330.2°C + 145.3°C}{45 \text{ days}}$$

$$= 11.3°C$$

We plot a point γ, with coordinates $t_{av} = 11.3°C$ and $n = 45$ days, onto Figure 1.56. From this point we trace a dashed line parallel to the multiannual line of average temperatures from February 25, up to the intersection with phenological curve 8. Thus $n = 73$ days, and pubescent ticks will appear on February 26 + 73 days = May 9. The foresight period of this forecast is 28 days (from April 11 to May 9).

Spring pasturing of livestock in the Hissar Valley (Tadzhikistan) usually begins at the end of February and finishes in April and May, the same field being used repeatedly during this period as pasture plants grow back. Let us take, for our calculations, the following initial dates for the pasturing of cattle: February 25, March 15, April 15, and May 15. These dates, therefore, will be the dates when the nymphs fall off the host's body, and when the ticks begin their development from nymph to imago. Therefore, we can calculate phenoclimatic forecasts with prolonged foresight for all the pasturing dates: February 25 + 80 days = May 16 (this was done above); March 15 + 67 days = May 21; April 15 + 50 days = June 4; May 15 + 39 days = June 23.

Let us pay attention to one important circumstance: the difference between the pasturings of February 25 and March 15 is 18 days, and this gives a difference of 5 days (May 16 and 21) between the dates of the imagoes' appearance; the difference between the pasturings of March 15 and April 15 is 31 days, with a difference of only 14 days between the dates of the imagoes' appearance (May 21 and June 4). Such a phenomenon has already been met repeatedly; it is a result of the phenological law, as follows: in the second half of summer, when temperatures increase, the development of biological individuals, which began their development later, begins to approach that of the individuals which began to develop earlier, at lower temperatures.

The emergence of imagoes produced by the repeated pasturing takes place almost simultaneously with the previous emergence; therefore, it is clear that the ticks will accumulate on the pasture and on livestock in great numbers all at once. This throws light on the following result of observations by Galuzo (1947) in the Hissar Valley: he writes that pubescent ticks appear on livestock in May and June in great numbers all at the same time. In June, the number of ticks on the bodies of the hosts reaches its maximum. We see here that the actual data coincide with the results of calculations.

We take the dates obtained above of the emergence of imagoes from nymphs to be the dates when pubescent *H. detritum* ticks attack livestock and the dates when livestock can become infected with theileriasis (in this situation not only the daytime temperatures, but also the 24-hour average temperatures, during the period from May 16 to June 23 in the Hissar Valley, are higher than 16° to 22°C, which is necessary for the imago to begin to migrate from the soil surface onto the herbage surface for the purpose of attacking livestock).

It is known that the pubescent ticks, which attack the animals, feed on them for 14 days, on the average. It is here that the impregnation of the females occurs. The impregnated females fall off again into herbage, where they lay their eggs, on temperature-dependent dates. In Figure 1.56 this dependence is represented by experimantal developmental curve 9, which was constructed by the author on the basis of data from Galuzo and is transferred here from Figure 1.36.

It should be noted that the analogous curve for the northern tick *I. ricinus* is located much lower than the phenocurve for the southern tick *H. detritum*; this indicates that the heat requirements of *I. ricinus* are less than those of *H.* detritum (Fig. 3.11).

Using curve 9 and the date when the ticks were actually observed or calculated leaving the host's body again, it is easy to obtain the date of the beginning of egg laying. For example, if the ticks attacked the animals on May 16, and left the host's bodies again on May 16 + 14 days = May 30 (here 14 days is the period when the ticks feed on the animals), then the beginning of egg laying in Dushanbe, corresponding to pasturing of February 25, will be May 30 + 14 days = June 13, where $n = 14$ days in the ordinate of the intersection of the average temperature line from May 30 with phenological curve 9.

The duration of this egg laying is also temperature dependent. Next we take experimental curve 6, which represents the duration of the egg laying. This curve is transferred onto Figure 1.56 from Figure 1.36. Thus we find that the egg laying, which began on June 13, will last until June 13 + 24 days = July 7, with an average temperature during the entire process of egg laying of $t_{av} = 25.3°C$ (this is the abscissa of the intersection of the average temperature line of June 13 with developmental curve 6).

From Galuzo we know that, at a temperature of 10°C, a female lays, on the average, 11 eggs per 10 mg of its initial weight, and at 30°C it lays, on the average, 90 eggs (the influence of temperatures on the intensity of egg laying can also be seen in Fig. 6.7). Hence we can calculate that at 25.3°C, one female will lay about 55 eggs per 10 mg of its initial weight.

If we make similar calculations for all the data corresponding to different dates of pasturing, we can see that *H. detritum's* egg laying in the conditions of the Hissar Valley (Dushanbe) lasts longer during the period from June 13 to August 13. According to Galuzo's observations in the Hissar Valley, the egg laying occurs during June, July, and August. Once again we see agreement between the calculations and the actual data.

According to phenocurve 7 (which, like curve 9, is located above the corresponding phenocurve for northern ticks), larvae will emerge from the first eggs (corresponding to pasturing on February 25) on June 13 + 32 days = July 15. The last eggs corresponding to pasturing on February 25 will give larvae on July 7 + 30 days = August 6. Here $n = 32$ and $n = 30$ are the ordinates of the intersections of phenological curve 7 with the average temperature lines from June 13 and July 7, respectively.

The duration of the development of the eggs (to larvae) is slightly influenced by atmospheric humidity. Galuzo (1947) gives a table presenting the relationship of the duration of the eggs' development to the temperature with several gradations of relative humidity. Using these data, the author of this book constructed phenological developmental curve 7 (Fig. 1.56) mentioned above in two sections: the high-temperature section is based on data for low relative atmospheric humidity, whereas the low-temperature section is based on data for high atmospheric humidity, as usually occurs. Therefore, when we work with curve 7, the atmospheric humidity is automatically taken into account. However, it is possible to construct different temperature curves for *H. detritum's* development from egg to larva, at different humidities. This has been done in Figure 1.36, and the curves have been transferred, for example, onto the PTN for Yavan (Fig. 1.50) or for Garm (Fig. 1.57). In these figures we use that curve (for *H. detritum*) which corresponds to the expected relative atmospheric humidity.

The date when the larvae begin to fasten onto livestock is found on the basis of the autumn date when the daytime air temperatures regularly fall below 15°C (this is the temperature criterion indicated in the literature). In Dushanbe this will be approximately mid-October.

The same date, for the Hissar Valley, was taken by Galuzo as the beginning of the larvae's attack on the livestock.

It is interesting that the same October date is obtained as a result of the calculation relating to the mass or final stages of development. The egg laying of August 16, which is the last that can produce larvae, expressed in Figure 1.56 by a meteorological line that is almost tangent to phenocurve 7, indicates that these larvae will emerge on August 16 + 75 days = October 30 (where $n = 75$ days corresponds to the intersection of the average-temperature line from August 16 with developmental curve 7). The date of October 30 is close to the autumn frosts, and therefore close to the time when the larvae lose their ability to fasten onto a warm-blooded host because of the fall in temperatures. If the larva appeared later than this, they would inevitably perish.

It is known that a larva that has fastened onto the host turns into a nymph in 8 to 12 days. This nymph is parasitic on the host during the entire autumn-winter period, and falls off in the spring of the following year during the first pasturing. After this the entire developmental cycle repeats itself.

Calculations identical to all of the above can be made for other pasturing dates for other regions of Tadzhikistan (Table 6.5).

Some practical conclusions that can be drawn from the calculations, in particular from those in Table 6.5, are as follows:

1. In normal years the pasture where the first pasturing occurred on February 25 must be considered especially dangerous regarding infection by *H. detritum* ticks from May 16 in the Hissar Valley and from May 8 in the Vakhsh and Yavan Valleys. If the first pasturing is on March 15, the dates of infection will be different (Table 6.5, columns 2 and 3).

Thus late (May) pasturing of cattle in the valleys is not only relatively inefficient, but is also apparently dangerous.

2. The dates given in column 3 are also relevant for the antitick treatment of cattle (e.g., dipping, using T. Ya. Vannovsky's method, 1965).

3. In the pasture where the first pasturing was on February 25, control measures against the basic mass of larvae must be begun on July 15 in the Hissar Valley, and on July 2 in the Vakhsh Valley, and so on (column 11). In Garm, larvae appear 38 days later than in the Vakhsh Valley (with pasturing on May 15).

It should be taken into consideration that, according to the calculations, larvae and pubescent ticks accumulate on the pasture in great numbers owing to the phenological principle: the falling off of the nymphs during pasturings in February and March is followed by almost simultaneous emergence of the larvae (column 11 or 12).

4. Phenotemperature nomograms make it possible to calculate the

TABLE 6.5. Development of *Hyalomma detritum* ticks in accordance with the dates of the host's pasturing in various climatic zones of Tadzhikistan (average data). Results of the calculations on PTNs are compared with the available actual data

Zone	Dates of the Pasturing of Livestock and the Falling Off of Nymphs	Calculated Dates of the Appearance of Imagoes, the Attack of Livestock, and the Possible Infection of Animals with Theileriasis	Actual Dates of the Appearance of Imagoes (after Galuzo)	End of the feeding of Imagoes on host, and falling off into pasture
1	2	3	4	5
Hissar Valley (Dushanbe area)	25 II 15 III 15 IV 15 V	16 V ⎫ 21 V ⎪ 4 VI ⎬ 23 VI ⎭	May June	30 V 4 VI 18 VI 7 VII
Vakhsh Valley (Kurgan-Tyube area)	25 II 15 III 15 IV 15 V	8 V ⎫ 13 V ⎪ 27 V ⎬ 16 VI ⎭	No data	22 V 27 V 10 VI 30 VI
Yavan	25 II	8 V–10 V	No data	22 V
Highland pastures (Garm)	15 V	2 VII	No data	16 VII

Zone	Calculated Dates of Egg Layings		Actual Dates of Egg Layings (after Galuzo)	Average Temperature at the Time of Egg Laying (in °C)
	Beginning	End		
1	6	7	8	9
Hissar Valley (Dushanbe area)	13 VI 18 VI 1 VII 20 VII	7 VII ⎫ 12 VII ⎪ 24 VII ⎬ 13 VIII ⎭	June, July, Aug.	25.3 25.8 26.5 25.9
Vakhsh Valley (Kurgan-Tyube area)	4 VI 9 VI 23 VI 13 VII	27 VI ⎫ 1 VII ⎪ 14 VII ⎬ 3 VIII ⎭	No data	26.6 27.1 28.2 28.3
Yavan	5 VI	27 VI	No data	27.2
Highland pastures (Garm)	29 VII	25 VIII	No data	22.5

TABLE 6.5. *(Continued)*

Zone	Number of Eggs Laid per 10 mg of Weight (after calculations)	Calculated Dates of the hatching of Larvae		The Last Egg Laying of the Year that is Capable (owing to succeeding heat conditions) of Producing Larvae (after calculations)
		Beginning	End	
1	10	11	12	13
Hissar Valley	55	15 VII	6 VIII	
(Dushanbe area)	60	18 VII	11 VIII	16 VIII
	62	30 VII	27 VIII	
	60	21 VIII	15 X	
Vakhsh Valley	62	2 VII	21 VII	
(Kurgan-Tyube	67	6 VII	26 VII	
area)	73	18 VII	9 VIII	23 VIII
	74	8 VIII	3 IX	
Yavan	67	1 VII	20 VII	4 IX
Highland pastures (Garm)	38	15 IX (from eggs laid on 29 VII)	From eggs laid on 25 VIII, larvae will not emerge	10 VIII

Zone	The Last Larvae that Can Emerge in the Year, Given the Heat Resources of the Region (after calculations)	Beginning of the Attack of Larvae on a Host (according to daytime temperatures lower than 15°C)	Beginning of the Attack of Larvae on a Host (according to observations by Galuzo)	Actual Multiannual Average Date of the First Autumn Frost in the Air
1	14	15	16	17
Hissar Valley (Dushanbe area)	30 X	Middle of Oct.	Middle of Oct. (and continuation in Nov.)	3 XI
Vakhsh Valley (Kurgan-Tyube area)	11 XI	Middle–end of Oct.	No data	29 X
Yavan	17 XI	Beginning of Nov.	No data	17 XI
Highland pastures (Garm)	3 XI	Beginning of Oct.	No data	24 X

extreme dates when the ticks occur in a pasturing season. For example, in most years in the Yavan region the period from the middle of March to the beginning of May may be free from pubescent *H. detritum* ticks; in Dushanbe the period free from pubescent ticks may shift to some 15 to 30 days later. In some years there is no tick-free period. This makes it clear why there are cases of theileriasis in livestock in winter (infection can occur even out-of-doors).

The middle of summer is also, for the most part, free from pubescent ticks (which could conceivably appear if pasturing took place at the beginning of summer), since the pasture is sterilized by the high summer temperatures.

The way to take into account those dates when the temperatures and humidity are critical for the ticks' existence was described in Section 2.3.

5. In Garm (1319 m above sea level) there are no optimal conditions for the development of *H. detritum* ticks because of the limited heat resources. For example, from the eggs that are laid at the end of the egg-laying period (column 7) larvae do not emerge (column 12). Apparently, the height of 1500 m in the Garm pastures (the pastures are located higher than the Garm meteorological station) is the vertical boundary of the distribution of *H. detritum* ticks, and therefore of theileriasis, in Tadzhikistan. In actual fact, in the pastures in Garm livestock virtually does not suffer from theileriasis. Thus driving the cattle to the mountains is one of the means of protection against the disease. However, we should take into consideration the possibility that nymphs may be carried into the mountains, and can develop there up to the stage of the pubescent ticks, which can infect livestock, especially in warm years.

In conclusion, we should point out that since the phenological curves for the ticks' development as used here are obtained from observations in a controlled temperature chamber, all of the above calculations relate to biotopes that are isothermal (in their 24-hour average temperatures) with the atmosphere (although the daytime and night-time temperatures in the biotope may be different from those in the atmosphere).

6.6. PROGNOSTIC ESTIMATION OF NATURAL ENDEMIC CONDITIONS FOR THE APPEARANCE AND DEVELOPMENT OF MALARIA IN POTENTIAL REGIONS AND IN FORMER MALARIAL REGIONS

In this section we continue the discussion of vector-borne, invasive diseases. We shall discuss an arthropodous vector of an invasive disease, which forms a complex with the pathogen.

As is known, about half of the Earth's population lives in malarial regions. Cases of this disease can definitely occur in newly irrigated regions, where the population has no immunity, for example, in the Yavan and Dangara Valleys in Tadzhikistan, USSR (these valleys are presented in Fig. 4.2., south-southwest of the points representing Yavan and Dangara). There are recurrences of malaria in Azerbaidzhan (USSR) and in many regions in other countries where malaria had seemed to be eradicated.

The method described below for constructing the parameter of malaria communicability can be transferred to the field of veterinary parasitology. In addition, a number of problems that are discussed in this section will probably interest agricultrual entomologists.

6.6a. Some Biological Curves, and the Developmental Dates and Number of Generations of Malaria Carrier *Anopheles superpictus* in Tadzhikistan.

On the basis of observations by various authors, the author of this book constructed phenological curves for the malaria mosquitoes *Anopheles superpictus* and *Anopheles maculipennis* and for the malaria pathogen *Plasmodium vivax* (Figs. 1.38-1.40). Such phenological curves, as we already know, present to a certain extent a biological "passport" of an organism, and characterize its heat requirements or its reaction to the heat factor.

These biological curves are then transferred onto the meteorological heat resources nets for a number of geographical regions in the USSR, the first being the PTN for Yavan (Fig. 1.51, since the previous PTN for Yavan in Fig. 1.50 is already overloaded with biological curves).

Besides these phenological curves, for further calculations a graph will be necessary, representing the decrease in the intensity of single (for one gonotrophic cycle) egglayings by the female mosquito during its lifetime. Here we see a resemblance to the sexual productivity of ticks, which can be seen in Figure 6.6.

It has also been found that the number of eggs in one egg laying of a mosquito in all proability increases as the ambient temperatures increase. This increase is small within the temperature range of 10° to 16°C, and is greater within the range of 16° to 23°C, as with ticks. This can be seen in Figure 6.7.

Using all these initial data, we first calculated the dates of the beginning and the end of the egg layings for each generation for Yavan. The beginning of the egg laying by wintering females was found by adding the length of one gonotrophic cycle of the mosquito female to the spring date when the 24-hour average temperatures regularly exceed 6° to 7°C (this is the date when the mass flight of the mosquitoes from their places of hibernation begins). The length of the gonotrophic cycle of the wintering generation is located on the PTN for Yavan (Fig.

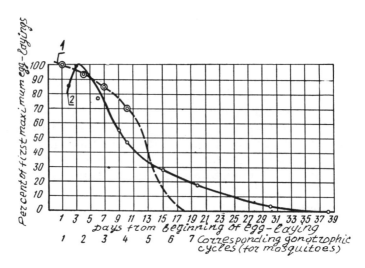

FIGURE 6.6. Productivity of egg layings. Curve 1: mosquito *Anopheles mac. messeae* (after O. A. Sibiryakova and M. A. Gobova, 1957); curve 2: livestock tick *Ixodes ricinus L.* (after E. M. Kheisin, 1954) at various stages of the egg-laying period, with $t = 20°C$ (egg-laying spectra).

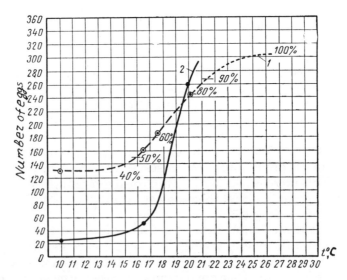

FIGURE 6.7. Productivity of a single egg laying of mosquito *A. mac. messeae*: curve 1, using Detinova's data (1936) which were bound with air temperatures by Podolsky. The 24-hour productivity of egg layings of livestock tick *I. ricinus L.*: curve 2, obtained using E. M. Kheisin's data (1954).

384

1.51) at the intersection of the average temperature line dated by the flight from the places of hibernation and biological curve 2. Thus the beginning of mass egg laying by the wintering females will be February 16 + 18 days = March 6, where February 16 is the beginning of the mass flight after wintering (February 16 is taken from Figure 1.51 as the date where the line of average temperatures begins from the mark 6°C on the horizontal axis). The date for the beginning of flight after wintering—February 16—is a real date, since mass bloodsucking by *A. superpictus* mosquitoes, even in regions of Middle Asia that are colder than Yavan, is observed as early as the end of February or the beginning of March. We should also note the data by G. A. Pravikov and L. I. Popov (1957), which show that in Iolotan and Kushka — regions with temperatures close to those in Yavan — the wintering *A. superpictus* females become active from the second 10 days of February. According to Chinayev, in Tashkent the first gonotrophic cycle of malaria mosquitoes lasts 15 days (according to my calculations it lasts 18 days), and the initial egg layings occur in the first 10 days of March (according to the calculations, on March 6).

Knowing that only two of the wintering females' gonotrophic cycles produce significant results (E. S. Kalmikov, 1959), we calculate, using the date of March 6 and the same phenocurve 2, that the end of the first egg laying will be on March 6 + 14 days = March 20 (Fig. 1.51).

On the basis of the intersection of the average temperature lines from March 6 and 20 with phenological developmental curve 1 for *A. superpictus* from egg to egg, we find the corresponding dates of the second egg laying (i.e., egg laying by the first generation): March 6 + 64 days = May 9; March 20 + 51 days = May 10.

V. V. Almazova (1959) observed, at a place 25 km away from Yavan (where it is somewhat cooler than in Yavan), the flight of the first generation of *A. superpictus* in the middle of May. The dates of May 10 and 11 are quoted by other authors, too.

As can be seen, because of the phenological principle, mosquitoes that began their development on March 20 almost catch up with mosquitoes that began their development on March 6. *Therefore, even if we had taken March 5, and not February 16, as the date of A. superpictus' mass flight from places of hibernation, there still would not have been any appreciable difference in the dates for egg laying by the first generation.*

The last females that began to lay eggs on May 10 will continue to lay eggs during several gonotrophic cycles. The duration of the egg laying and its end can be determined on the basis of the lifetime of the mosquitoes that laid eggs on May 10, that is, that formed wings on May 7 (since the length of one gonotrophic cycle in the first half of May is 3 days, as can be seen from biological curve 2 in Fig. 1.51). This lifetime, and the date of the females' death, are given by the intersection of the average temperature line from May 7 with biological curve 3*: May 7 + 17 days = May 24; t_{av} = 20.6°C, where n = 17 days and t_{av} = 20.6°C are, respectively, the ordinate and absicissa of the intersection of the average temperature line from May 7 with curve 3, which represents the lifetime of old females (Fig. 1.51). Therefore, on the whole, the second egg laying will begin on May 9 and finish on May 24.

According to observations by Pravikov and Popov (1957), in Iolotan imagoes of the first generation appear during the period between the end of the first and the beginning of the third 10-day period of May.

Clearly, all the calculations above were made on the basis of multiannual data. However, if the initial date of the forecast (flight after wintering or the first egg laying, etc.) were the result of an actual observation in a particular year, then any of the above calculations may become what we call a phenoclimatic forecast with prolonged foresight. In fact, the actual initial date of the forecast is the integral of all influences of place and time.

Nevertheless, if it is necessary, we can easily take into account the long-term weather forecast for a particular year. Let us assume that we know from the long-term weather forecast that in March and April the temperature is expected to be 1°C higher than the norm. Then in the previous problem (where we defined the date of the second egg laying on the basis of a first egg laying occurring on March 6), the second egg laying, which interests us here, occurs not on May 9, but on May 2: March 6 + 57 days. Here n = 57 days is the ordinate of intersection of phenological curve 1 with the average temperature line from March 6 in Figure 1.51, that is not with the multiannual line, but with the parallel line that is located 1°C to the right of the multiannual one in accordance with the scale. This is a phenosynoptical forecast.

In the same way as was used in previous chapters for various biological objects, we can also calculate a phenoforecast with shortened foresight for the second egg laying of A. superpictus using the actual temperatures for part of the forecast period.

Let us return, however, to defining the dates of the egg layings of A. superpictus for all subsequent generations in multiannual terms, that is, by means of phenoclimatic forecasts with prolonged foresight. It turns out that the third mass egg laying begins on May 30 (on the basis of the beginning of the second egg laying on May 9 and on the basis of curve 1 in Fig. 1.51) and finishes on June 17: on the basis of the end of the second egg laying on May 24, and on the basis of curve 1, we find June 10 to be the date of the third egg laying and, corresponding to this, June 8 to be the date of the appearance of wings in mosquitoes which will be alive, according to curve 3, up to June 17. The third mass egg lay-

---

*Curve 3 shows that the lifetime of mosquitoes that already have wings decreases as the temperature increases. The same can be seen for other poikilotherms, as well as for viruses (Fig. 3.12).

ing proceeds at an average temperature of about 26°C, over the entire period from May 30 to June 17. The fourth mass egg laying lasts from June 14 to July 8, and so on. To clarify this see Table 6.6.

We proceed with the calculations until the beginning of mass egg laying by the next generation passes the boundary of the mass entry of mosquitoes into the diapause. The date of this entry into the diapause can also be calculated (see below).

Thus it is possible to calculate not only the developmental dates of individual generations, but also the overall number of generations of the insect (and not only on the basis of multiannual average temperatures, but also on the basis of temperatures for each concrete year). For example, A. superpictus in Yavan (where it is hot) gives, according to calculations, 10 to 11 generations. This agrees with the data from N. K. Shipitzina (1957).

However, the drying up of reservoirs can upset the calculations for some generations of mosquitoes. The ways to account for this factor, as well as other critical factors, were given in the description of the mathematical method (see Section 2.3).

6.6b. Calculating the Dates when Insects Go into Diapause and Hibernation.

The method of calculating the diapause is of particular biological importance.

In Section 4.1h it was already mentioned that to calculate the date when the mass facultative diapause occurs for a given insect in a given natural climatic region consists of finding the date of the last mass egg laying of the year that can be followed by the mass entry of insects into the stage most apt for hibernation. This stage, for A. superpictus and A. maculipennis mosquitoes, is the flight of the imagoes; for cotton bollworm, it is pupation; and for the Colorado potato beetle, it is again the imago.

A sharp decrease in the mosquitoes' egg layings usually occurs, in natural circumstances, as a result of the appearance of young females that are unable to lay eggs. The time when such females appear is taken as the date when the mosquitoes enter the diapause. It is possible that at this time the old females also sharply reduce their egg layings.

The dates when the mosquitoes' egg layings decrease sharply in nature (as a result of the old and the young females being unable to lay eggs) are apparently related to the time necessary for the development of the imago before hibernation. The offspring of biological individuals or biological species that continued to lay eggs after their critical date perished, and did not continue the species. Those species and individuals that stopped egg laying too early lost their advantages in the competition and thus disappeared.

TABLE 6.6.   Order of calculations for egg laying dates of *Anopheles superpictus*

Flight Date from Places of Hibernation	First Egg Laying			Second Egg Laying			
	Gonotrophic Cycle	Dates	Phenocurve	Gonotrophic Cycle	Dates	Preceding Wing Formation	Phenocurve
16 II	1	16 II + 18 = 6 III	2	1 . . .	6 III + 64 = 9 V		1
	2	6 III + 14 = 20 III	2	1 . . .	20 III + 51 = 10 V	7V	1
				Last	7 V + 17 = 24 V		3

What is this critical date for the cessation of egg laying, even though winter begins at a different time every year? Apparently, in the process of their historical development, insects adapted to the multiannual average dates of the onset of autumn and winter. These multiannual dates, or the dates closest to them, are those which in fact occur most often in each natural climatic region. This is why many authors note the astonishing constancy, year after year (within certain limits), in the dates when mosquitoes enter the diapause in a given region. Of course, the insects' nervous system, in its "prevision" of the multiannual temperature trend, must be "guided" or, more accurately, stimulated by some kind of advance signal. A signal for the decrease in temperatures may be the decrease in day length, which begins in autumn. The "stimulus" may be that remarkable sense of time (known by the name of "physiological clock") which is peculiar to many animals (and even to plants). It is being studied intensively now by the new branch of cybernetics-bionics.*

Of course, all of this, for the present, is a working hypothesis, but we shall now verify whether this hypothesis gives real results. Thus, in accordance with the hypothesis described above, we find in Figure 1.51

---

*Bionics [bio + (electro)nics] deals with the application of principles of activity found in living systems, and principles of biological processes, to engineering problems. I believe that, vice versa, the use of methods from the exact and engineering sciences for the purpose of explaining processes found in living nature also relates to the field of bionics, or at any rate to biophysics in general. From this point of view this author's work is devoted to bionics and biophysics.

on the PTN for Yavan (Fig. 1.51).

	Third Egg Laying				Fourth Egg Laying		
Gonotrophic Cycle	Dates	Preceding Wing Formation	Phenocurve	Gonotrophic Cycle	Dates	Preceding Wing Formation	Phenocurve
1	9 V + 21 = 30 V		1	1	30 V + 15 = 14 VI		1
1				1			etc.
1	24 V + 17 = 10 VI	8 VI	1	1			
Last	8 VI + 9 = 17 VI		3	1	17 VI + 13 = 30 VI	28 VI	1
				Last	28 VI + 10 = 8 VII		3

the multiannual average temperature line which only touches, but does not intersect, phenocurve 4 for the mosquitoes' development from mass egg laying to the mass appearance of imagoes (of course, for the cotton bollworm we would have to construct a curve for the development from egg to pupa). This will be the line from September 26 (it is shown by a dashed line in Figure 1.51). Therefore, on September 26 the last mass egg layings, which are able to produce imagoes, take place in the field in Yavan, and the mass entry into the diapause occurs.

The line of average temperatures from September 26 touches developmental curve 4 at a point with ordinate $n = 52$ days. Therefore, the last mass emergence of imagoes of the year occurs on November 17: September 26 + 52 days.

It is known that the multiannual date of the first autumn frosts in Yavan is November 17. Therefore, the biological genus is adapted so that the last mass emergence of imagoes occurs just as the first autumn frosts begin, and consequently they immediately go into hibernation (in all of the above, we refer to the mass beginning of hibernation). The agreement of the data is quite striking!

The average temperature over the period from the mosquitoes' last mass egg laying in the field to the last mass appearance of imagoes is $t_{av} = 14.8°C$ (see the abscissa of the point where the two lines touch); the multiannual 24-hour average temperature for September 26 is 21.2°C; for November 17 it is 8.5°C (in Fig. 1.51, on the horizontal axis, see the points where the average temperature lines from September 26 and November 17 begin). As can be seen from the calculations, the entry into the diapause (September 26) occurs when the temperatures are still high; this is also well known from many observations. The entry into hibernation (November 17) occurs at the generally known temperature of 7° to 9°C. From Table 6.7 we see that all these temperature criteria are quite stable for very different climactic zones.

Having the PTNs available for other regions of the USSR (Figs. 1.56, 1.99, and 1.108), we can, therefore, make identical calculations for different geographical areas: from Middle Asia to Moscow and from Moscow to Amur (Table 6.7).

It can be seen from Table 6.7 that the calculated dates and temperatures are confirmed by observations, although the difference in the dates when the diapause occurs in different geographical regions may naturally be more than a month and a half (Yavan-Zavitinsk).

Next, in spite of the fact that Zavitinsk is located south of Moscow, the diapause in Zavitinsk occurs a little earlier, according to the calculations. This corresponds to the known facts that diapause in the Far East occurs earlier than in the west (and in Middle Asia, which is relatively east, diapause occurs earlier than in the Caucasus, which is relatively west) at the same latitude.

It can be seen from Table 6.7 that the temperature criteria accompanying the process (which occurs at different times) of the transition to diapause and hibernation are quite constant. This constancy is also confirmed by A. Ya. Storozheva (1957) and N. K. Shipitzina (1957). These criteria are as follows: mass diapause of *A. maculipennis* and *A. superpictus* occurs in most of the USSR territory at a time (in summer or in autumn) when the 24-hour average temperatures of the outdoor air fall below 18° to 21°C; the average temperature over the period from the cessation of mass egg layings in nature to the last mass wing formation is, almost everywhere, 13°-15°C; the mass onset of hibernation occurs when, in autumn, the 24-hour average temperatures of the outdoor air fall below 7° to 9°C (Table 6.7).

Let us test the theory of the diapause described above with another biological object—the Colorado potato beetle. As was described in Section 2.3, special experiments in Pruzhani showed that the critical day length for the beetle's maturation for egg laying is 17 hours 15 minutes. When the day is shorter than this, all beetles are ready to go into diapause without laying eggs. In Pruzhani such a day length occurs after July 21.

In Kursk (the geographical latitude of which is close to the latitude of Pruzhani), the critical day length of 17 hours 15 minutes occurs, according to the astronomical reference books, after July 23. This is shown in Section 2.3, and in Chapter 5, and is expressed by hatched area I on the PTN for Kursk (Fig. 1.93).

Since the egg laying by young beetles occurs, approximately, 10 days later after their appearance, then the last eggs of the Colorado potato beetle in the field in Kursk must be laid on July 23 + 10 days = August 2.

Since the date when the egg layings decrease sharply in nature is considered to be the moment when the poikilotherms' diapause begins, August 2 will be the date when the Colorado potato beetle goes into diapause in the conditions of Kursk. The basis of this calculation is, as can be seen, the *"light" theory of the diapause.*

On the other hand, according to the heat hypothesis for the diapause described above, we find the line tangential to phenocurve 12 from egg to imago in Figure 1.93. This tangent is the line of average temperatures dated August 4 and 5. Therefore, on August 4 the last eggs that are able to give imagoes will be laid—the only stage in which the Colorado potato beetle is able to survive hibernation. In other words, August 4 is the date of the diapause according to the *heat hypothesis.*

As can be seen, the light and the heat hypotheses of diapause give us similar dates (August 2 and August 4). These hypotheses are connected to each other because the initial cause of the diapause is the heat factor, which makes itself felt by means of the light signal.

6.6c. The Number of Individual Insect Carriers in Relation to Overlapping Generations.

Having calculated for Yavan the dates of the beginning and the end of the egg layings for each generation (Table 6.6), as well as the date of cessation (up to the next spring) of egg layings in nature in relation to the diapause (Table 6.7), and having the experimental curves for the sexual productivity of female mosquitoes (Figs. 6.6 and 6.7), we can make calculations concerning the numbers of the insects.

When doing this, we proceed from the following fact (which is confirmed by the calculation): on the way from egg to imago so many individuals perish that in the end one female produces little more than one female of the next generation which is able to survive up to middle age. Thus the increase (known from observations) in the numbers of insects during a season must be explained not only by the "fission" of a female in the process of propagation, but also by the overlapping of generations, especially in the south (however, this overlap is related indirectly to the "fission").

An individual gono-active female of the malaria mosquito lays eggs during several gonotrophic cycles, with decreasing intensity from cycle to cycle, as is shown in Figure 6.6 by the experimental dotted line. The peak of this line, representing the quantity of eggs laid in the first gonotrophic cycle, depends to a high degree on the temperature, and is determined by the dotted curve in Figure 6.7. Knowing (from previous calculations) the average temperatures at which individual generations of mosquitoes develop, we can take from Figure 6.7 the data about the quantity of eggs laid by an individual female from a given generation at the beginning of its gonotrophic activity. But this quantity of

TABLE 6.7 Dates when malaria mosquitoes enter diapause and hibernation, as well as temperature criteria accompanying these processes, in various regions of the USSR, after calculations on PTNs and actual data.

			Calculated Data			
Location, Height above Sea Level	Date of Mass Diapause	Date of Mass Entry into Hibernation (at Imago Stage)	24-Hour Average Temperature of Outdoor Air at Time of Mass Entry into Diapause (in °C)	24-Hour Average Temperature at Time of Mass Entry into Hibernation (in °C)	Average Temperature for the Period from the Last Mass Egg Laying to the Last Mass Appearance of Imagoes (in °C)	Actual Data
1	2	3	4	5	6	7
			Anopheles Superpictus			
Yavan, 664 m (Tadzhiki-stan)	26 IX	17 XI	21.2	8.5	14.8	Mass entry into diapause in 1956, 25 km from Yavan (where it is somewhat cooler than in Yavan) occurs on 21 IX with 24-hour average temperature of about 21°C (Almazova, 1959). 17 XI is multiannual average date of the first autumn frosts in outdoor air of Yavan. Actual data of O. F. Buyanova (1959) for Nurek (25 km from Yavan) in 1955, as well as other observations, are also close to the calculated data.

Dushanbe, 803 m (Tad-zhikistan)	11 IX	13 XI	20.1	8	13.7	Mass entry into diapause is close to 11 IX. For instance, after E. S. Kalmikov (1959), in the Dushanbe suburbs by 18 IX, 60% of females were in a state of gonotrophic dissociation; the first females in this state were observed by Kalmikov on 26 VIII, and the last on 26 IX; then 50% on $$\frac{26\ \text{VIII} + 26\ \text{IX}}{2} = 10\ \text{IX.}$$ The first autumn frosts in the air are November 3 to 8. According to observations in Dushanbe by E. P. Lupova (1954), females with inactive ovaries constitute, in the first 10 days of September, 30.8%; in the second 10 days, 52.5%; etc.

Anopheles maculipennis

Moscow (Pavlovsky Posad), 150 m	12 VIII	26 IX	≈17	≈8	≈13	Mass diapause: at the beginning of the second 10 days of August (N. K. Shipitzina). The first autumn frosts are in the air on 25 IX.
Zavitinsk, Amur district, 400 m	9 VIII	29 IX	≈20	7	14.2	According to the maps and cartograms constructed by N. K. Shipitzina (1957), in the neighborhood of Blagoveshchensk (where it is a little warmer than in Zavitinsk), the mass cessation of blood sucking occurs, according to the average multiannual data, on 7 VIII, and mass diapause occurs on 10 VIII. The first autumn frosts in the air are on 28 IX.

eggs must be divided by 10, since, according to V. N. Beklemishev, only 7 to 10% of the individuals that went through the first larva stage survived up to the stage of wing formation. Let us call them surviving eggs (mosquitoes with wings will hatch from them later) or "effective egg layings."

If, for each day of a prolonged period of a generation's egg laying, we take into account one female that lays eggs for the first time, we shall obtain a series of elementary lines, each of them similar to the dotted line in Figure 6.6. Egg layings by individual females will be totaled each day, and as a result the diagram of egg layings by a given generation will be expressed by a figure that is similar to the trapezium in Figure 6.8. From this figure we see that the third mass egg laying began on May 30 (see Table 6.6 and the upper section of the first line in Fig. 6.8) and finished on June 17 (see Table 6.6 and the lower section of the last line), that is, the egg laying lasts for the period shown by the calculation whose result is given in Section 6.6a and in Table 6.6. It is indicated in the calculation that the average temperature over the entire period of the third egg laying is 26°C. From the dotted line in Figure 6.7 we see that at such a temperature, the productivity of a single egg laying by a malaria mosquito at the beginning of its gonotrophic activity is 300 eggs; that gives 30 surviving eggs (300 ÷ 10). That is why each line in Figure 6.8 begins with exactly this value.

Thus each line of the series of curves in Figure 6.8 shows the law of the decrease in egg layings by an individual female from the first gonotrophic cycle, which gave 30 surviving eggs, to the last gonotrophic

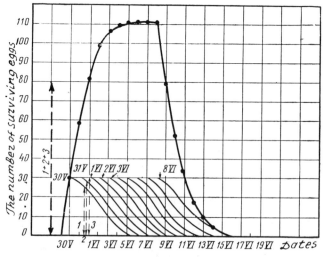

FIGURE 6.8. Diagram for the third effective egg laying of *A. superpictus* in Yavan (according to calculations).

cycle in the female's life, which gave almost no eggs at all. The entire gonotrophic activity of an individual female lasts for the period calculated in Section 6.6a and in Table 6.6. For example, a female that begins its mass egg laying on June 8 continues it during almost its entire lifetime up to June 17 (see the last curve of the series of curves and the last line in Table 6.6, in the columns related to the third egg laying).

Naturally, it is clear that females that belong to the same generation, but appear on different days, begin their egg layings at different times. Therefore, on any day of the egg laying period of a given generation, a female laying eggs for the first time can always be found together with an older female, a predecessor that has not yet finished its egg layings (but from whose previous eggs larvae have already hatched), and so on. It is clear that, in such a case, the productivity of the female that lays eggs for the first time on a particular day will be totaled together with the quantity of eggs laid by the female predecessor. For example, in Figure 6.8 it is shown that 30 surviving eggs, laid on June 1 by a female laying eggs for the first time (see dotted arrow 3), are totaled with 27 eggs laid on June 1 by a female predecessor (dotted arrow 2) and with 23 eggs laid on June 1 by the predecessor of the predecessor (all three dotted arrows, of course, coincide and refer to June 1, but for the sake of visual clarity, they are drawn separately). Thus the total of all the surviving eggs laid on June 1 in Yavan is $30 + 27 + 23 = 80$ (see the large dotted arrow $1 + 2 + 3$ in Fig. 6.8). In the same manner we total the eggs laid on May 31, June 2, June 3, and so on. The results obtained are plotted in Figure 6.8 in the form of big points joined by a thick line, which resembles a trapezium.

Such trapezia should be constructed for all the generations of the insect. The procedure for construction can be simplified, using the following practical advice. The height of the trapezium is defined by the initial quantity of the surviving eggs. If this initial quantity is 30 (as it will always be at average temperatures higher than 23°-24°C), then this height will be equal to approximately 110 eggs (as in Fig. 6.8). The length of the trapezium's base corresponds to the distance between the date preceding the beginning of all the egg layings by a given generation and the date of the finishing of all the egg layings by this generation. The length of the upper trapezium's base depends on the inclination of the elementary lines. But the upper base always begins from the date that is close to the end of the first elementary line and finishes at the date of the beginning of the last elementary line. Therefore, the construction of trapezia for the succeeding generation should begin with plotting the last and the first elementary lines representing the egg laying by an individual female. When doing this, the elementary curves can be simplified by being represented by straight lines (with an inclination expressing the length of the egg laying by an individual female).

For example, the diagram of the fourth egg laying looks as shown in Figure 6.9. This diagram is simplified in the following way. It can be seen from Table 6.6 that the fourth egg laying finishes on July 8 (see the last line). Since we intend to replace the elementary curves by straight lines, we take July 6. This egg laying is finished by a female that formed wings on June 28, and that began egg laying on June 30 (see the second to the last line in Table 6.6). In Figure 6.9 we mark the points that correspond to July 6 and June 30. We join these points by a straight line. Next, in Table 6.6 we see that the fourth egg laying began on June 14 (see the first line). A point corresponding to this date is also to be marked in Figure 6.9. From this point we draw a straight line parallel to the line June 30–July 6. Between these extreme lines we draw a number of parallel straight lines for each day. Then we draw the upper base at the level of about 110 eggs. The way to find the beginning and the end of the upper and lower bases has been described above.

As can be seen, the diagram of the fourth egg laying is wider than that of the third egg laying. Each subsequent diagram is wider than the preceding one. Owing to this, as well as to the increased closeness of generations to each other caused by the rise in temperatures, the gap between the generations first closes up and then the generations overlap each other. This involves a sharp increase in the number of carriers.

Actually, the "trapezia," that is, the diagrams for egg layings of all generations, are put onto the same figure. If the generations overlap each other, the diagrams are totaled, as is shown in Figure 6.10. For

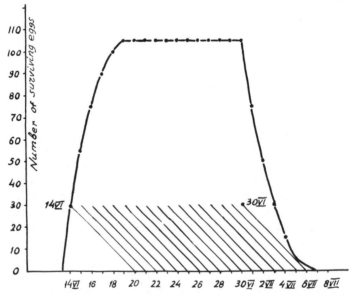

FIGURE 6.9.   Diagram for the fourth effective egg laying of A. *superpictus* in Yavan (according to calculations).

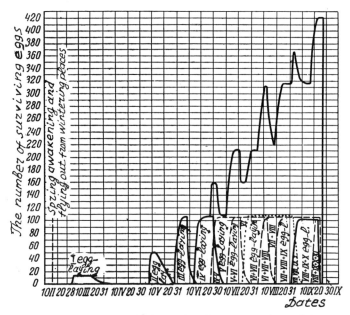

FIGURE 6.10. Totaling the egg layings of various generations of *A. superpictus* in Yavan (according to calculations).

example, on June 30 we have a superposition of the diagram for the fourth egg laying on the diagram for the fifth egg laying at the interval equal to 80 eggs; then the total will be equal to 160 eggs. That is marked in Figure 6.10 by the upward extension of the totaling curve over the date of June 30.

As can be seen, the theoretical line of the increase in egg layings is suddenly broken and brought to zero by the vertical line that corresponds to the date of mass diapause in Yavan—September 26. It can be seen in the same figure that by the time the mosquitoes go into diapause, that is, by the end of the season, in nature four generations are accumulated which coexist in an identical developmental stage (the eighth, ninth, tenth, and eleventh egg layings, which begin the generations of the same number). Remember that, according to V. N. Beklemishev (1944), by the end of the season not less than three generations should coexist. The number of generations coexisting at different developmental stages must be higher than this.

The values marked on the vertical axis of Figure 6.10 cannot be considered as the quantity of mosquitoes caught over one day, nor as the general quantity of mosquitoes in the conditions of Yavan. However, the quantity of surviving eggs taken from the totaling curve on one date or another can be considered as a certain equivalent (perhaps, the smallest equivalent) of the abundance of mosquitoes in nature. This equivalent would have to be multiplied by a certain constant, or even

variable, coefficient, in order to find out the absolute quantity of mosquitoes in nature in a given region; but this coefficient is unknown for
the time being. For example, the final date of September 26 corresponds
to 420 surviving eggs, whereas the date of June 30 corresponds to only
160 surviving eggs (Fig. 6.10); therefore, in the conditions of Yavan,
the quantity of insects on September 26 is at least three times greater
than their quantity on June 30 (with a similar availability of reservoirs
during the entire season). In the given calculating system , the height
of a trapezium diagram for each generation is of minor importance for
the formation of such a ratio: in Figure 6.10 it can be clearly seen that
the height of all elementary trapezia, beginning from the third egg
laying, is equal. The key factor is the quantity of coinciding trapezia,
and this depends on the trapezia's width.*

It should be said that the calculation described above for the number
of eggs was begun with one wintering female. For example, the first egg
laying by the wintering female consists, according to the calculation in
Section 6.6a, of two gonotrophic cycles, that is, two single egg layings
on March 6 and 20, each of them producing 13 surviving eggs (see Fig.
6.10, the first egg laying). The figure of 13 eggs was obtained in the
following way: the temperature on March 6 is 9.0°C and on March 20
it is 10.6°C (In the PTN in Fig. 1.51, see the point on the horizontal
axis from which the lines of March 6 and 20 rise). According to Figure
6.7, at such temperatures the quantity of eggs is 130; 13 surviving ones
occur in each egg laying, that is, in all there are 26 surviving eggs. Let
us assume that from 26 eggs will emerge 13 females and 13 males. We
assume that of the 13 females, two survive to middle age—one female
from each single egg laying; let this and some other exaggerations be
a compensation for the underestimation caused by the equal height of
the trapezia after the second egg laying. These two females are able to
have five gonotrophic cycles each and to form a diagram of the second
egg laying that is shown in Figure 6.10. Constructing the diagram for
the third egg laying and for the following ones has been already described at the beginning of this section (6.6c). Therefore, the values
designated on the vertical axis in Figure 6.10 can also be considered as
the minimum number of offspring given by one wintering female over
a 24-hour period (the offspring are composed of several coexisting
generations). During the entire season one wintering female, apparently, gives offspring equal to the integral of the totaling curve on
Figure 6.10, that is, to the area under this curve (i.e., the part of the
offspring with wings, which is able to survive up to the fourth gonotrophic cycle and more).

For each peak of egg layings there is a corresponding peak of the

*The width of the trapezia is unequal, and reflects the "fission" of a female in the process
of reproduction; in the second half of summer, it also reflects the phenological principle
mentioned above.

number of females, which occurs earlier by the time interval equal to one gonotrophic cycle, that is, in the conditions of Yavan 2 to 4 days earlier. Thus, having moved the entire totaling curve of egg layings to the left by 2 to 4 days, we obtain a diagram of the number of females and compare it with the actual observations (Fig. 6.11).

As can be seen in Figure 6.11, the actual data confirm the calculation, although the acutal curves are displaced somewhat to the right, since they relate to regions colder than Yavan, for which these calculations are made (in Yavan there were no observations).

Moreover, we see that the theoretical curve shows nonsystematical zigzags, as often occurs in nature. But, referring to Figure 6.10, we note that these zigzags are not casual: each upward flight of the totaling

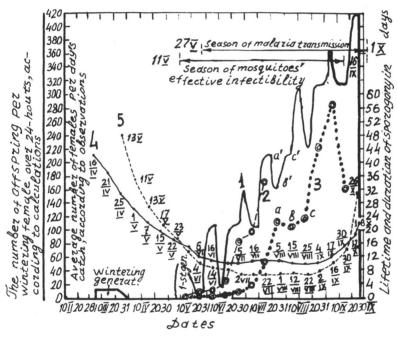

FIGURE 6.11. Diagram for the number of *A. superpictus* in Yavan, using calculations on the PTN, in comparison with actual observations in the neighboring (but somewhat cooler) areas, as well as the lifetime of old females and the duration of sporogony. Curve 1: solid thick line—results of theoretical calculations for the number of mosquitoes; curve 2: coarse dotted line—actual number of *A. superpictus* (taken each 10-day period) in Kurgan-Tyube (56 km from Yavan) in May-July 1950 (after E. S. Kalmikov and A. Ya. Lysenko, 1959); curve 3: line with points—actual number of *A. superpictus* females (taken each 10-day period) in the Komsomolabadsky district (90 km from Yavan) in May-September 1955 (after L. Ya. Ilyashenko, 1959); curve 4: solid thin curve—lifetime of old *A. superpictus* females (composing 2-3% of a population) in relation to the dates of wing formation (the dates along the line are the dates of death); curve 5: fine dotted line—sporogony duration of *Plasmodium vivax* in the mosquito in relation to the dates of the beginning of sporogony (the dates along the line are dates when sporogony ends).

curve is bound up with the beginning of the next generation. Hence it follows that if frequent measurements of the number of insects are available, then calculating even a large number of generations and their dates in nature is not impossible; it is only necessary to follow the "upward flights" of the curve representing the numbers. But this is a subject for separate study.

Conclusions from the calculations made above are as follows.

1.   One of the factors causing the increase in the numbers of insects from the beginning to the end of a season is, apparently, the progressively rising overlap and totaling of generations (this is the case for mosquitoes, e.g., if reservoirs are available).

2.   As is known, the number of many (but, of course, not all) insect species begins to rise especially intensively in the second half of a year. This occurs, concurrent with other causes, because the overlap of generations begins, even in the south, in the second half of a year, when temperatures are high enough and are favorable for the quick development of generations and for reducing the gaps between the generations.

3.   Other things being equal, the chances of malaria transmission increase as the number of carriers increases. If this is correct, then the development of malaria with short incubation in the second half of a year apparently must also be explained by the overlap of the mosquitoes' generations in the second half of a year.

4.   If the increase of malaria carriers is caused, to a certain degree, by the totaling of generations, and this totaling proceeds most intensively at high temperatures, then one of the causes of the extensive distribution of malaria in the south is also the intensive overlap in this area of many coexisting mosquito generations.

5.   The known summer depressions in the number of mosquitoes are explained not only by the drying up of the reservoirs, and by the oppressive (for mosquitoes) conditions of high temperature and dry air, but also by the circumstance that the depressions, naturally, occur between two overlaps of adjoining generations.*

The above calculations for the number of insects are made on the basis of multiannual average temperatures (to these temperatures the developmental dates of generations are related, as well as the values for t_{av} of the processes, calculated in Section 6.6a). Of course, however, the number of insects can also be calculated for each concrete year using the temperatures in that year.

It should be noted that the intensive increase in the number of insects at the overlap of generations was observed in nature by N. L. Shchotkin (1960), B. M. Yakunin et al. (1979), and others.

*This is especially clear on the graphs for Tashauz in Turkmenia, which are not given here.

6.6d. Malaria Mosquito and Malaria Plasmodium, and their Interrelation. Construction of Epidemiograms.

We discussed above those components of the parameter of malaria communicability which relate to the vital activity of the mosquito—the carrier. Let us focus on the components that concern the development of the plasmodia—the pathogens of malaria—in the body of the mosquito after it feeds on a sick person.

Sporogony (the process of the development of the plasmodia in the mosquito's body), when begun at different times of the year, takes a longer or shorter time in relation to the temperature: in spring and autumn it takes longer than in summer. On the other hand, the lifetime of the mosquito also changes in relation to the season: females that developed wings in spring and autumn live longer than those that developed wings in summer.

As is known, only those mosquitoes which survive to the end of sporogony or longer are dangerous. Therefore, in order to judge the epidemiological danger of mosquitoes, or, more exactly, the *season of the mosquitoes' effective infectibility* by the pathogens of malaria, it is necessary to compare the duration of sporogony with the mosquito's lifetime in different months. This will enable us to find the spring and autumn dates of wing formation, following which the mosquito can survive to the end of sporogony. In other words, this comparison will make it possible to realize the principle of correlation ("counterpoint"—according to the terminology of Sh. D. Moshkovsky, 1950).

For example, on the PTN for Yavan (Fig. 1.51) using the dates of wing formation and the beginning of sporogony once at the 10-day period, we can find the corresponding lifetime and duration of sporogony. That is shown by the intersections of the average temperature lines corresponding to the dates of wing formation and beginning of sporogony with biological curves 3 and 5. The results obtained from this intersection are put onto a graph (Fig. 6.11), where the dates of wing formation and beginning of sporogony for the mosquito are designated on the horizontal axis, and the lifetime and the length of sporogony in days on the vertical axis. The points of these results are joined, forming two lines that intersect each other twice (in spring and in autumn); their salient part is directed to the horizontal axis. These two intersections show that the most long-lived females (their quantity in a population is 2-3%; phenocurve 3 relates to them) survive up to the end of sporogony. The section of the curves between the intersections shows that the mosquito outlives sporogony; the sections of the curves before the spring intersection and after the autumn intersection show that the mosquito does not survive to the end of sporogony.

For Yavan the curve of the mosquito's lifetime intersects the curve of sporogony at the abscissae close to May 11 and September 16. This means that mosquitoes that get wings on May 11 and September 16 survive to the end of sporogony, and those that get wings between these dates even outlive sporogony. All this can be clearly seen in Figure 6.11. For example, the female *A. superpictus* that develops wings on March 25 will live 31 days, that is to March 25 + 31 days = April 25. This latter period and date are designated on curve 4 by the point with coordinates March 25 (on the horizontal axis) and 31 days (on the vertical axis; in the given case, the vertical axis should be seen on the right-hand side of Fig. 6.11). Actually, the line of average temperatures from the initial date of March 25 intersects phenocurve 3, representing the mosquito's lifetime (Fig. 1.51) at exactly $n = 31$ days.

One to two days after the appearance of wings, the mosquito can feed on blood from a person ill with malaria. Let us assume that this occurs on March 26. Therefore, March 26 is the beginning of the sporogony of *Plasmodium vivax* in the mosquito's body. In this case, the duration of sporogony, using curve 5 (Fig. 1.51), will be $n = 48$ days, and the sporogony will finish on March 26 + 48 days = May 13. This date and this period are shown on curve 5 in Figure 6.11 by a corresponding point (which is displaced slightly to the right in comparison with the corresponding point of the mosquito's death). As can be seen, the mosquito will die on April 25 and the sporogony will finish only on May 13. Therefore, such a mosquito is not dangerous for healthy people. On the contrary, the mosquito that develops wings, for example, on July 25 (Fig. 6.11) will live up to August 5; the sporogony that began on July 25 or 26 will finish on August 1. This means that the infected female will live 4 days after the end of sporogony (August 5 – August 1). Four days are sufficient for two gonotrophic cycles, and therefore for the mosquito to suck blood twice, possibly infecting two people.

Thus *the season of the mosquitoes' effective infectibility* lasts in Yavan for 128 days—from May 11 to September 16. But this is correct only if during this period mosquitoes will be available. Therefore it is necessary to realize one "counterpoint" more—to put the seasonal trend of lifetime and of the sporogony's duration onto the background with the seasonal trend of the mosquitoes' numbers (this trend was calculated by us above). That has been done in Figure 6.11. It can be seen from this figure that during the entire season of the mosquitoes' effective infectibility in Yavan, there can in fact be mosquitoes (in accordance with the temperature conditions), although they will not exist in equal quantities throughout this season.

The PTN makes it possible to detail the calculation for the season of the mosquitoes' effective infectibility: to calculate the quantity of gonotrophic cycles for which a female has time, over her whole life, over one sporogony cycle, and after sporogony. For this purpose, we select on the PTN an average temperature line that corresponds to the date of wing formation in mosquitoes or to the date of the begining of sporogony; then, on the basis of the intersection of this line with biological curve 2 (Fig. 1.51), we define the duration of the first gonotrophic cycle. On the basis of the end of the first cycle we define, in the same way, the length of the second gonotrophic cycle and so forth, until the whole lifetime or the whole length of sporogony which corresponds to the selected starting time is exhausted. If the length of one gono-trophic cycle changes little during the lifetime and during sporogony, then, in order to find the quantity of gonotrophic cycles, the lifetime or the length of sporogony is divided by the length of one gonotrophic cycle. The difference between the quantity of gonotrophic cycles over the mosquito's whole life and that over one sporogony cycle gives (within the limits of a single malaria season) the *quantity of epidemio-logically dangerous cycles.*

It is possible to give a graphic representation of the calculated sea-sonal trend in the quantity of gonotrophic cycles over the mosquito's whole life, over a sporogony cycle, and after sporogony (Fig. 6.12).

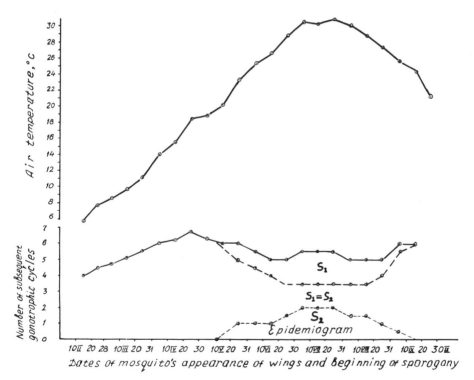

FIGURE 6.12. The quantity of gonotrophic cycles made by the most long-lived (2-3% of a population) *A. superpictus* females in Yavan (Tadzhikistan) over the entire life (thin solid line), over the sporogony period (coarse dotted line), and after the sporogony, which began after the first flight, i.e., the number of epidemiologically dangerous cycles (fine dotted line), in relation to various (in a year) dates of wing formation of the females and in relation to various dates of the beginning of sporogony. These are results of the calculations made on the PTN. Solid thick line: seasonal trend of air temperatures in Yavan on the basis of multiannual data.

As can be seen from Figure 6.12, in the conditions of Yavan the number of gonotrophic cycles over the lifetime of a female increases as the temperature increases only in the case of females which develop wings before May. Although the temperature after this continues to rise, the quantity of gonotrophic cycles begins to decrease somewhat. This is explained as follows. If the mosquito's lifetime is divided by the length of one gonotrophic cycle, the result of division will be the quantity of gonotrophic cycles during the mosquito's life. This quantity increases first because, with the temperature rise, the length of one gonotrophic cycle decreases more quickly than the length of the mosquito's life. When the temperature continues to rise, the length of one gonotrophic cycle decreases less quickly than the mosquito's life. As a result, the quantity of gonotrophic cycles during the mosquito's life begins to decrease and not increase as it did before. The specific character of the temperature trend in a given region is also important.

The quantity of gonotrophic cycles of a female over one sporogony cycle, according to the author's calculations, decreases as temperatures increase (at the end of spring—six cycles; at the hottest time—three to four cycles; nearer to September, again six cycles), in full accordance with the known data of P. G. Sergiyev, M. F. Shlenova, and Sh. D. Moshkovsky, about the correlation of the duration of sporogony with the length of time taken to digest the blood at various temperatures. When the sporogony finishes, the females become epidemiologically dangerous, at the fourth to sixth gonotrophic cycle from the beginning of the appearance of wings (Fig. 6.12).

The number of gonotrophic cycles of a female after the end of sporogony, that is, the number of epidemiologically dangerous cycles, increases, according to the calculations, as temperatures increase. Thus the extent of malaria in the south and in warm years is greater than in the north and in cold years; this fact is accurately proved theoretically (especially if we add the effect of the overlap of generations, which was discussed above). The fact is that the seasonal trend of the quantity of epidemiologically dangerous gonotrophic cycles (after sporogony) was not yet calculated (or such calculations are extremely few); this seasonal trend is a basis for malaria's distribution.

Knowing the dates and the length of the season of the mosquitoes' effective infectibility, and the distribution of the quantity of epidemiologically dangerous gonotrophic cycles according to various dates in this season, we can draw an *epidemiogram*. It is to be drawn on the following axes: ordinate—quantity of dangerous cycles; abscissa—dates of wing formation or dates when sporogony begins. That has been done in Figure 6.12 and, separately, in Figure 6.13.

Such an epidemiogram characterizes the potential or the real epidemiological capacity of the season of effective infectibility (more exactly, the season of malaria transmission or even the malaria season) caused by an *individual mosquito female*. The larger the area (integral)

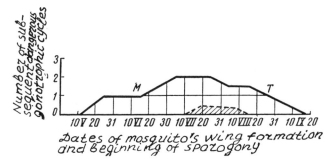

FIGURE 6.13. Epidemiograms for Yavan, 664 m above sea level (solid line), and for Khorog, 2077 m above sea level (dotted line), in Tadzhikistan, for an individual female of malaria mosquito.

of the epidemiogram for a given region, the larger the capacity of the season:

$$S = \int_{\tau_1}^{\tau_2} f(\tau) \cdot d\tau \tag{23}$$

where S is the epidemiological capacity; $f(\tau) = y$ is the equation of the epidemiogram's curve, that is, the dependence of the quantity of epidemiologically dangerous gonotrophic cycles y on the time τ in the season; τ_1 and τ_2 are the beginning and the end of the season of the mosquitoes' effective infectibility, respectively.

The area can be defined easily with the use of proper apparatus or by counting the squares. For example, it can be seen in Figure 6.13 that the potential epidemiological capacity of a season in Yavan is almost 11 times as large as the capacity in the Khorog highlands. Apparently, Khorog's height of 2077 m above sea level is close to the vertical boundary of malaria's distribution in Tadzhikistan. Hence it becomes clear why malaria did not and does not occur every year in Khorog. But at the same time a very large potential capacity in Yavan, on the whole, should be noted. It is, for example, twice as large as the capacity in Dushanbe. Therefore, Yavan was intentionally chosen for malariological calculations, since there it is necessary to prepare anti-malarial arrangements carefully because of the fact that the valley has been supplied with water.

Of course, the area of the epidemiogram is a relative representation of the epidemiological capacity of the season. But nevertheless, this area shows that the epidemiological capacity depends both on the duration of the season (length of the epidemiogram) and on the quantity of dangerous blood suckings (width of the epidemiogram), and that this capacity can be expressed by a product of the length and the average width (see Fig. 6.13).

On the basis of the data in Figure 6.13 and other calculations, a conclusion can be made that the number of epidemiologically dangerous gonotrophic cycles (i.e., the number of dangerous bloodsuckings), on the whole, is not large: even in very hot Yavan at the hottest time there are only two cycles; in Dushanbe there is one; and in Khorog there is one half of a cycle.

Maybe the erroneous belief that a female can infect only once is explained by the fact that, after the end of sporogony, very few gonotrophic cycles remain, and therefore very few epidemiologically dangerous stings remain in the life of even a long-living female in such a hot region as Yavan. And in general, without the admission that the abundance of mosquitoes is the most important cause of malaria's distribution, it would not be clear why malaria is so widespread on the face of the earth. Not only is man protected from malaria by medical science and practice (to a greater or smaller degree in relation to social conditions), but man has a lot of, as it were, "natural protections" from malaria, among them the following: of all mosquitoes only *Anopheles* transmits malaria; of all *Anopheles* only females are dangerous (males feed on the nectar of plants); among the females only the long-living ones are dangerous, that is, those which survive up to the end of sporogony; of the four to seven stings made by a long-living female (see the solid thin line in Fig. 6.12) only the last one or two stings are dangerous, and those only if the female succeeded in sucking the blood of a man and not that of livestock (which females prefer), and provided that this man is ill, and not healthy; the female must suck blood soon after wing formation (otherwise the sporogony will be late); finally—the mosquito must feed during the season of mosquitoes' effective infectibility, and this season in many regions is short or nonexistent. Briefly, V. N. Beklemishev (1957) was right when he wrote that the number of populations of a carrier in a particular breeding ground is one of the key factors in malaria epidemiology. That has been confirmed by this author's calculations given above.

Taking into account the number of females converts the epidemiogram for an individual female into a seasonal trend of the parameter of malaria communicability. In order to do such converting, it is necessary to shift the epidemiogram by the distance corresponding to the period of the infection's turnover. This will be discussed below.

In order to calculate the dates of a season of the mosquitoes' effective infectibility and the epidemiological capacity of this season, it is not necessary to calculate the whole seasonal trend of lifetime, sporogony, and so on, as was described above. To find the "counterpoint" quickly and to find out when a female mosquito has to develop wings in order to survive up to the end of sporogony, in other words, to find the boundaries of the season of the mosquitoes' effective infectibility, it is necessary to solve a system of three or even four equations.

1. An equation representing the relationship of the lifetime of a female that has developed wings to the temperature (this equation is represented in Fig. 1.51 by biological curve 3).

2. An equation of the length of sporogony in relation to temperatures (this equation is represented in Fig. 1.51 by phenological developmental curve 5).

3. A spring equation and autumn equation of heat resources, expressed by corresponding lines of average temperatures on the net of Figure 1.51. Moreover, the results of the calculations must be presented against a background of the diagram of the mosquitoes' abundancy (the diagram has also been calculated).

Neither the method of totals of effective (as well as active) temperatures, nor the method of totals of developmental rates are able to solve in essence a system of several equations; not to mention that each of these equations is empirically complicated and that a temperature total is not constant for the full length of temperature curves representing the duration of sporogony or lifetime. The method proposed here solves this complicated system in the following way: on the intersection of biological curves 3 and 5 (Fig. 1.51), the length of sporogony equals, exactly, the mosquito's lifetime. Through the mentioned intersection run two lines of average temperatures: the line from May 9 and the line from September 18, representing dates when females develop wings and when sporogonies begin; these dates cause the simultaneous end of both processes. But since a mosquito feeds not at the moment of wing formation, but 1 or 2 days later, then in order for the end of sporogony to coincide with its death, the mosquito must get wings not on May 9 and September 18, but on May 11 and September 16, when the duration of sporogony is somewhat shorter than the mosquito's lifetime. May 11 and September 16 will be, therefore, the beginning and the end of the season of the mosquitoes' effective infectibility (in addition, during this whole period there is a sufficient quantity of mosquitoes).

As is known, the method used in most cases in the USSR for determining the dates of the season of the mosquitoes' effective infectibility (a basis of the malaria epidemiology) proceeds, obviously, from the circumstances as follows: the duration of sporogony and the lifetime of a mosquito with wings are close to a certain degree if the sporogony and the life began or finished at a 24-hour average temperature of 16°C, taking into account, of course, the temperature rise after the beginning or before the end of the season (Moshkovsky et al., 1951; Shipitzina, 1957; Kalmikov, 1959). Hence according to the method mentioned, it is accepted that the season of the mosquitoes' effective infectibility in a given region begins on the date when 24-hour average air temperatures exceed 16°C, and finishes on the date when the last sporogony cycle *begins*; this last sporogony cycle must finish on the autumn date when

24-hour average air temperatures fall below the criterion of 16°C. Sometimes, the abundance of mosquitoes is taken into account. Individual attempts have also been made to take into account empirically the number of mosquitoes that have survived up to the end of the first sporogony cycle. If there are no mosquitoes, the beginning of the season of effective infectibility moves forward. That disturbs the principle based on the assumption that the season of the mosquitoes' effective infectibility begins on the date when 24-hour average temperatures exceed 16°C (E. S. Kalmikov, 1959). Mosquitoes may not exist in spring (if it is cold) or in summer (if reservoirs dry up); in autumn mosquitoes almost always exist. Therefore, taking into account the abundance of mosquitoes does not affect the determination of the date in autumn when the season of effective infectibility finishes: this determining, based on the criterion of 16°C, does not change.

As a result of the application of the Podolsky method, it becomes clear that if sporogony and life begin in spring or finish in autumn at 16°C (but proceed, therefore, at higher temperatures), the mosquito does not survive to the end of sporogony in actual fact, either at the beginning of the season or at the end. For example, in Yavan the mosquito's lifetime and sporogony become equal when these processes begin or finish at 19° to 19.5°C (taking into account, of course, higher temperatures between the beginning and the end), and last not 40 days (Shipitzina, 1957), but only 16 to 17 days: on the PTN for Yavan (Fig. 1.51) we can see that the intersection of biological curves 3 and 5 gives both the lifetime and duration of sporogony equal to 16 to 17 days (on the vertical axis). The average temperatures line running through this intersection, is close to May 11 (it is the date that will be, as we already know, the beginning of the season of the mosquitoes' effective infectibility). The line of May 11 begins on the horizontal axis at $t_{av} = 19.5°C$, which is the 24-hour average temperature on May 11. We obtain the same for the end of the season: September 16 + 16 days = October 2 where September 16 is the starting date of the second meteorological line running through the intersection of phenocurves 3 and 5. The line of average temperatures from October 2 rises from the horizontal axis at $t_{av} = 19.4°C$. Therefore, the last cycle of life and sporogony (which are equal) will also finish at a 24-hour average temperature of 19.4°C to 19.5°C (but, generally speaking, each region has its own temperature and duration of these periods).

Thus, according to the Podolsky method, the season of the mosquitoes' effective infectibility begins on May 11 and finishes on September 16, lasting 128 days.

But the calculation, made using the generally accepted method in the USSR with the criterion of 16°C, gives the beginning of the season of the mosquitoes' effective infectibility on April 16 for Yavan and the end of the season on September 23, with a duration of 160 days. These dates can be obtained from the formula by Sh. D. Moshkovsky (1951), which is well known to malariologists. But it is even more simple to use the PTN for Yavan (Fig. 1.51): from the point corresponding to 16°C on the horizontal axis rises the spring line of average temperatures of April 16—this is the beginning of the season. The last sporogony cycle in the season must begin so that it finishes at a 24-hour average temperature of +16°C. The autumn line rising from the mark 16°C on the horizontal axis is the line of October 15, as can be seen on the same PTN. Therefore, it is necessary, with the help of curve 5, to find the date of the beginning of the sporogony that finishes on October 15. This is done by means of the method of retrospective forecasting described in the chapter about phytophenology, and more simply, this can be done by two or three attempts to select a starting date. One of the attempts will show that the beginning of sporogony should be on September 23: the average temperature line from September 23 intersects phenocurve 5, representing the sporogony, at $n = 22$ days. Hence the end of sporogony will be September 23 + 22 = October 15.

Therefore, according to the old method, the season of the mosquitoes' effective infectibility finishes on September 23 (it should be remembered that the end of a season of the mosquitoes' effective infectibility is not the end of the last sporogony cycle—October 15, but its beginning—September 23); the duration of the season is September 23 – April 16 = 160 days.

Thus, the season of the mosquitoes' effective infectibility calculated with the old method is 32 days longer than it is in reality, as is shown by the Podolsky method (160 – 128). How much superfluous means were spent for protection against malaria? And at what cost to the environment (DDT was used against mosquitoes, in addition to being used against cotton pests)?

Moreover, for the regions on the terrestrial globe, where the smallest value of a 24-hour average temperature does not fall below 16°C and where there are enough mosquitoes all year round, the method, which is generally used for calculating the season of the mosquitoes' effective infectibility, is useless, since for such regions this method will always automatically give a season of effective infectibility equal to 365 days, an obvious exaggeration.

A calculation made with the Podolsky method according to outdoor temperatures of the air was described above. It is known that the temperature in houses (where pubescent malaria mosquitoes spend a considerable part of their lives) is approximately 1°C higher than the outdoor temperature in the warm half of a year on the 24-hour average. Then we have to imagine that the intersection of curves 3 and 5 occurs not at the average temperature lines from May 9 to 11 and September 16 to 18, but at the lines located to the left by 1°C according to the scale, namely, May 3 and September 23. This last result is quite clear: since it is warmer in houses, the season begins earlier and finishes later than the season calculated using outdoor temperatures.

Of course, in the same way, *forecasts of the season of the mosquitoes' effective infectibility* should be made for each concrete year that has its own temperature.

Last, if it is necessary to judge the real dates of a season *post factum*, then segments of several lines of actual average temperatures, which are necessary for calculations, must be drawn.

The epidemiogram is calculated in a simple way as follows: it is absolutely clear that the number of epidemiologically dangerous gonotophic cycles that correspond to the beginning of the season of the mosquitoes' effective infectibility in Yavan on May 11 is equal to zero (since the female's life and sporogony, which begin almost simultaneously at the beginning of the season of the effective infectibility, finish simultaneously). Females that develop wings in Yavan, on May 20, for example, outlive the sporogony by 1.5 days (in Fig. 1.51 at the place where the average temperature line of May 20 runs, curve 3 representing the lifetime lies 1.5 days above the sporogony curve 5). In order to find out how many epidemiologically dangerous gonotrophic cycles will be made during these 1.5 days, it is necessary to know the duration of one such cycle at the end of May/beginning of June. From biological curve 2 we see that at the beginning of June the duration of one gonotrophic cycle equals two days. More exactly, it can be done in the following way: the average temperature line from May 20 intersects the developmental curve 5 at $n = 13$ days; therefore, the sporogony finishes and dangerous gonotrophic cycles begin on May 20 + 13 days = June 2. The line from June 2 intersects curve 2, representing the duration of a gonotrophic cycle, at the ordinate $n = 2$ days. Therefore, during 1.5 days 75% of a gonotrophic cycle will occur [(1.5 ÷ 2)100]. A part of this 75% cycle will be the starting stage of the gonotrophic cycle, during which the infecting sting can occur. In the same way we find females that develop wings on May 31 and June 10 will have one epidemiologically dangerous cycle, females that develop wings on June 20 will have 1.3 dangerous cycles, on July 10, two cycles, and so on up to the end of the season of the effective infectibility—September 16. At the end of the season there will again be zero epidemiologically dangerous gonotrophic cycles. We put the obtained results, as before, onto the graph of the epidemiogram (Fig. 6.13).

In view of the question about correlations of the duration of sporogony with the mosquito's lifetime, there appears to be one more explanation of "anophelism without malaria," its heat interpretation: if in some region there are malaria mosquitoes, but they do not survive up to the end of sporogony, there will be no malaria. It is necessary to construct the curve of lifetime in relation to the temperature, and not for *A. superpictus* but for *A. maculipennis*. Then it can be seen at what temperatures this curve intersects the curve of the sporogony duration, and it can be judged in what climatic situation and in which latitudes those places are located which are notable for "anophelism without malaria." In the same way the northern boundary and the height boundary of malaria distribution can be indicated (although, in such an interpretation, the limits of the usual conception of this question are somewhat extended). The previous explanation of the phenomenon of "anophelism without malaria" consists, as is known, of the theory that liquidation of diseases in such places proceeds quicker than the distribution of the disease (R. Ross, 1911; Sh. D. Moshkovsky, 1950).

The possibility of calculating the number of gonotrophic cycles over a particular period was illustrated above. The calculated physiological ages of anopheles females (the physiological age is determined by the number of gonotrophic cycles that have already been made by a female) coincide well with the actual ages stated during dissections. Therefore, a partial mathematical addition to the old method (Mer—Polovodova—Detinova) for determining the physiological age of anopheles females appears feasible. Let us attempt to calculate the number of gonotrophic cycles made by the oldest females of *Anopheles superpictus* (e.g., those composing 2-3% of a population) in the conditions of Yavan on May 26. This date is close to the date of death of old females, which is why we pose the question of the number of gonotrophic cycles made by a female during her whole life. By means of a retrospective phenological calculation on the PTN with biological curve 3 (Fig. 1.51), we define the date of wing formation in females that die on May 26: it will be May 9 (May 9 + 17 days = May 26). On the basis of curve 2 giving the length of one gonotrophic cycle, we ascertain that during the period from May 9 to 26, the duration of the gonotrophic cycle changes little and equals 2.7 days on the average. Therefore, during 17 days of life, the female will have 6.3 gonotrophic cycles ($17 \div 2.7$).

It is possible to calculate the dates and the length of the season of malaria transmission, having available the above calculated dates of the mosquitoes' effective infectibility. As is known, the *season of malaria transmission* begins at the moment when the first real sporogony cycle finishes (the real cycle implies a cycle whose duration is equal to or less than the lifetime of the female mosquito, and during this cycle, of course, there must be mosquitoes). In other words, the season of malaria transmission begins at the moment when mosquitoes become infectious for healthy people for the first time in the year. The end of transmission is the end of the last real sporogony cycle in the year (the transmission season is shifted in time).

Therefore, in order to calculate the beginning of the season of malaria transmission in Yavan, it is first necessary to find in Figure 1.51 the ordinate of the intersection of the average temperature line from May 11 (that corresponds to the beginning of the mosquitoes' effective infectibility) with biological curve 5, representing the duration of sporogony of *Plasmodium vivax*. This ordinate is $n = 16$ days. Next, the obtained ordinate is to be added to May 11: May 11 + 16 days = May 27.

This will be the beginning of the season of malaria transmission. In the same manner we find the average temperature line from September 16 (that corresponds to the end of the season of the mosquitoes' effective infectibility). The intersection of this line with biological curve 5 gives $n_1 = 15$ days. Therefore, the end of the transmission season will be September 16 + 15 days = October 1. The length of the season of malaria transmission proves to be almost equal to the length of the season of the mosquitoes' effective infectibility—127 days.

If we compare the season of the mosquitoes' effective infectibility and the season of malaria transmission with the previously calculated dates of the mosquitoes' development (Figs. 6.10 and 6.11), it can be seen that in Yavan the season of effective infectibility is begun by mosquitoes of the first generation which make the second egg laying. They are also the first to transmit malaria. The season of effective infectibility is finished by females of the ninth generation; they are also the last that transmit malaria, if the temperature in the houses is close to the outdoor temperature and if mass transmission is meant.

The described calculation of the season of malaria transmission (communicability) is made on the basis of multiannual temperatures. If necessary, a similar calculation can be made on the basis of concrete temperatures of a given year (indoors or outdoors), and a forecast can be calculated, as well.

After all the above, it becomes clear that the method for calculating plasmodia's development can be used for some other hemosporidia—pathogens of serious blood parasitoses in cattle, horses, and sheep.

6.6e. Malaria Mosquito, Plasmodium, and Man. Seasonal Trend of the Parameter of Malaria Communicability as a Characteristic of the Epidemiological Capacity, Taking into Account the Numbers of the Carrier.

If to the duration of the parasite's incubation in the mosquito (i.e., to the duration of the sporogony) we add the duration of the incubation in man (15 days for *Plasmodium vivax* of Southern variety), we obtain the *turnover period of the infection.*

If we add the period of the parasite's incubation in man to the beginning and to the end of the season of malaria transmission, we obtain the dates of the beginning and the end of display of new malaria cases, that is, we obtain the boundaries of the *malaria season.* For example, in Yavan the malaria with short incubation has the dates of a malaria season as follows: May 27 + 15 days = June 11 and October 1 + 15 days = October 16, where May 27 and October 1 are the beginning and the end of the season of malaria transmission respectively, and June 11 and October 16 are the beginning and the end of the malaria season, respectively.

Now we have the basic data for calculating the parameter of malaria communicability.

It is known that the *parameter of communicability* α is a certain measure of the possibility ("facility") of the pathogen to be transmitted in given conditions (Moshkovsky, 1950). In our case (as regarding malaria) and in other similar cases, the seasonal trend of the parameter of communicability is also thought of in terms of epidemiological capacity of the season when taking into account the number of carriers.

Since the possibilities of transferring malaria, other things being equal, are directly proportional to the numbers of the carrier, then the seasonal trend of the parameter of malaria communicability, first of all, has to "trace" the seasonal trend of the carrier's numbers. The trend of numbers is represented in Figure 6.11. But it is necessary to shift the peaks of the numbers to the right by the distance corresponding to the turnover period of the infection, that is, by the period of the parasite's incubation in the mosquito and in man $(Q_a + Q_h)$.

Of course, when doing it, only those carriers are taken into account which are within the boundaries of the season of the mosquitoes' effective infectibility. And this season, being shifted by Q_a, gives the season of malaria transmission, that is, the length of the *contagious period i*.

The trend of the parameter of malaria transmission depends not only on the number (A) of female mosquitoes, and on Q_a, Q_h, and i, but also on a number of other components among which the most important are, apparently, the following: the mosquito's survival up to the date of the sporozoite's ripenings (S), the lifetime of the mosquito after the sporozoite's maturation in its body (V), and activity of females—frequency of gonotrophic cycles (U). The parameter α is proportional to these values. It is clear that the epidemiogram for an individual female gives a complex representation of the components V, S, and U (Fig. 6.13). When constructing the sesonal trend of the parameter of malaria communicability, all that remains is to shift to the right (by the period $Q_a + Q_h$) the peaks of the carrier's numbers taken from Figure 6.11; then it is necessary to double these peaks in the places where the quantity of epidemiologically dangerous blood suctions is doubled, in accordance with Figure 6.13. This must be done because the level of danger that the epidemiologically dangerous females pose doubles, and therefore the parameter of malaria communicability also doubles, if the females suck the blood of healthy people twice (when having sporozoites in salivary glands). For example, the heights of the communicability parameter will be equal to the heights of the carrier's numbers as long as female mosquitoes have no more than one epidemiologically dangerous cycle. As can be seen in Figure 6.13, females that develop wings from May 11 to June 16 and after August 23 have up to one dangerous cycle (to the end of life). With respect to epidemiology, the initial stage of the gonotrophic cycle that is important involves feeding on the body of a healthy man by a female mosquito that has ripe sporozoites. Therefore, 1.3 gonotrophic cycles as shown in Figure 6.13 will be considered as having two dangerous suctions. Then, when constructing the parameter, the heights representing the number of mosquitoes are doubled from June 16 to August 23: in Figure 6.13 see points M and T which limit this period; then see the vertical line segments designated by the same letters in Figure 6.14 on the coarse dotted curve representing the numbers of the carrier. The coarse dotted curve of the carrier's number in Figure 6.14 is copied from Figure 6.11 (line 1).

In Figure 6.14 point M' on the solid curve of the communicability parameter has the same height as the corresponding vertical line segment M on the curve giving the carrier's number (to the left of point M'); in addition, the parameter's height H equals the height of numbers H, and $\delta' = \delta$. But from line segment M to line segment T, the height of the drawing representing the numbers is doubled (when constructing the parameter of communicability), so that the parametric point T', for instance, has a height that exceeds twice the height of the corresponding vertical line segment T on the curve of numbers (just as the parameter's height $2H_2$ corresponds to the height of numbers H_2, the parameter's height $2H_1$ corresponds to the height of numbers H_1). After T (on the right of it) the size of the drawing representing the numbers becomes equal to the size of the parameter's drawing as it was before M (on the left of M),(e.g., the parameter's height H_3 equals the height of numbers' H_3, and Δ' corresponds to Δ at the end of the season of effective infectibility).

Therefore, the parameter of communicability is expressed by the units that are relative: for example, as a unit of the parameter, the height of the numbers' peak, which begins the season of the mosquitoes' effective infectibility, is taken (see the horizontal dotted line in Fig. 6.14).

Thus when constructing the parameter of malaria communicability, seven components of this parameter have been taken into account (A, Q_a, Q_h, i, V, S, U), *seven of ten mentioned by malariologists* (Moshkovsky, 1950).

Of course, when constructing the parameter, factual data about the carrier's numbers can also be used.

Let us transfer the calculated trend of the parameter of malaria communicability in Yavan from Figure 6.14 to Figure 6.15. While doing this, let us somewhat change the correlation (used in Fig. 6.14) of the vertical and horizontal scales. Then let us compare the calculated trend of the parameter for Yavan with the actual trend of the rate of malaria morbidity and of the carrying of parasites in neighboring regions.

As can be seen in Figure 6.15, the calculations are confirmed by the actual data. This can also be seen from the following description made by A. S. Monchadsky and A. A. Shtakelberg (1943) about the actual trend of the malaria morbidity rate in Tadzhikistan. They wrote that

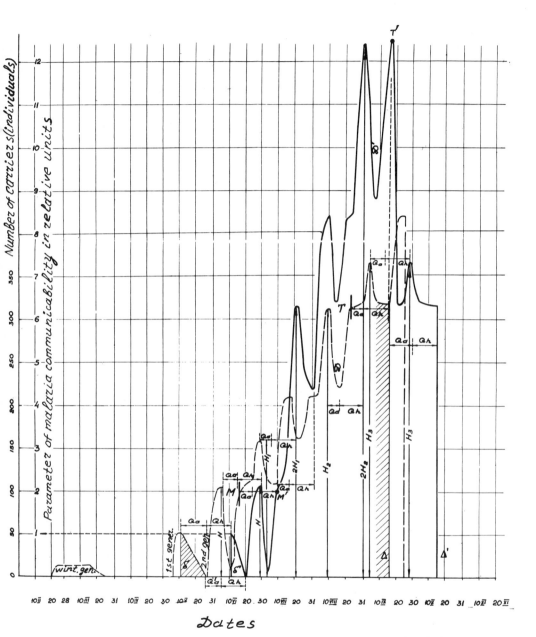

FIGURE 6.14. Constructing the seasonal trend of the parameter of malaria communicability in Yavan α (obtained as a complex value of the natural preconditions for the malaria epidemiology). Coarse dotted line: the trend of the number of carriers taken from Figure 6.11 or 6.10; solid thick line: trend of the communicability parameter. The matched areas are boundaries of the season of the mosquitoes' effective infectibility.

the curve, giving malaria cases in the regions where *Anopheles superpictus* is the primary or single carrier, is very characteristic: the curve is notable for a sharp rise in July and August, continuing up to September, and then a fall, sometimes accompanied by a small rise. Let us compare this small rise with the right shoulder of the parameter's curve in Figure 6.15. The shoulder was formed as a result of a continued increase in the carrier's numbers while the number of dangerous suctions decreased sharply.

There is another way to estimate the epidemiological capacity of the season: estimating with the use of graphs giving the seasonal trend of the number of epidemiologically dangerous females of malaria mosquitoes (T. S. Detinova). The advantages of the method described above of estimating using the epidemiogram and the seasonal trend in the communicability parameter, in comparison with the T. S. Detinova's method, are as follows.

1. The epidemiogram and the parameter can be constructed for regions where there is neither a given illness nor a carrier for the present. Therefore, they can reflect the potential epidemiological capacity.

For constructing the edipemiogram and parameters of communicability, curves are used that represent the duration of life, sporogony, gonotrophic cycle, and so on; these curves are a rather stable "passport" for an organism and can be constructed, for regular use, on the basis of observations in a controlled temperature chamber or in the field, not necessarily in the years and at the region for which the epidemiological capacity is estimated.

2. The epidemiological characteristic of a bioclimatic region must reflect the average multiannual situation. In order to derive such a characteristic with Detinova's method, it is necessary to have a multiannual series of observations in a given region for the structure of populations with respect to physiological ages. Such multiannual series are either not available at all or are available for rare regions only. While on the contrary, the calculating epidemiograms and parameters of communicability with the use of multiannual heat resources net (which can be constructed for any region where there is a meteorological station) immediately gives a multiannual characteristic.

It is clear that epidemiograms and communicability parameters can be constructed for each individual year (on the basis of actual temperatures) by means of moving the lines of the heat resources net by the distance corresponding to the value of the temperature anomaly, as was shown above repeatedly.

3. Using a long-range weather forecast, it is possible to construct a forecasting epidemiogram or communicability parameter for a forth-

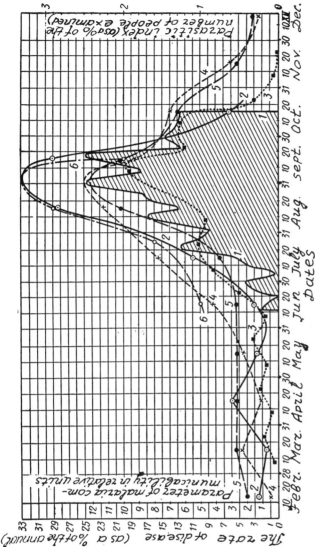

FIGURE 6.15. Seasonal trend of the potential endemic parameter of communicability of 3-day malaria with short incubation in Yavan (Tadzhikistan), on the basis of calculations, in comparison with actual data about the morbidity rate and parasite carrying. Curve 1: calculated parameter based on average multiannual data (parameter transferred from Fig. 6.14; the area under the curve of the parameter is hatched). Curves 2, 3, 4, and 5 are constructed by using actual data on the morbidity rate, respectively: for the Samarkand district in Uzbekistan (1942) (after V. M. Ramennikova); for the entire Tadzhikistan (1958–1962) (after M. P. Atrakhov); for the Vanch district in Tadzhikistan (1943) (after B. I. Isetov); for Dagestan (1933) (after Sh. D. Moshkovsky and others). Curve 6: parasitic index for the Rushan district in Tadzhikistan (1953) using actual data provided by B. I. Isetov.

417

coming season in a given region. The old method does not give such a possibility.

4. The graph giving the seasonal trend of the number of epidemiologically dangerous females has not been interpreted as an area integral either by Detinova or by others up to now (it was not interpreted as an area integral before this author's report at the Moscow Institute of Medical Parasitology in 1964 or before the first edition of this book in 1967). Without such an interpretation, the quantity of epidemiologically dangerous females on a given date is not indicative of the epidemiological capacity, if in one region after this date females suck blood once, and in another region they accomplish three dangerous suctions.

5. It is known that the number of epidemiologically dangerous females on a given date, according to Detinova, is determined by comparing the actual, ascertained (by means of the dissection of females) physiological ages in a population with the calculated number of gonotrophic cycles made by a female over one sporogony cycle. If a sporogony cycle is reckoned from a given date "forward," and the physiological age on this initial date is reckoned "backward," that is, concerning females that develop wings before the given date, then the calculated number of gonotrophic cycles over the sporogony period becomes uncomparable with the actual ascertained age. The Podolsky method does not have this shortcoming.

In a number of cases (especially when the number of the carrier is approximately equal in different years in different regions) the comparative epidemiological capacity of a region and of a year can be expressed by the epidemiogram (Fig. 6.13) without constructing the seasonal trend of the communicability parameter.

On the basis of calculations that concern mosquitoes with wings, the dates for imagocidal work in Yavan were calculated. Using phenological curves of larvae's development (see curves 6 and 7 in Fig. 1.51), the dates of larvicidal work are also calculated. On the basis of the calculations given above, it is also possible to plan medicinal-prophylactic arrangements. But all these data are not given here.

6.6f. Malariological Calculations for the Moscow Suburbs.

The majority of the malariological calculations given above were made for the Yavan Valley. We supplement them with calculations for the Moscow suburbs and compare the latter with actual observations. This will be especially interesting because all malariological calculations will be made this time for the geographical zone that differs sharply from Middle Asia.

We will describe the results of malariological calculations, carried out by listeners of the seminars on the application of the Padolsky method for phenological forecasting and bioclimatic estimations in the field of parasitology. Seminars and lectures were conducted by the author of this book at the Institute of Medical Parasitology and Tropical Medicine in Moscow in December 1963 and January 1964.

The calculations were made with the help of the PTN for the Moscow suburbs (Fig. 1.99); the heat resources net was constructed on the basis of multiannual observations at the Pavlovsky Posad meteorological station; the phenological developmental curve for *Anopheles maculipennis*, from egg to imago, was based on the actual data of N. K. Shipitzina; the biological curve of the duration of one gonotrophic cycle, — based on observations made by M. F. Shlenova and A. Ya. Storozheva with the addition of a small period of time for the search of food and for the dropping of eggs; the curve of the sporogony duration was constructed on the basis of data provided by V. P. Nikolayev. The technique for solving the problems described below was given above.

Problems of the First Type. Phenological forecasts with prolonged foresight. On the basis of the date of May 2, 1945, when wintering females *A. maculipennis* began to lay eggs in the Moscow suburbs, and on the basis of the phenological developmental curve 1 (Fig. 1.99), the beginning of the flight of the first generation was calculated as June 13. The actual beginning of the flight of the first generation in the Moscow suburbs in 1945 was observed by N. K. Shipitzina on June 10. The error of the forecast was 3 days. It should be said that the observations of 1945 were used neither for constructing the phenological curve nor for constructing the heat resources net (verification on the independent material).

On the basis of the actual date of June 30, 1950, when the third mass egg laying took place (made by the second generation), the beginning of the mass flight of the third generation was calculated as July 18. That coincides with the actual flight date. As can be seen, the errors of the forecasts do not increase from generation to generation, as sometimes happens when calculations are made using the temperature totals method.

Problems of the Second Type. Phenoclimatic forecast with shortened foresight, taking into account the actual temperatures over a part of the forecast period. The beginning of the first egg laying (made by wintering females) in 1948 was observed on May 1. We must determine the beginning of the mass flight of the first generation, if the actual temperatures over the period from May 1 to 10, 1948 are known. We calculate the actual average temperature over this period in the Moscow suburbs. It turns out to be equal to

14.7°C. We put point b with coordinates 14.7°C and 10 days onto the nomogram (Fig. 1.99) (point a is an acutal 24-hour average temperature on May 1, 1948). From point b we trace a dashed line parallel to the multiannual line of average temperatures dated May 1, and so on. As a result we obtain the date of the beginning of the first generation's mass flight: May 25. The actual date was May 29. The error of the forecast is +4 days. Because of the 6°C anomaly in 1948, the phenoclimatic forecast with prolonged foresight gives an error of −16 days.

Problems of the Third Type.   Phenosynoptical forecast, taking into account a long-term weather forecast. If the long-term weather forecast for May 1948 indicates the anomaly of +2°C, then in the preceding problem the error of the phenoforecast for the unusual year of 1948 proves to be −9 days.

Problems of the Fourth Type.   Phenoclimatic forecast with prolonged foresight, taking into account the temperature differences between the inhabitation places of the organism and the outdoor air. The calculation on the PTN for the Moscow suburbs shows that the sporogony of *Pl. vivax*, which begins on July 5, should finish on August 6 at outdoor temperatures, and on July 30 at cow-shed temperatures, if the latter are 1°C higher than the outdoor temperatures. Actually, the average temperature line dated July 5, intersects biological curve 4 at $n = 32$ days. The end of sporogony will be July 5 + 32 days = August 6. Moving the average temperature line dated July 5 to the right by 1°C, we see that the new line intersects biological curve 4 at $n = 25$ days, when the end of sporogony in a house will be July 5 + 25 days = July 30.

Problems of the Fifth Type.   Retrospective phenological calculation. On the basis of the date when an imago begins to fly, July 14, we find that the eggs from which these imagoes emerged were laid on June 26 (the line of average temperatures from June 26 intersects phenological curve 1 at $n = 18$ days, when the appearance of wings will occur: June 26 + 18 days = July 14). This was the phenomenon that N. K. Shipitzina observed in the Moscow suburbs in 1951.

Problems of the Sixth Type.   Bioclimatic. It is known that the mass flight of mosquitoes from wintering places coincides with the time when temperatures exceed +6° and +7°C. On the PTN shown in Figure 1.99, we see that in the conditions of the Moscow suburbs this criterion for mass flight is met approximately on April 24. Then the first egg laying, that begins the first generation, will be, according to phenocurve 3, on April 24 + 18 days = June 12. All remaining calculations are made on the basis of phenocurve 2. As a final result we obtain, according to the multiannual data in the Moscow suburbs, four generations of *A. maculipennis;* this coincides entirely with the facts given by N. K. Shipitzina. It is important that the last possible egg laying, according to these calculations, relates to the date of August 10. This is close to the date of mass diapause (see the next problem).

Problems of the Seventh Type.   Forecasting the dates of the mass diapause of *A. maculipennis*. Looking for the tangent to developmental curve 1 of *A. maculipennis* from egg to imago (the tangent *ef* is parallel to the average temperature lines on Fig. 1.99), we find that the mass diapause in the Moscow suburbs (southwest of the forest zone) begins on August 12 or 13. This is reconfirmed by the data of N. K. Shipitzina.

Problems of the Eighth Type.   Concerning the mosquitoes' gonotrophic activity.

On the PTN shown in Figure 1.99, it was calculated that, on the average, for multiannual indoor temperatures during one sporogony cycle that begins on July 5, there will be five gonotrophic cycles, and, therefore, at the sixth gonotrophic cycle females will become epidemiologically dangerous: we find the duration of the sporogony (25 days) in Figure 1.99 at the intersection of the average temperature line from July 5 (the line shifted 1°C to the right because indoor temperatures are 1°C warmer) with biological curve 4. Then at the intersection of the average temperature line from July 5 with biological curve 3, we find that the duration of the first gonotrophic cycle is 5 days. Therefore, the second cycle will begin on July 10 (July 5 + 5 days). At the intersection of the average temperature line from July 10 with curve 3, we find the duration of the second cycle, and so on, up to the moment when 25 days will be exhausted. But since the duration of the gonotrophic cycles changes little in this case, it is possible to calculate the number of gonotrophic cycles all at once: 25 days: 5 days = 5 gonotrophic cycles.

Problems of the Ninth Type.   The same calculation as above, but on the basis of actual temperatures over a given year.

The PTN shows that in 1954, for example, because of very high temperatures, females could become epidemiologically dangerous during the fourth gonotrophic cycle: in 1954 the segment *cd* of the actual average temperature line from July 5 intersects biological curve 4 at the duration of sporogony, $n = 11$ days, and at $t_{av} = 24.8°C$ (Fig. 1.99). At such a temperature the gonotrophic cycle lasts a little more than 3 days. Therefore, 11 days:3 days = 3.7 gonotrophic cycles. As can be seen, this can happen not only in Africa (Detinova) and in Middle Asia, but sometimes also in the Moscow suburbs.

Problems of the Tenth and Eleventh Type.   Concerning the calculation of the dates of the season of the mosquitoes' effective infectibility and the dates of the malaria transmission season, as well as concerning the construction of an epidemiogram. These problems are discussed in enough detail above.

Problems of the Twelfth Type.   The calculations of given developmental stage/temperature combinations in the development of organisms. For example, when must the egg laying of A. *maculipennis* occur, so that in the natural conditions of the Moscow suburbs the developmental period from egg to the appearance of wings proceeds at the average temperature of 12.2°C over this period? Is such a developmental stage/temperature combination possible in the conditions of the Moscow suburbs?

On the PTN for the Moscow suburbs, shown in Figure 1.99, phenological curve 1 for A. *maculipennis's* development from egg to the appearance of wings is intersected at $t_{av} = 12.2°C$ by two average temperature lines—from April 29 and August 12 (the line from August 12 is tangent to phenocurve 1). Therefore, the required developmental stage/temperature combination can occur, in the majority of years, only twice a year: at the time of egg laying on April 29 and on August 12 (the mass egg laying on August 12 is the last in the year; it verges on the mosquitoes' going into diapause, as is already known, see the seventh problem).

6.7. PHENOLOGICAL FORECASTING USED FOR THE CONTROL OF OTHER VECTOR-BORNE DISEASES: DERMAL LEISHMANIASIS AND PLAGUE

It is known that leishmaniases are a group of parasitical diseases caused by intracellular parasites in man and some animals. These parasites are protozoa from the *Leishmania* genus. We shall discuss only the dermal leishmaniasis—the tropical ulcer or Borovsky disease, primarily of the zoonous type.

The pathogen of the dermal leishmaniasis of the zoonous type is *Leishmania tropica major;* the carrier is the *Phlebotomus papatasii* mosquito.

G. E. Gozodova, Z. I. Martinova (1967, 1968), and A. V. Kondrashin (1967, 1970) used Podolsky's phenotemperature nomograms for calculating the quantity of generations of mosquitoes in the southern

Tadzhikistan and in the Pamirs. It turned out that in the regions where calculations showed one generation of mosquitoes there were almost no cases of dermal leishmaniasis. In the regions where calculations showed two to three generations, the dermal leishmaniasis was distributed widely; even in Kalay-Humb, with two generations of mosquitoes, 16.5% of the population was affected by the dermal leishmaniasis.

Thus the calculation of the quantity of generations of mosquitoes, on PTNs for various arid landscapes, made it possible to classify these landscapes according to the steadiness of the leishmaniasis epidemiological situation, and even to judge the morbidity rate of the population. The corresponding medical-geographical maps were composed.

It seems that with the use of PTNs, it would also be possible to calculate the seasonal trend in the rate of sickness caused by leishmaniasis in man and animals. That would be useful for prophylaxis and medical treatment.

Second (after the section about malaria), we make sure that the carrier and its numbers are the probable key factor of the intensity and distribution of vector-borne diseases.

Finally, as a reminder, mosquitoes are carriers not only of leishmaniases (dermal and visceral), but also of mosquito fever, of bartonellosis, and perhaps of viral neuroinfections.

. . .

It is known that plague is the most dangerous disease to man and animals. The pulmonary form of plague in man in recent times resulted in 100% death of ill people. People also fall ill with bubonic plague.

Natural breeding grounds of plague lie in a very wide belt surrounding the entire terrestrial globe, from 32° south latitude to 49°, and even to 51°, north latitude.

The pathogen of plague is the bacterium *Pasteurella pestis;* the carriers are fleas of different species from the order *Aphaniptera.*

For successful plague control, it is necessary to annihilate its carrier; to annihilate the carrier, it is necessary to know its phenology in different spatial breeding grounds. Here the new method of phenoforecasting proved to be useful.

Padolsky's method was applied in this field of medical parasitology for the first time by V. N. Kunitzky, V. M. Volkov, Z. F. Lelikova, and O. S. Agundina (1974). They used the new method for calculating the number of generations of *Xenopsylla skrjabini* fleas parasitizing on rodents *Rhombomys Opimus Licht* in the northern section of the Caspian lowland. It is interesting that the net of heat resources for these calculations (on the PTN) was based on the temperatures taken directly in burrow of rodents. Dates of the development of fleas, obtained on the PTN, proved to be close to the actual dates in most cases. For instance, in the vicinity of the town of Guryev in 1971, the calculated dates of the maximum emergence of fleas (which correspond to various dates of the maximum reproduction) were July 2 and 15, August 28, and September 21. The actual dates of the maximum emergence of fleas were July 2 and 28, August 26, and September 20, respectively. In one case, however, we see a considerable error.

Kunitzky along with the coauthors came to the conclusion, made on the basis of the PTN, that in the Caspian lowland *X. skrjabini* develops the first generation completely, the second generation incompletely, and third generation partially.

It is interesting that from this work one more illustration of the general phenological principle is obtained: fleas that have emerged from later eggs develop along with the fleas that came from earlier eggs (this principle, for different organisms, was mentioned in previous chapters, but Chapter 8 will be specifically devoted to this principle).

Concerning the reproduction rhythm of other species of fleas—*Xenopsylla gerbilli* — which also parasitize on the rodents *Rhombomys Opimus Licht*, but in Southern Balkhash Territory (Balkhash is a lake in Kazakhstan) two articles were published by N. T. Kunitzkaya, V. N. Kunitzky, D. M. Gauzshtein, N. M. Savelova, and I. V. Morozova (1974, 1977). In contrast to the work cited about *X. skrjabini*, the heat resources net is represented here by the ground temperatures recorded at the meteorological station at a depth of 60 and 40 cm, where *X. gerbilli* mainly become localized during the preimago developmental stages (Fig. 1.63). Constructing the nets on the basis of the biotope's temperatures is logical and advantageous when the controlled temperature chamber's phenocurves are used (see Section 2.4). The experimental temperature phenocurves are put onto the net of Figure 1.63 for two graduations of relative humidity in the soil air. On such a PTN it is possible to take into account several factors at once.

For instance, one of the *peaks* in egg layings by fleas was observed on May 6. When can we expect the corresponding *peak* in appearance of imagoes at a depth of 60 cm at a relative humidity of 91% of the soil air? The line of average temperatures from May 6 on the complete net (Fig. 1.63) intersects phenocurve 2 at $n = 54$ days. Then the imago's peak that interests us will be May 6 + 54 = June 29. In fact, it was observed on June 25, 1970 (Table 6.8). With a foresight period of almost two months, the error of the forecast is not big.

If the given relative humidity is 75%, it is necessary to look for the intersection of the meteorological line with phenocurve 1; then $n = 63$ days and the date for the imago is May 6 + 63 = July 8.

For a depth of 40 cm, where in spring, summer, and autumn it is warmer than at the depth of 60 cm (this can be seen when comparing the segments of the meteorological lines for 40 cm with the corresponding lines of the complete net for 60 cm), when the relative humidity is 75%, we have $n = 58$ days and May 6 + 58 = July 3.

We proceed now with the complex of calculations. Unfortunately, we have no data about the lifetime of *X. gerbilli* fleas which parasitize on rodents *Rhombomys opimus Licht*. However, if we assume that the biological curve, giving the lifetime of suslik and marmot fleas outside the host's body, is similar to the curve for.*X. gerbilli*, then *X. gerbilli* fleas that appeared on June 29 (see the first problem above) will be alive up to June 29 + 293 days = April 18. Here $n = 293$ is the ordinate of the intersection of the meteorological line from June 29 (the complete net shown in Fig. 1.63) with phenocurve 3. This phenocurve is constructed by the author of this book using the data of I. G. Ioff (1941). Thus the suslik fleas and marmot fleas can hibernate in burrows and even preserve the plague microbes in their bodies until the next season (it should be noted that rat fleas *Xonopsyla cheopis*, in the state of being infected, have a shorter life—by as much as 50 days).

It should be explained why the net in Figure 1.63 consists of smooth lines, although it is based not on the average multiannual temperatures, but on data accumulated over the year 1970 alone. The cause is the regular smoothness of changes in the soil temperatures, especially at a depth of 60 cm.

Thus we bring together in Table 6.8 the actual data and the results of calculations made by N. T. Kunitzkaya et al.

Phenoforecasts for 1970, as can be seen in Table 6.8, are good, but they must be regarded as phenosynoptically reliable forecasts (see Section 3.1e), since they are based on the temperatures of 1970, which were known in advance (the PTN is constructed only on the basis of

TABLE 6.8 Peaks (maxima) of egg layings and of the appearance of *Xenopsylla gerbilli* young fleas in Southern Balkhash Territory in 1970.

Actual Peaks of Egg Layings		Peaks of Imago Appearance			
			Actual Data		Errors of
Dates	Percentage of Females Laying Eggs	Dates Calculated on PTN	Dates	Percentage of Young Fleas	Pheno-forecasts (in days)
15 IV	30	27 VI	—	—	—
6 V	14	29 VI	25 VI	56	−4
7 VI	46	18 VII	19 VII	28	+1
15 VI	60	2 VIII	6 VIII	64	+4
15 VII	56	23 VIII	21 VIII	30	−2
27 VII	22	6 IX	7 IX	40	+1
12 VIII	11	22 IX	21 IX	47	−1
27 VIII	14	13 X	12 X	19	−1

Average error without distinction of plus or minus   /2/

these temperatures). If we had calculated on the multiannual net, the successfulness should have been less. However, B. M. Yakunin's work which is discussed below, is based only on the multiannual net of average temperatures, and nevertheless, the calculations proved to be almost of the same exactness.

The question about the quantity of the generations of fleas is very important. There was no united opinion about this because of the overlap of generations. For instance, *X. gerbilli* and *X. skrjabini* fleas were considered to produce from one generation (Kiryakova, 1970) to three to five generations (Darskaya, 1970; Zolotaryova, Afanasyeva, 1969). With the help of the PTN, Kunitzkaya found that the effective egg layings by *X. gerbilli* in Southern Balkhash Territory could be from March 26 to September 10, 1970. At this period *X. gerbilli* could develop one generation completely, the second generation incompletely, and the third generation partially. The third generation begins to appear on October 12 (see the last line in Table 6.8).

The next work was devoted to determining the quantity of generations of *Xenopsylla skrjabini* fleas in Mangyshlak, in Eastern Caspian Territory, a zone of transitional deserts (B. M. Yakunin, N. A. Chernova, and N. T. Kunitzkaya, 1979). In Figure 1.64 a PTN is given that was calculated by Yakunin et al. and was somewhat supplemented by this author.

Yakunin showed that the relative air humidity influences the species' developmental rate to a lesser degree than the temperature does. Apparently, the relative air humidity in soil is close, almost always, to the humidity that is optimal for fleas; even in the center of the Kara-Kum Desert, it does not decrease lower than 60%.

On the basis of actual dates of egg layings, the corresponding dates of the appearance of imagoes were calculated on the PTN by Yakunin et al. The exactness of such calculations proved to be 1 to 3 days. The quantity of generations according to the calculations is four. The second, third, and fourth generations parasitize at the end of summer and at the beginning of autumn. The third and fourth generations hibernate in torpor. In spring of the next year they propagate and in June they give the beginning of the first summer generation. All this was shown by Yakunin with the help of the PTN.

The cause of the high number of fleas in autumn was explained by Yakunin as follows: three generations overlap each other and the generation's lifetime in the imago stage increases. This explanation is similar to this author's conclusions regarding mosquitoes (see the previous section).

All in all, Yakunin et al. came to the conclusion that the Podolsky method for phenological forecasting proves to be quite correct (according to Yakunin's expression) and *can be recommended for epizootological inspection and for the study of the natural breeding grounds of plague.*

One other work, of the works known to this author, belongs to M. A. Samurov (1977). It concerns forecasting the number of *Xenopsylla conformis* fleas in the Volga/Urals sand territory. Samurov has only begun to use the new method of forecasting, but he considers this method to be a good aid in the practical appraisal of the state of the population of fleas.

In conclusion, it should be remembered that fleas transfer not only microbes of plague, but also pathogens of rat spotted fever, melioidosis, and other serious infectious human diseases.

6.8. THE POSSIBILITIES OF THE APPLICATION OF THE BIOMATHEMATICAL METHOD IN CALCULATING THE DEVELOPMENTAL DATES OF VIRUSES AND CONSERVING THEIR ACTIVITY (YELLOW FEVER AND FOOT-AND-MOUTH DISEASE)

As a result of the fact that some branches of science are still isolated, very important comparisons sometimes escape the investigators' field of vision. Let us compare curves 2 and 3 in Figure 1.38. In spite of a great difference between the organisms of pathogens (the malaria plasmodium and yellow fever virus), their temperature developmental curves coincide along a great distance. This can be accounted for by the fact that for both pathogens there is a single carrier—the mosquito; the carrier forms such a close biological connection with the pathogens that it (the carrier) requires from the different pathogens similar periods of development in the carrier's body.

For example, filter-passer of dengue fever and of the yellow fever is transferred by the *Aëdes aegypti* mosquito. It is not mere chance that these pathogens are also close to each other in the relationship between their development and temperature. For instance, at the temperature of +24°C a female mosquito becomes infectious 8 to 11 days after sucking the blood of a man ill with dengue fever; this is almost the same amount of days as the yellow fever incubation lasts in the same mosquito at the same temperature.

Such an approach (graphic or nomographic) to the joint study of different organisms will help to explain a number of interesting and important phenomena. For example, why do some infections advance in a parallel fashion in their epidemiological and epizootological display? Can this phenomenon be explained by the fact that they have similar carriers?

If we transfer the curve giving the development of the yellow fever virus from Figure 1.38 onto the heat resources net of a particular region, for example, onto the PTN for Yavan (Fig. 1.51, developmental curve 11), it is possible to do various calculations of the developmental dates of this virus. Let us assume that an *Aëdes* mosquito has sucked the blood of a man ill with yellow fever on March 26. When will this mosquito become dangerous for healthy people in the climatic conditions of Yavan?

We find in Figure 1.51 the average temperature line from March 26, and we follow it up to the intersection with curve 11, giving the development (in the mosquito's body) of the yellow fever virus, and find the ordinate of the intersection to be $n = 50$ days. Then the virus will be "ripe" in the mosquito on May 15 (March 26 + 50 days), that is, only 1 day later than the plasmodium in the same conditions.

The calculation of a phenoforecast on the basis of actual temperatures over a given year, or on the basis of temperatures from a long-range weather forecast, was given above repeatedly for other organisms.

There is no yellow fever in the USSR, but there are, as in other regions of the world, its carriers-*Aëdes* mosquitoes. As is known, mortality from yellow fever reaches 90% and above.

Thus some calculations, which concern the development of viral diseases (in animals and plants) and their transmission, can be made in a similar way as was described for malaria, since the proposed method makes it possible to do calculations not only for the pathogen, but also for its carrier. Moreover, some calculations made for malaria plasmodium can be transferred to the field of forecasting yellow fever (maybe for dengue fever also) without any essential changes in quantities, since the developmental curves for the *Plasmodium vivax* and for the yellow fever virus almost coincide.

The calculation of dates of the plasmodium's development was considerably complicated by the limited lifetime of the carriers after wing formation. Since the carriers of yellow fever (and of dengue fever) live as long as three to five months, then the probability that the virus will develop up to its infectious stage in the carrier's body is not limited by the mosquito's lifetime.

. . .

We focus in more detail on some peculiarities of the foot-and-mouth disease virus and on forecasting the dates of inactivation of the virus in the environment. This work was done by A. A. Slepov and the author of this book in Kursk, in CBEZ (1976).

Foot-and-mouth disease is dangerous, very infectious, and widely spread in the world. This disease affects cattle and some other animals; people can also fall ill with this disease. The longer the activity of the virus is preserved in the environment, the more intensive is the distribution of the infection from the breeding grounds. That is why there is an interest in this area.

The relationship of foot-and-mouth disease virus to heat from the standpoint of the inactivation of the virus on different substrates was studied by Slepov in 1974. The general principle of the field experiment was suggested by the author of this book: similar to the cases where crops were sown at different times of the year, and where oncospheres of the beef tapeworm, as well as eggs of fasciolae and molluscs, were put into the environment at different times of the year (see the appropriate sections in this book), objects that were contaminated by the foot-and-mouth disease virus were put into the environment each month for an entire year, and in the summer, twice a month. This made it possible to subject the pathogen to the effect of different temperatures in natural conditions: in winter, to low temperatures, and in summer, to high temperatures.

The research was carried out on the A_{22}-type foot-and-mouth disease virus extracted from the air tests in the breeding ground of the infection. The virus was reproduced on the primary culture of kidney tissue of newborn rabbits. The virus suspension was put onto small pieces of wood, paper, cotton cloth, and small bundles of hay and straw. The objects contaminated by the virus were located on earth samples in Petri dishes. The latter were placed on a layer of earth in a metal box. The box was put into the ground so that the objects contaminated by the virus would be on the level of the upper soil layer at the temperature and humidity peculiar to the soil. The entire system was placed in the shade.

When doing the experiments, the strict arrangements were made to guarantee the absolute isolation of the virus-containing material (the territory was isolated with a metal net on each side and from above; the entrance to the biotopes was closed and sealed up).

In all, 14 experiments were made. The virus indication was carried out each week, and during some periods every two days. For this purpose each sample (substratum contaminated by the virus), in separate test tubes, was inundated by 2.0 ml of nutrient medium. The nutrient solution was filtered 30 minutes later; the filtrate, in 0.2 ml portions, was inoculated in the cell culture of rabbit kidney in test tubes. The test tubes were incubated for 5 days at a temperature of 37°C, being examined every day for the availability of a cytopathic effect. In the case of a negative result, the procedure was repeated as many as six times, to ensure that the virus was inactivated.

The entire period from putting the virus into the environment up to the day of its inactivation, for the initial date of each experiment and for each substratum, was related to the air temperatures. In order to do this, the usual 24-hour average temperatures of the local meteorological station, over all the days of the mentioned period of the given experiment, were totaled (the first day of the period is included by the account, but the last is not). The obtained total, Σt, is divided by the number of days (i.e., by the number of constituents), and the average temperature over the period, t_{av}, is obtained:

$$t_{av} = \frac{\Sigma t}{n}$$

The points with coordinates t_{av} and n were put onto Figure 3.12 (the black points are for the virus on the straw substratum) and onto Figures 6.16 to 6.18 (for other substrata). As a result, the aggregate of points for each substratum is obtained. Then a regression line is drawn. It can be calculated on the basis of all the laws of mathematical statistics, but we can use the method of "geometric alignment of points": a smooth curve is drawn by hand so that each point or group of points above the curve, as far as possible, will find its "counterbalance" below the curve; the entire number of points above the curve should be approximately equal to the number of points below the curve, without counting those lying on the curve. Such a curve is phenological or biological. In short, we proceed in the same way as was described at the beginning of this book for other organisms.

When analyzing phenocurve 1 in Figure 3.12 or in Figures 6.16 to 6.18, it can be seen that the virus "lives" longer with a temperature decrease. Apparently, this is not only the preservation of the virus in anabiosis; in any event, on the phenocurves no break or leap is noticeable at the places of transition from positive temperatures to negative. What we are probably seeing here is a particular display of the general principle for many organisms of different organizational level. In certain cases, their lifetime is longer, the lower is the temperature of the environment. For instance, in Figure 3.12 we see that foot-and-mouth disease viruses, oncospheres of beef tapeworm, fleas, and malaria mosquitoes live longer at low temperatures.

We see in Figure 3.12 that each organism has its own curve. From the previous chapter it is known that the phenological curve is rather stable for a given organism and characterizes its genotype, sometimes

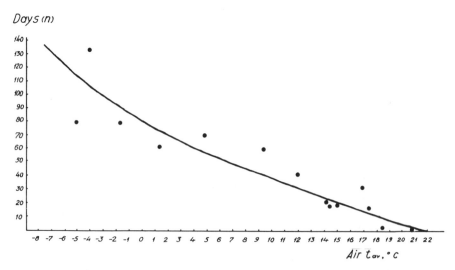

FIGURE 6.16. Phenocurve giving the duration of activity (lifetime) of the A_{22}-type virus of foot-and-mouth disease in the environment: in shade, on paper substrate.

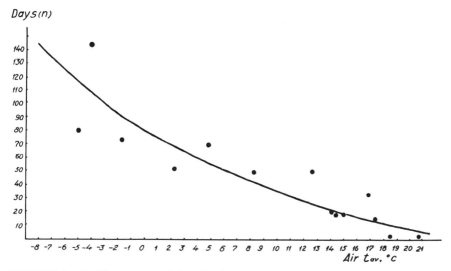

FIGURE 6.17. Phenocurve giving the duration of activity (lifetime) of the A$_{22}$-type virus of foot-and-mouth disease in the environment: in shade, on cotton cloth.

with phenotypical features (note that both phenolines for beef tapeworm-N2 and N3-would coincide if they were based on the temperatures of biotopes, and not on air temperatures; see below).

Finally, in Figure 1.93 we ascertain that the kind of substrate (object), the surface of which is contaminated with the virus, is not of great importance for the survival of the virus (however, the virus lives

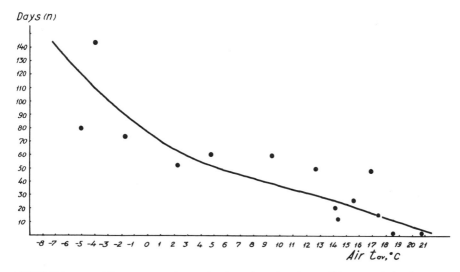

FIGURE 6.18. Phenocurve giving the duration of activity (lifetime) of the A$_{22}$-type virus of foot-and-mouth disease in the environment: in shade, on wood.

somewhat longer on wood than on straw in a considerable part of the temperature range). Apparently, the heat properties of a substrate (its heat conductivity and heat capacity) are important when the virus gets into the interior.

In order to determine the lifetime (activity duration) of the virus in the environment, it is necessary to compare its reaction to heat with the region's heat resources. That is, as we already know well, it is necessary to superimpose the biological phenocurves onto the meteorological net of one or another geographical region. This is done in Figures 1.93, 1.106, and 1.107.

Let us assume that the foot-and-mouth disease breeding ground is activated in the Kursk district in a concrete year on October 15. The question is, what is the length of the period during which the A_{22} virus in the breeding ground, and in the surrounding territory, will be infectious and, therefore, will threaten to infect animals? Let the virus-containing material be straw litter which is located in shadow.

The solution is as follows: since the process began on October 15, we are interested, first of all, in the heat resources of the Kursk district beginning with October 15. Therefore, on the PTN for Kursk (Fig. 1.93) we find the average temperature line from the initial date of October 15. Then we look for the intersection of this meteorological line with phenological curve 1 for straw. From the intersection we move the pencil horizontally to the left up to the vertical axis of the graph, where we find $n = 90$ days. That will be the period that interests us.

If wood is contaminated by the virus, then we look for the intersection of the average temperature line dated October 15 with phenocurve 2. This intersection corresponds to $n = 105$ days.

As can be seen, when forecasting the method makes it possible to take into account not only the heat factor, but also the nontemperature factors: in the given case, properties of the contaminated objects. If temperature phenocurves 1 and 2 concerned not objects, but, for instance, different graduations of air humidity or of day length, then in a similar way temperature and humidity would be taken into account, or temperature and duration of daylight, or all three factors. Such curves are difficult to obtain in the field conditions, but they are not difficult in polythermohygrostats or in chambers of artificial climate.

If the key factor proved to be not the temperature, but the relative air humidity or solar radiation (although the temperature is, as a matter of fact, an indicator of the solar radiation stress), then it would be possible to construct a net of average humidity over any periods or of average radiation. However, onto such a net the phenological curve must be put which binds the period of virus preservation with the humidity or with the solar radiation.

In spite of the fact that the heat resources net is based on multiannual data, the phenoforecast is fixed for a given year and farm by a concrete initial date. For example, if in another year or at another farm the breeding ground was activated, say, on June 20 (but not on October 15, as was assumed in the previous problem), then for straw and wood we have 11 and 15 days, respectively (not 90 and 105 days as in the previous problem). The line of average temperatures from June 20 is interpolated on the net between the lines dated June 15 and 25.

Such phenoforecasts are called forecasts with prolonged foresight. Average errors of forecasts with prolonged foresight seldom exceed the limits of 10 to 15% out of the foresight period. Such a satisfactory exactness accounts for the importance of the initial date of the forecast and for the biological "inertia" of an organism's development or of the vital activity.

Nevertheless, the result can be made more accurate by a forecast with shortened foresight, which takes into account the actual temperatures over a part of the forecasted period. For example, in the first problem, the natural inactivation of the foot-and-mouth disease virus on straw, according to the phenoforecast with prolonged foresight, should occur 90 days after October 15, that is, on January 13; this is a preliminary forecast. A more accurate forecast can be given, say, on November 25, provided that the actual 24-hour average air temperatures over the period from October 15 to November 25, that is, 41 days, are acquired from the local meteorological station. These actual temperatures are then totaled (taking into consideration positive and negative temperatures) and divided by the number of days. As a result we obtain for t_{av}, for instance, $-0.1°$. Point m, with coordinates $t_{av} = -0.1°$ and $n = 41$ days, is put into Figure 1.93. If in that year the temperatures were similar to the average multiannual temperatures, then point m should fall obligatory on the multiannual line of average temperatures dated October 15. But instead it falls to the left of this line, indicating that the actual temperatures in a given year are lower than the multiannual norm (by almost $1.5°$). From point m we trace a dashed line parallel to the multiannual line from October 15. This dashed line intersects phenocurve 1 at the ordinate $n = 97$ days, but not 90 days as it was before; therefore, the inactivation moment should be expected 7 days later, that is, on January 20. Here not only is the biological "inertia" used, but also the meteorological inertia (see Sections 3.1a and 3.1d for plants).*

Thus, the phenoforecasting method does not need the long-term weather forecast, which, unfortunately, is not always successful. And this is one of the advantages of the new method of phenological forecasts and bioclimatic estimations in comparison with the traditional methods (although for the foot-and-mouth disease virus there are no traditional methods).

*Although the narration is told in such a way that a specialist, having insufficient time, could find the information, necessary for phenoforecasting, in the section concerning his speciality in this book, it is recommended that one not "lock" himself into his particular field of knowledge. For results of such "locking," see the beginning of Section 6.8.

As to the long-term temperature forecast, it can be easily taken into account, in principle, by a phenosynoptical forecast. For example, in the first problem with the initial date of October 15, it is known that at the end of autumn and at the beginning of winter the temperatures are expected to be higher than the norm by 1°. Then the multiannual line of average temperatures dated October 15 is simply transferred horizontally to the right by 1°, according to the scale. In this case n will be equal to 82 days.

The heat resources net can work in a radius of 100 to 200 km (and sometimes more) for the flat country; it is only necessary to move the appropriate lines of average temperatures by the value of the spatial difference between air temperatures of the "base" point, for which the net is constructed, and the point for which this "base" net is applied. For this purpose, we act in the same way as was done for the phenosynoptical forecast.

More distant regions need their own nets. But the same phenocurves can be superimposed onto various nets in more or less similar climatic regions, since phenocurves are rather stable when characterizing some of the "passport" data of an organism. This has been shown repeatedly in this book.

Finally, concerning the accounting of the microclimate of biotopes on a PTN, no special consideration of the inhabitation medium is needed, if both the heat resources net and the phenocurve are based on air temperatures. When using air temperatures at the standard height of 2 m (Stevenson screen), the heat reactions of an organism rise or fall exactly as the biotope's heat resources rise or fall. The microclimate is taken into account here indirectly and automatically. For example, in Figure 3.12 two phenological lines for the beef tapeworm can be seen. They show that at an average air temperature, for instance, of 15°C, oncospheres will live 76 days on sunlit soil cleaned from verdure, and 129 days in grass in the shade. That is understandable because the temperature on a sunlit bare surface is higher than in the shaded grass, though the air temperature may be the same. It is clear that these two phenolines, being transferred onto a heat resources net, will give different phenoforecasts for various biotopes.

Any special consideration of the biotope's microclimate is also not needed, if the phenocurve is based on temperatures of the biotope, and the net is based on temperatures of the atmosphere, however, 24-hour average temperatures of the biotope are equal (isothermal) to the corresponding temperatures of the atmosphere. In this case even a thermostat's curves can be superimposed onto the net. A more detailed discussion can be found in Sections 2.4 and 6.3 of this book.

The above work has a hopeful continuation: if the same curve is allowed to be superimposed onto heat resources nets for various geographical regions, then having two or three such curves available for large naturally-climatic zones (in order to take into account the phenotypical "imprints" on the genotype) makes it possible to obtain, on PTNs for various regions, the dates of inactivation of the virus in one or another season. These dates are then put onto a geographical map, and isophenes of identical inactivation dates are drawn for each season. This is done in the same way as was described for the piorcer, the Colorado potato beetle, and the fall webworm.

On the basis of a PTN it is possible to construct phenoforecasting calendars, which are simple to use, for practical workers. This calendar can be constructed for various regions, in the way discussed above. In the initial column of such a calendar an actual date, when the foot-and-mouth disease breeding ground is activated, is found; in one of the neighboring columns (in relation to the actual temperature anomaly over a part of the forecast period), the corresponding date of inactivation of the virus is found.

CHAPTER SEVEN

Application of the Phenological Method in Microbiology. Forecasting the Duration of the Bactericidal Phase of Milk. Forecasting the Duration of Butter Formation in Cream.

This preliminary investigation was made by the author of this book together with V. G. Gizatulin in 1967; it was continued by the author of this book in 1971.

It is known that as a result of the activity of lactic-acid bacteria and kefir bacteria the fermentation of milk occurs. In this process the milk sugar molecules turn into lactic-acid molecules.

But freshly obtained milk has bactericidal properties that protect milk from fermentation for some period of time. When bacteria propagate in sufficient quantity, the bactericidal phase finishes and intensive fermentation begins.

Since the propagation of bacteria depends on the initial bacterial pollution of milk and on its temperature, then the duration of the bactericidal phase of milk depends on these conditions (of course, at a certain constant level of the milk's bactericidal properties, at a certain initial pH level of acidity, etc.).

The dependence of the duration of the milk bactericidal phase on temperatures, at different degrees of cleanness of milk, was studied, for instance, by A. F. Voitkevich and R. B. Davidov. Their data are represented graphically by the author of this book in Figure 7.1 (lines I, II, and III).

Apparently, these relations were studied in laboratory conditions at the temperature constant during each experiment. However, in natural conditions the temperature changes sharply during a 24-hour period, and therefore the duration of the milk bactericidal phase, in addition, will depend on the hour of milking and on the 24-hour temperature trend.

It is possible to solve this latter problem by the mathematical method of phenoforecasts; but the net of heat resources must be based not on the annual, but on the 24-hour temperature trend of the air.

In climatic reference books there are average multiannual air temperatures, by the hour, for each individual month (these multiannual hourly temperatures are the results of a systematic treatment of thermograph tapes). Using one such reference book, we constructed nets of average air temperatures (average over periods measurable in hours) in a 24-hour cycle for various months in the town of Grodno in Byelorussia (Fig. 7.1).

Such 24-hour nets are constructed in the same way as the annual nets, but instead of dates (10-day periods), hours of a 24-hour period are used.

For example, the average multiannual air temperature in July at 1 o'clock is equal to 15.1°C; at 2 o'clock to 14.6°; at 3 o'clock to 14.2°C, and so on. Then the line of average temperatures from the initial hour (1 o'clock) is calculated as follows:

$$t_{av,1\text{-}1} = 15.1°C$$

(in Fig. 7.1 see point a with coordinates $t_{av} = 15.1°$, $n = 0$ hours); $t_{av,1\text{-}2} = (15.1°C + 14.6°C)/2 = 14.8°C$ (in Fig. 7.1 see point b with coordinates $t_{av} = 14.8°C$, $n = 1$ hour); $t_{av,1\text{-}3} = (15.1°C + 14.6°C + 14.2°C)/3 = 14.6°C$ (point c with coordinates $t_{av} = 14.6°C$, $n = 2$ hours), and so on.

In the same way we act for the initial hour of 2 o'clock; $t_{av,2\text{-}2} = 14.6°C$; $t_{av,2\text{-}3} = (14.6°C + 14.2°C)/2 = 14.4°C$, and so on.

As a result, a net of heat resources appears (Fig. 7.1).

It is clear that the average multiannual 24-hour temperature will be equal over the period from 1 o'clock to 24 o'clock (i.e., at $n = 23$ hours), or from 2 o'clock of one date to 1 o'clock of the next date, or from 3 o'clock to 2 o'clock, and so on. Therefore, all the lines of average temperatures intersect each other at one point d at $n = 23$ hours. But the average temperature over the period from 1 o'clock of one date to 10 o'clock of the next date will differ from the average temperature over the period from 2 o'clock to 11 o'clock, or from 3 o'clock to 12 o'clock, since night temperature is not equivalent to day temperature. Therefore, at $n > 23$ hours, the average temperature lines diverge again (Fig. 7.1).

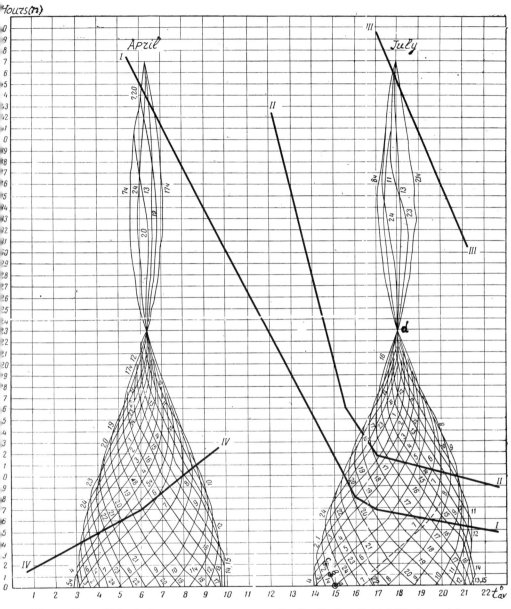

Figure 7.1. Biological-temperature nomogram for the Grodno district (Byelorussia) with biological curves. Curve I: duration of the bactericidal phase of milk obtained in regular conditions; curve II: the same as curve I, but for aseptic milk; curve III: the same as curve I, but for pasteurized milk; curve IV: duration of butter formation in cream.

Since the biological lines giving the duration of the milk bactericidal phase are constructed with the same coordinates (t_{av}, n hours) as the meteorological lines of the heat resources net, both kinds of lines can be combined on a single biological-temperature nomogram, as is done in Figure 7.1.

Let us assume that milking was done in July at 5 o'clock in the morning in the Grodno district. What will the bactericidal stage of ordinary milk be if its initial temperature is equal to the air temperature, and, later on, if the temperature of the milk changes as the air temperature changes? In order to answer this question, we proceed as before. We find on the nomogram for July the average temperature line designated by the initial forecasting moment, that is, by 5 o'clock; then we trace it up to the intersection with biological curve I; from the intersection we move the pencil horizontally to the left—up to the vertical axis where we read $n_I = 6.5$ hours. This is the duration of the bactericidal stage of milk which was obtained in regular conditions.

If we trace the same average temperature line up to the intersection with biological line II, we obtain the duration of the bactericidal stage of aseptic (cleaner) milk: $n_{II} = 11$ hours.

Finally, the intersection of the line designated by 5 o'clock with biological curve III indicates the duration of the bactericidal stage for pasteurized milk: $n_{III} \approx 46$ hours.

At a 10-o'clock milking, when the temperature of the environment is higher, the duration of the bactericidal stage decreases in comparison with the 5-o'clock milking: $n_I = 5.5$ hours; $n_{II} = 10$ hours; $n_{III} = 45.5$ hours.

If the air temperature on a given day of a given year turns out to be higher than the multiannual norm for this day, for instance, by 2°C, then for the 5-o'clock milking, we trace up to the intersection with curve I not the multiannual average temperature line from 5 o'clock, but the parallel dotted line located 2°C further to the right of the multiannual line. Then $n_I = 6$ hours. As can be seen, the bactericidal stage has become somewhat shorter in comparison with the norm.

In order to judge in the morning whether the coming day or part of it will be warmer or colder than the norm and by how many degrees, the factual morning air temperature should be measured (at a generally accepted height of 2 m) and compared with the multiannual temperature norm. The norm is read at the point from which one or another average temperature line rises from the horizontal axis of the Figure 7.1. For instance, the line marked by 5 o'clock on the July nomogram begins at 14°C; this will be the norm. If the actual temperature is 16°C, then the day begins by exceeding the norm by 2°C; this difference, we can assume, will be approximately the same during the whole day (in most cases).

In order to determine the temperature anomaly, the weather forecast can be also compared with the norm.

The same nomograms, as for July, must be constructed for other months, too. For example, in Figure 7.1, as well as the nomogram for July, we see a nomogram for April. From this April nomogram it follows that with milking at 5 o'clock, ordinary milk is preserved for 44 hours, but in July ordinary milk lasts only 6.5 hours, because temperatures in Grodno in April are on the level of 6°C. Not only with the 5-o'clock milking, but with any other time of milking, the duration of the bactericidal stage of ordinary milk in April reches 43 to 45 hours, as can be seen from the intersection of biological line I with the group of temperature curves at the top of the April nomogram.

All of this seems to be logical. Only two circumstances can give rise to doubt: first, we began all calculations with the suppositions that the initial temperature of milk was equal to the air temperature, and second, that the milk temperature changes similarly to the air temperature. The second circumstance is confirmed, provided that the milk is kept in relatively small containers and outside refrigerators (e.g., when being transported in ordinary cars or in special conduits which come down from distant mountain pastures).

In the first circumstance, the duration of the bactericidal stage calculated on the nomogram could be prolonged by 1 to 2 hours if the milk underwent a single artificial cooling after milking (especially in southern regions in summer). Or the calculated bactericidal stage could be shorted by the same 1 to 2 hours, if the temperature of the fresh milk (without cooling) differs considerably from the air temperature (especially in northern regions in the cold part of a year).

Such biological-temperature nomograms can be constructed for various months and climatic regions.* On these nomograms we can take various hours of milking, various temperature anomalies (by means of comparing air temperatures during milking with the temperature norm for each hour of milking, known from the nomograms; see above), and different levels of milk cleanliness. Then it is possible to compile calendars as in Table 7.1.

It would be rather simple to read such calendars, and they would be useful for workers on livestock farms, in distant pastures, and for others.

In calendars, as in nomograms, the mean solar time is given. The way to convert the time used by the population into the mean solar time can be found in the local meteorological station, and then one can

---

*For some regions it would be necessary to obtain in a laboratory biological curves for the region's own strains of lactic-acid bacterium, since there is information that northern strains have a lower optimum temperature of development.

TABLE 7.1. Fragment of the calendar giving the duration (in hours) of the bactericidal phase of milk for the Grodno district in Byelorussia.

Month	Time of Milking (hours)	Air Temperature during Milking (in °C)	Ordinary Milk	Aseptic Milk	Pasteurized Milk
		9	14.5	41.0	
		11	9.5	14.0	~57
		13	7.0	11.5	50
	5	14 (norm)	6.5	11.0	46
		15	6.0	10.5	43
		17	5.5	9.5	35
		19	5.0	8.5	~19
July					
		11	11.5	41.0	
		13	7.5	12.0	~57
		15	6.5	10.5	50
	7	16 (norm)	6.0	10.0	46
		17	5.5	9.5	43
		19	5.0	8.5	34
		21	4.5	7.5	~13

replace the mean solar hours of milking in the calendars by usual hours.

Forecasting the duration of butter formation in cream is also possible, as we will now show.

Of course, there will be an inverse relation here: the lower the temperature, the quicker the formation of butter will occur. A line of such a relationship (IV), obtained, obviously, from laboratory data, is given in Figure 7.1.

If the transformation of cream to butter proceeds in natural conditions (probably expedient only at the relatively cold parts of the year), then the forecast is given very simply. Let us assume that cream began to transform to butter at 5 o'clock in April in Grodno. Then we trace the average temperature line marked by 5 o'clock on the April nomogram (Fig. 7.1) up to the intersection with line IV. This intersection corresponds to $n = 6$ hours. Therefore, the cream transforms to butter in 6 hours.

It should be noted that the temperature totals method is unable to work with such a line as line IV (inverse relation to the temperatures).

Of course, many things in this chapter concern distant pastures in developing countries, especially if these pastures have no suitable equipment (e.g., refrigerators).

CHAPTER EIGHT

The General Phenological Principle

In the previous chapters the author noted that crops sown later begin their development in the first half of summer at higher temperatures, and therefore develop with greater speed than crops sown earlier, thus gradually approaching the dates of developmental stages in early crops. However, in the second half of summer and in autumn, the developmental stages of crops sown late coincide with lower temperatures, and in comparison to the same stages in crops sown early, proceed slower than the latter; as a result, the difference between the dates of identical stages in crops sown late and in crops sown early increases again from stage to stage. In Table 8.1 the results of field experiments are given.

As can be seen in Table 8.1, in all years and in various versions of sowing dates, there is an unchangeable decrease of the intervals between versions from stage to stage in the spring/summer period, usually up to July and August inclusive; and in the autumn period these intervals increase. For example, in the third upper line we see that the 34-day difference between the April and May sowing dates in 1949 diminished to 31 days by sprouting, 19 days by budding, and 16 days by flowering, since the latest stage among those being discussed— the flowering stage—even with May sowing proceeds not later than July 30.

But when comparing the May sowing with the June sowing in the same year, 1949, we can first see that the difference decreased (28, 24, 22 days) until July 24, and then that it increased again (up to 36 days) until September 4. Comparison of the June sowing of 1949 with the July sowing shows that the turning point comes during the earlier stage: in the previous example (May sowing–June sowing) between budding and flowering, and now between sprouting and budding (30, 6, 42 days). But, as before, the turning point occurs in summer (after August 4).

TABLE 8.1. Phenological observations of development of cotton variety 108-F in the field experiments with sowing at different times of the year.

		Date				
Year	Versions of Sowing Dates	Sowing	Sprouting	Budding	Flowering	Opening of First Bolls
1	2	3	4	5	6	7
1949	April sowing	21 IV	4 V	14 VI	14 VII	16 IX
	May sowing	25 V	4 VI	2 VII	30 VII	Did not occur
	Difference between April and May versions (in days)	34	31	19	16	—
	June sowing	22 VI	28 VI	24 VII	4 IX	Did not occur
	Difference between May and June versions (in days)	28	24	22	36	—
	July sowing	22 VII	4 VIII	4 IX	Did not occur	Did not occur
	Difference between June and July versions (in days)	30	6	42	—	—
1950	March sowing	27 III	30 IV	8 VI	6 VII	28 VIII
	April sowing	21 IV	6 V	10 VI	11 VII	11 IX
	Difference between March and April versions (in days)	25	6	2	5	14
	May sowing	17 V	25 V	20 VI	23 VII	5 X
	Difference between April and May versions (in days)	26	19	10	12	24

TABLE 8.1. *(Continued)*

Year	Versions of Sowing Dates	Date				
		Sowing	Sprouting	Budding	Flowering	Opening of First Bolls
1	2	3	4	5	6	7
1951	March sowing	21 III	8 V	10 VI	16 VII	18 IX
	April sowing	21 IV	10 V	10 VI	20 VII	24 IX
	Difference between March and April versions (in days)	31	2	0	4	6
	May sowing	15 V	21 V	21 VI	26 VII	5 X
	Difference between April and May versions (in days)	24	11	11	6	11
1952	March sowing	24 III	17 IV	31 V	6 VII	5 IX
	April sowing	23 IV	6 V	10 VI	14 VII	23 IX
	Difference between March and April versions (in days)	30	19	10	8	18
1953	April sowing	15 IV	29 IV	4 VI	5 VII	5 IX
	May sowing	15 V	24 V	26 VI	25 VII	20 X
	Difference between April and May versions (in days)	30	25	22	20	45
	July sowing	22 VII	27 VII	24 VIII	15 X	Did not occur
	Difference between May and July versions (in days)	68	64	59	82	—

TABLE 8.1. (*Continued*)

Year	Versions of Sowing Dates	Date				
		Sowing	Sprouting	Budding	Flowering	Opening of First Bolls
1	2	3	4	5	6	7
1953	August sowing	17 VIII	23 VIII	24 X	Did not occur	Did not occur
	Difference between July and August versions (in days)	26	27	61	—	—
1954	April sowing	18 IV	29 IV	3 VI	6 VII	9 IX
	May sowing	25 V	31 V	29 VI	3 VIII	Did not occur
	Difference between April and May versions (in days)	37	32	26	28	—
	June sowing	28 VI	3 VII	28 VII	31 VIII	Did not occur
	Difference between May and June versions (in days)	34	33	29	28	—
	July sowing	20 VII	27 VII	24 VIII	8 X	Did not occur
	Difference between June and July versions (in days)	22	24	27	38	—
	August sowing	18 VIII	27 VIII	6 X	Did not occur	Did not occur
	Difference between July and August versions (in days)	29	31	43	—	—

Similar situations were also observed in other years: the difference between crops sown in June and July in 1954 proves to be the smallest soon after sowing (i.e., in July, as before); and for crops sown in April and May the turning point occurs only during flowering (but also at the end of July–beginning of August). Therefore, through regulating sowing dates this turning point can be shifted to the desired developmental stage.

Thus the turning point does not necessarily coincide with the budding stage of cotton (although the budding stage can be critical in other aspects). And this turning point happens not because each subsequent stage from sowing to budding proceeds at increasing values of the lower temperature threshold B (and therefore the developmental stages of crops sown earlier are forced to "wait" for the same stages of crops sown later, i.e., to wait until the temperature of the environment rises up to the necessary "tension") as it was explained by T. D. Lysenko (1928, 1949, 1952), but because of the reasons explained at the very beginning of this chapter. As to the disparity of the environmental temperatures with the lower developmental thresholds, direct analysis of the results of observations shows that, beginning from the end of April, the average temperatures in the cotton-growing regions of Tadzhikistan are much higher, even in comparison with the very high value of B given by Lysenko for cotton variety 508 on the eve of budding.

Therefore, the decrease of the intervals between the same stages in crops sown at different times in spring and summer and the increase of these intervals in autumn is accounted for by the correlation of the developmental rate of an organism and the heat resources of the environment. Then the PTN must reflect the objective law. Let us take two sowing dates and make corresponding calculations on the Dushanbe PTN (Table 8.2).

As can be seen in Table 8.2, the PTN confirms entirely the above conclusions from the field experiments: until July the intervals (differences) decrease regularly from stage to stage (30, 18, 9, and 7 days), and in autumn they increase again (12 days).

An especially great increase of intervals occurs with late-sown crops: even a 2-day interval during the flowering period increases up to 8 days by the ripening period of cotton. Therefore it is advisable to strive for accelerated development of crops at the flowering stage. There is also a certain practical use for another effect of the phenological principle: the uselessness, and sometimes even harmfulness, of sowing cotton too early in spring in the USSR, because crops sown later practically overtake in their development crops sown too early. It is also interesting that the PTN confirms entirely the fact known from practice: crops sown too early sprout later than crops sown late, since

TABLE 8.2. Developmental dates of cotton variety 108-F sown at different times of a year. Results of the calculations are taken from the PTN for Dushanbe.

	Date				
Version of Sowing Dates	Sowing	Mass Sprouting	Mass Budding	Mass Flowering	Mass Opening of the First Bolls (According to the Observation Method of the Hydrometeorological Service)
April version	1 IV	23 IV	4 VI	8 VII	8 IX
May version	1 V	11 V	13 VI	15 VII	20 IX
Difference between the April and the May versions (in days)	30	18	9	7	12

the seeds lose germinating energy because of being in cool soil for too long a time.

Therefore, the PTN helps to interpret and to explain the phenomena correctly. Moreover, the PTN makes it possible to calculate sowing dates that should cause the dates of developmental stages in given climatic conditions to be close. This is important for agricultural crop breeding.

For example, it is known (D. V. Ter-Avanesyan, 1954) that, when crossing cotton intravarietally, the best results are obtained with cross-pollination of "parent" couples that are remote from each other in their sowing dates. For this purpose it is necessary to have simultaneous or almost simultaneous flowering of crops sown on different dates. The PTN or phenological calendars make it possible to calculate the different sowing dates that cause close flowering dates. For example, in the phenological calendar for budding/flowering for Dushanbe (Table 3.9), we take two close dates of mass flowering, say July 7 and 12 (zero temperature anomaly). Then mass flowering on July 7 on one field can coincide with the beginning of flowering (which becomes mass on July 12) on another field. In this calendar (Table 3.9) it can be seen that the flowering of July 7 corresponds to the mass budding of June 4, and the flowering of July 12 corresponds to the budding of June 10. In the calendar for the sprouting/budding period (Table 3.8) we find that the budding of June 4 corresponds to the mass sprouting of April 22 and the budding of June 10 corresponds to the sprouting of May 5. In the calendar for the sowing/sprouting period (Table 3.7) we find that the sprouting of April 22 corresponds to the sowing of March 26, and the sprouting of May 5 corresponds to the sowing of April 23 (with zero temperature anomaly and normal agrotechniques). As can *be seen, simultaneous cotton flowering, necessary for crop breeders and seed growers, can be reached in most years at highly different sowing dates*—March 26 and April 23. In addition, according to the calendar for the sowing/sprouting period, it turns out that mass sprouting on April 22 is the earliest sprouting, among the possible sproutings in the climate of Dushanbe in most years. But the transfer of the obtained pair of sowing dates forward does not result in such simultaneous flowering, as can be seen from the same calendars. Therefore, a crop breeder working in the Hissar Valley in Tadzhikistan must make special arrangements so that the sprouts of April 22 will not suffer from heavy showers and hail.

The PTN also makes it possible to foresee the results in cases where it is difficult to judge *a posteriori:* the phenological principle gets complicated when we deal with the plants that develop at above-optimal temperatures (i.e., at the rising branch of the phenological curve), or when we deal with the regions with *a double maximum in the annual temperature trend.*

Of course, the phenological principle discussed is not exhausted by the world of plants. This principle is also to be observed in the vast world of poikilotherm animals. Some aspects of this latter problem have already been discussed in previous sections and chapters. For instance, according to the calculations on PTNs, pubescent ticks *Hyalomma detritum* are accumulated in great numbers on pastures owing to the phenological principle, since individuals that began their development in late spring overtake the development of earlier individuals. And the actual data of I. G. Galuzo (1947) entirely prove this accumulation of individuals. In the section about helminthes, the same could be seen with respect to the invasion larvae of the *H. contortus.* In the section devoted to the malariological calculations for Yavan, attention was paid to the fact that the females that flew out from wintering places at very different times, later on gave offspring whose emergence dates differed from each other much less. The manifestation of the phenological principle described above with respect to the protozoa *Plasmodium vivax* can be found on the PTN for Yavan (see Fig. 1.51, biological curve 5, and Table 8.3).

A more distinct picture among all of the above was observed in the survivability of beef tapeworm's oncospheres. In that case the difference of four months between the dates when eggs got into the environment, diminished by up to one month at the moment of the eggs' death (Section 6.3).

The phenological principle can also be seen with respect to fleas—carriers of plague bacteria (Section 6.7).

TABLE 8.3. The revelation of the phenological principle in the development of the malaria plasmodium.

Version	Beginning of Sporogony	Duration of Sporogony (in n days)	End of Sporogony
May	15 V	14	29 V
June	15 VI	8	23 VI
Difference (in days)	**31**	—	**25**
August	15 VIII	7	22 VIII
September	15 IX	14	29 IX
Difference (in days)	**31**	—	**38**

Thus we take the most widespread situation. This is the annual temperature trend with one maximum (out-of-equator latitudes), and the development of biological organisms at below-optimal and optimal temperatures. Then the *phenological principle* can be formulated, on the multiannual scale, as follows: *at different dates of the initial developmental stage in plants or poikilothermic organisms in a given region, the difference between the dates of the subsequent developmental stages decreases in the spring–summer period when temperatures rise, and increases in autumn when temperatures fall, provided that other conditions, which are not equal during the season, do not interfere.*

CHAPTER NINE

Criticism Against the Podolsky Method, and the Dissemination of this New Method

1. In different times various authors pointed out the shortcoming that is common to both the Podolsky method and the temperature totals method. Both methods take into account only one factor in the development of plants and poikilothermic organisms — heat. Now this shortcoming has been overcome in principle in the new method (Section 2.2 and a number of other places in this book). The matter depends on vast application of multifactoral forecasting in practice. Unfortunately, most people prefer to use only the main, single-factoral relationship of the developmental period to the temperature. This is explained not only by the difficulty in obtaining a multifactoral relationship in the field experiment, but also by the fact that situations in which the nontemperature factors influence the developmental *dates* to a practically significant degree are not common. It happens even more seldom that the key factor for developmental dates is not the temperature. For example, a hydrophilous biological organism may stop its development and even perish at humid deficiency; but if such an organism is in a developmental state, then its developmental rate is determined, mainly, by the temperature. Nevertheless, the author believes that situations in which developmental periods depend as much on nontemperature factors as on the main factors, must be searched for diligently, and the new method's effectiveness in such situations must be checked. Humidity resource nets, for instance, have already been constructed by the author (Figs. 1.131-1.133). The matter depends on phenocurves that bind developmental dates with nontemperature factors.

2. The necessity was pointed out (by Member of the USSR Academy of Medical Sciences P. G. Sergiyev) of minding the critical values

of the factors that determine the development of biological organisms. The author took into consideration this important point, too (Section 2.3).

3.   Professors L. N. Babushkin and A. S. Danilevsky noted that it is difficult to perform agroclimatic mapping with the proposed method. This branch of experimental-mathematical phenology was worked by the author shortly after (Section 4.2b).

4.   Z. A. Mishchenko (1962, pp. 78-79) noted that the Podolsky method, like the temperature totals method, does not reflect the influence of the amplitude of 24-hour temperature fluctuations on the development of plants. Danilevsky expressed the similar opinion, but he noticed that *biological curves based on natural experiments, being put onto the nomogram, eliminate this shortcoming (but not entirely).* Member of the USSR Academy of Medical Sciences Sh. D. Moshkovsky generalized the shortcoming mentioned as follows: various lines of average temperatures on the PTN can intersect the phenological curve, giving the development of a biological organism, at one point and give, therefore, a single solution to the problem—n days and t_{av}. However, the single average temperature t_{av} at the intersection point is formed, in the mentioned cases, by various series of individual temperature values which can influence the biological organism differently.

The author agrees with this criticism and will make efforts to overcome this shortcoming, which is common for both the new method and the temperature totals method. Nevertheless, it is necessary to take into consideration the following:

a. It is known that the 24-hour temperature fluctuations exert some specific influence on the developmental rate of plants and poikilo-thermic organisms, mainly in spring and autumn, when both night temperatures and 24-hour average temperatures fall below the biological threshold, but the development advances slightly, owing to daytime temperatures. However, if the relationship of the plant's interdevelopmental stages to the average temperatures over these periods is based on experiments with crops sown at different times in a certain region, then the low-temperature part of this relationship (presenting the spring and autumn development) cannot help taking into account the integral result of the developmental acceleration that occurs under the influence of high daytime temperatures, and the result of the delay in development due to the influence of low night temperatures. If these fluctuations in temperature did not influence the development, then the low-temperature sections of the phenocurves would go in somewhat different directions than was fixed by the experiments, for example in Figure 1.2. The same can be said about poikilotherms that produce several generations during a season: the low-temperature section of the phenocurves based on field observations for such organisms relates to the spring and autumn generations (Figs. 1.1, 1.34-1.36, and others).

A very clear example of taking into account, on the PTN, the influence of 24-hour temperature fluctuations upon development can be seen in Figure 1.34. Here the lower temperature threshold for the development of *H. contortus* geohelminth seems to be much lower than the limits mentioned in the literature, and is referred to as the constant temperature in controlled temperature chambers. This discrepancy is explained by the fact that in Figure 1.34 average temperatures over periods were taken which were obtained from the 24-hour average temperatures. And although, at a constant temperature of, for example, +4°C, the *H. contortus* does not develop, nevertheless, the spring (or autumn) 24-hour average temperature of +4°C relates to such high daytime temperatures on the soil surface in Tadzhikistan that the development from egg to third-age larva is completed in the field during 24 to 26 days (in Fig. 1.34 see the last left-hand points of observations). It becomes clear that if we put the phenological curve, giving the *H. contortus*'s development, onto the PTN, then the latter will take into account the influence of 24-hour temperature fluctuations on the dates of *H. contortus*'s development.

b. Taking into account the influence of the daytime temperature semiwave on the developmental rate of organisms may be important for relatively cold regions. But, if the temperature phenological curve is based on data from both cold and warm regions, then the curve's low-temperature section, "working" on the nomograms for cold regions, also takes into account the influence of the daytime semiwave in these regions.

c. The influence of critical, for the organism's life, temperature values refers also to the influences of the 24-hour temperature trend on the organism's development, since the autumn frost that put an end to vegetation of the plant is, as a rule, the night fall in temperature, and the high temperature that killed a tick or a helminth is the daytime temperature. The method for taking into account critical temperature fluctuations of this kind has already been developed by the author (Section 2.3).

d. We should acknowledge that a living organism does not react upon any fluctuation in temperatures; it reacts most frequently upon a certain integral result in the form of average temperature. It is good that the new method takes into account, first of all, and automatically, the structure of the temperature total (this total is inconstant and depends on the height of 24-hour average temperatures composing the total). This was the greatest error in the temperature totals method: *the error caused by insufficiently taking into account the swing in temperature fluctuations is a relatively small error.*

Further elaboration of the methods, used for taking into account the influence of temperature fluctuations, is conceived as being in the area of the "elasticity" of the low-temperature end of the phenological curve. It is necessary to know how to shift this end upward or downward, in relation to autumn and spring 24-hour fluctuations in temperatures, in a particular region, in a particular year.

5. V. A. Smirnov (the All-Union Plant-Growing Institute) noted the difficulty in constructing phenological temperature curves in cases where the points are scattered irregularly on a graph. Unfortunately, this difficulty is real, and it is common for both the new method and the temperature totals method. However, if Smirnov meant the poor quality of the phenological observations, in terms of the scattering of points, then it is necessary to press for improvement in this area. If another cause is meant, namely the absence of the relationship between the development of plants and the temperature, then we have to look for the primary relationship between the development and another seasonal factor. In the very rare cases, where there is no obvious relationship between the development and meteorological factors at all, or between the development and the seasonal factors in general (this is hardly probable), the new method, and especially the temperature totals method, or any other forecasting method is inapplicable.

In the author's opinion, the main shortcoming of the new method is the inapplicability directly for warm-blooded animals (although the phenological forecasting helps, indirectly, to take care of vegetable food for warm-blooded animals and to protect them from illnesses).

Finally, we should note that P. G. Sergiyev gave his approval to the fact that temperatures are written on the PTNs in the scale with zero degrees; but Sh. D. Moshokovsky and A. A. Shigolev questioned the correctness of such writing. The analysis of this correctness and the reply to critics have been given in Section 2.1, which described the properties of the PTN.

Nevertheless, almost all critics recognized considerable advantages of the new method. These advantages are corroborated by the graph that represents the extent to which the new method has been disseminated (Fig. 9.1).

Figure 9.1 includes only the works published in the USSR, and only those works of various authors which were available to the author of this book. The overall number of works is, apparently, much greater and many of them are unknown to this author. Information in abstracts published in the USSR and in other countries has also not been taken into account.

If the interest for the method had been exhausted, the curve in Figure 9.1, after a certain rise, would have run horizontally, parallel to the axis of abscissae. Hence the first derivative representing the number of references with respect to time, $\partial N/\partial \tau$, would be zero. It is not like this; on the contrary, the second derivatives, $\partial^2 N/\partial \tau^2$, appear after the books were published by the author of the method (points B and C in Fig. 9.1). It is interesting that the appearance of the second derivative, that is,

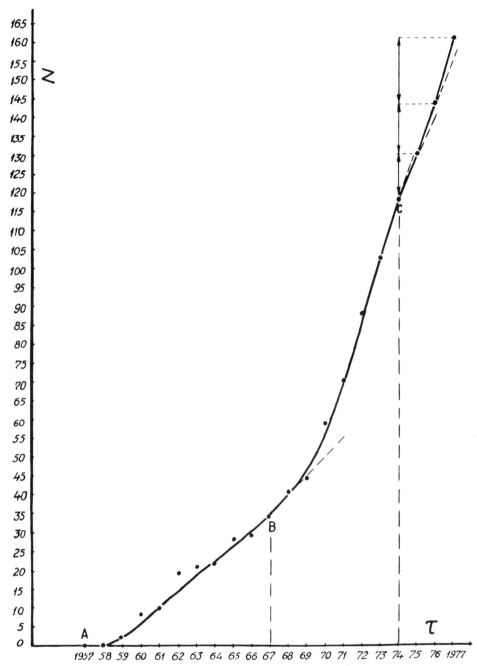

FIGURE 9.1. The curve of citation: the increasing total (N) of references and works by various authors (with the exception of works by Podolsky) published in the USSR and devoted to the new method of phenological forecasting, in the course of years (τ). Point A: the time when the first article by the author of this method was published (1957); Point B: the time when the first Russian edition of the author's book was published (1967); Point C: the second edition of the book (1974).

the increase of steepness of the quotation curve, does not coincide with points B and C, but begins approximately two years later. This is the result of the reader's "psychological" reaction to the novelty, and the time necessary for application of the method by other authors with their own data; there is also the time needed for the preparation of scientific works, and for the "incubation" of such works in editorial offices.

As to the abstracts published outside the USSR, there are descriptions of Podolsky's method in "Biological Abstracts": Vol. 34, 1959, Septemper-December, 9436; Vol. 50, No. 10, 1969, May, 51119; Vol. 60, No. 8, 1975, Oct., 44970; Vol. 61, No. 6, 1976, March, 30734; and others. In the "Biological Abstracts" there are also descriptions of the author's adherents' works: Vol. 65, No. 11, 1978, June, 63194; Vol 70, No. 6, 1980, September, 37795; and others.

Podolsky's article "A new method for phenological prognosis and agroclimatic appraisal using the heat factor" (*Doklady Akad. Nauk SSSR*, Bot. Sci. Sect., 121, 1958) has been translated entirely from Russian into English.

The method was also used in the educational literature as follows: in the United States: *Agricultural Meteorology* by J. Yu. Wang, First edition, 1963, Wisconsin University (pp. 124-129, 271, 399); Second edition, 1967, California University; Third edition, revised, 1972, California University (pp. 290, 353-354, 369). In the USSR: *Agricultural Meteorology* by V. I. Vitkevich, First edition, 1960 (translated into English in Jerusalem, 1963); Second edition 1965; *Agricultural Meteorology* by G. V. Rudnev, First edition 1964; Second edition, 1973; *Agricultural Meteorology* by Yu. I. Chirkov 1979; *Insect Phenology* by B. V. Dobrovolsky, 1969; *Biological Forecasting*, by A. M. Maurin and B. N. Tardov, 1975 (Latvian University), and others.

The Podolsky method was also included in the syllabus of the course "Forewarning and forecasting the propagation of agricultural pests and illnesses" for all the agricultural universities in the USSR (1975). In this syllabus Podolsky's book *Phenological Forecasting*, Second edition, was recommended as a textbook.

With the use of Podolsky's method, about 30 dissertations have been prepared and defended for scientific degrees in biology, geography, agriculture, pharmaceutics, medicine, and veterinary medicine (sciences are mentioned in descending order according to the quantity of dissertations on each science).*

---

*In the USSR synopses are published with the right of a manuscript (although they are typeset). This is why such synopses are not given in the reference section at the end of this book.

CHAPTER TEN

Further Development of the Theory and Practice of the Podolsky Method, and its Prospects

Totals of active and effective temperatures, totals of squared temperatures, totals of developmental rates, coefficients, indexes, and other artificial units used in agricultural meteorology, phenology, and in phenological forecasting, were being devised for centuries, since, apparently, no method could be found for directly solving natural problems—in particular, phenological problems.

The heat resources of a geographical region are expressed, most simply and naturally, by an ordinary annual temperature trend. On the other hand, an organism's reaction to heat is expressed by an empirical phenocurve, which is also quite natural. Therefore, the phenological problems must be solved directly on the annual temperature trend, using phenological curves. It seems that this was a "dream" of everybody who dealt with agrometeorology and phenoforecasting, from Réaumur (1735) to the author of this book. This problem has been solved, at last, using the Podolsky method (A. S. Podolsky and V. V. Denisov, 1975).

But, first, a fragment of the Kursk heat resources net must be detached from Figure 1.93, which is overloaded by curves. Then this fragment must be supplied by a few new (for this fragment) phenocurves (Fig. 10.1).

As can be seen in Figure 10.1, the average temperature lines are rectilinear within the limits of the segments composing these lines —for instance ab and $c'd'$ at the beginning of the lines. On the basis of the average temperature lines, for example, from the initial date of April 5 (5 IV), we can find the 24-hour average air temperature for April 5 on the axis of abscissae (point a). The average temperature over a 10-day period reckoned from the same initial date of April 5 will be $t^{\mathrm{av}}_{\text{5-15 IV}}$.

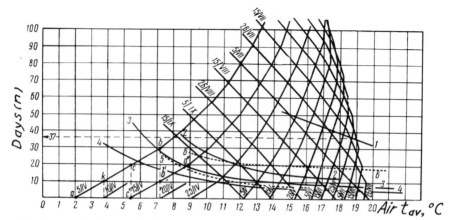

FIGURE 10.1. A PTN for Kursk. Phenological curves : curve 1: spring wheat variety Kharkovskaya 46 from mass heading to mass wax ripeness; curve 2: lily-of-the-valley (medicinal raw material) from the beginning of vegetation to the beginning of flowering; curve 3: the duration of the incubation period of wheat stem rust; curve 4: development of larvae of nematode *Dictyocaulus viviparus* to the third invasive stage; curve 5: winter rye from sprouting to tillering, with a moisture reserve of 35-55 mm in the arable layer (E. S. Ulanova, 1959); curve 6: the same as curve 5, but with a moisture reserve of 15 mm.

Since the temperature at the period between April 5 and 15 changes linearly, we can write the following:

$$t^{av}_{5\text{-}15\,IV} = \frac{t_{5\,IV} + t_{15\,IV}}{2} = t_{10\,IV}$$

Therefore, the average temperature over the period from April 5 to 15 will be equal to the temperature of the middle of this period, that is, to the temperature on April 10 (point k on line segment ab). Next, the average temperature over the 20-day period beginning from the same initial date of April 5, that is, over the period from April 5 to 25, will be $t^{av}_{5\text{-}25\,IV}$. The temperature over the period between April 5 and 25 will also change linearly, and therefore we can write

$$t^{av}_{5\text{-}25\,IV} = \frac{t_{5\,IV} + t_{15\,IV} + t_{25\,IV}}{3} = \frac{t_{5\,IV} + t_{25\,IV}}{2} = t_{15\,IV}$$

Thus, the average temperature over the 20-day period is equal to the temperature of the middle of the period between April 5 and 25 (point c on line segment ab).

Now we can reason, in the same manner, about the 30-day period from April 5 to May 5, and find

$$t^{av}_{5\,IV\text{-}5\,V} = \frac{t_{5\,IV} + t_{5\,V}}{2} = t_{20\,IV}$$

This will be point b on line segment ab.

If we calculate, in a similar way, the average temperature line $c'd'$ from another initial date, for example, from April 15, we can obtain the following: the 24-hour average temperature on April 15 is $t_{15\,IV}^{av}$ (point c'),

$$t_{15\text{-}25\,IV}^{av} = \frac{t_{15\,IV} + t_{25\,IV}}{2} = t_{20\,IV} \text{ (point } b')$$

$$t_{15\,IV\text{-}5\,V}^{av} = \frac{t_{15\,IV} + t_{5\,V}}{2} = t_{25\,IV} \text{ (point } d')$$

Points c and c' represent the same temperature on April 15, and points b and b' represent the temperature on April 20; therefore, segment $c'b'$ of the line dated April 15 is identical with segment cb on the average temperature line ab dated April 5. But this line segment cb is not located at the beginning of the line dated April 5. It can be shown, in an identical way, that the average temperature line dated April 25 would be identical with the line dated April 5, but with its subsequent segment above point b, if, of course, the average temperature lines were rectilinear the entire way along. Therefore, in this case all the average temperature lines (from the initial dates of April 15, 25, . . .) are represented by various segments of the same straight line ab (and of its continuation) dated April 5. Therefore, this line reflects the temperatures over various dates in the year ($t_{5\,IV}$, $t_{10\,IV}$, $t_{15\,IV}$, $t_{20\,IV}$, $t_{25\,IV}$, . . .). Hence line ab with the continuation represents the annual temperature trend in reverse axes in comparison with the regular presentation of the annual temperature trend (time is on the vertical axis and temperature is on the horizontal axis). Moreover, the temperatures, for example, $t_{5\,IV}$ and $t_{15\,IV}$ or $t_{15\,IV}$ and $t_{25\,IV}$ are separated by a space corresponding to a 10-day period, whereas on line ab and on its continuation (Fig. 10.1) these temperatures are separated by the extent corresponding to a 20-day period. Therefore, the annual temperature trend must be drawn in the double scale of the axis of dates in comparison with the scale of periods on the heat resources net or on the phenocurve (Fig. 10.2).

Thus the average temperature lines of the heat resources net and the annual temperature trend coincide within their rectilinear segments (A. S. Podolsky, 1967, pp. 18-19; 1974, p. 24.).

Each annual temperature trend (in spring-summer and summer-autumn periods) has rectilinear segments; therefore, it is possible to solve problems in phenological forecasting using the annual temperature trend directly, without resorting to constructing heat resources nets, if short interdevelopmental stages are under consideration. And, as before, we are free of the artificial notions about temperature totals and thresholds. Let us construct, as an example, an annual temperature trend with reverse axes on the basis of multiannual 10-day average temperatures of the air in Kursk: on the horizontal axis we put temperatures with a scale of $1°$ to 1 cm; on the vertical axis we put the dates

—midpoints of 10-day periods—for instance, April 5, 15, and 25, and May 5, 15, and 26, and so on, with a scale of 10 days to 2 cm (Fig. 10.2 has been given in smaller scales). Ten-day average temperatures are attributed to the middle of 10-day periods: the average temperature over the first 10 days of April is attributed to April 5, over the second 10 days to April 15 and so forth. Points with coordinates of date and temperature are joined by rectilinear segments, and the broken line, composed in this way, will be the annual temperature trend (Fig. 10.2).

Onto this annual trend we superimpose phenological curves 1 to 6

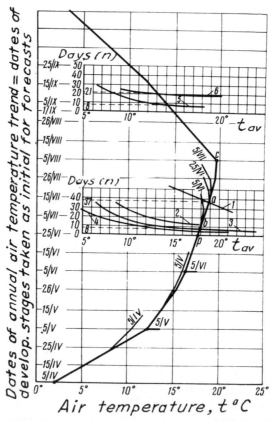

FIGURE 10.2. The solution of phenological problems on the annual temperature trend. Phenological curves: curve 1: spring wheat variety Kharkovskaya 46 from heading to wax ripeness; curve 2: lily-of-the-valley (medicinal raw material) from the beginning of vegetation to the beginning of flowering; curve 3: duration of the incubation period of wheat stem rust in relation to the average temperature over the period; curve 4: development of larvae of nematode *Dictyocaulus viviparus* to the third invasive stage; curve 5: winter rye from sprouting to tillering, with the moisture reserve of 35–55 mm in the arable layer; curve 6: the same as curve 5, but with the moisture reserve of 15 mm. Boldfaced line: the annual temperature trend in Kursk based on multiannual average data. Thin lines with dates: average temperature lines taken from the heat resources net.

plotted with the scale of periods n, which is half as large as the scale of dates of the annual trend: 10 days correspond to 1 cm (phenological curves for both Figures 10.1 and 10.2 are taken from the previous sections of this book).

Let us consider two simple problems:

1. It is necessary to determine the date when the wax ripeness of spring wheat variety Karkovskaya 46 occurs in Kursk, provided that the multiannual average date (or the date for a particular year) of mass heading is June 25.
2. It is necessary to calculate the duration of the incubation period for the agent of stem rust in wheats, provided the date of infection is June 25.

To solve these problems we must compare graphically the equations giving the organism's heat requirements, that is, curves 1 and 3, with the heat resources equation, which is represented not by a net, but directly by the annual temperature trend. For this purpose we superimpose phenocurves 1 and 3 onto the annual temperature trend, so that the zero ordinate of the phenological curve is brought in coincidence with the initial forecast date of June 25; we find this date on the annual trend. Temperatures of the phenological curve must coincide with the identical temperatures of the annual trend (Fig. 10.2). Then the ordinates of the points where the phenocurves intersect the annual temperature trend will give interdevelopmental stages on the phenocurves' graph. These periods are 37 days (point a) for the first problem, 8 days (point b) for the second problem. Therefore, wax ripeness of wheat will occur in Kursk on June 25 + 37 days = August 1, and the date of the end of the incubation period will be June 25 + 8 days = July 3.

The actual date of wax ripeness in spring wheat in Kursk is July 29. As can be seen, the forecast is a success, and, in addition, it is in agreement with the forecast made using the heat resources net (Fig. 10.1). The same can be said about the forecast for the stem rust agent.

If the initial date of the phenoforecast is earlier than June 25, the graph of phenological curves is shifted down the field of the annual trend graph, and vice versa. For this purpose it is advisable to draw the graph of phenocurves on tracing paper. *Still better is to construct a plastic plate on which the multiannual temperature trend is drawn, and another, smaller transparent plastic plate—"runner"—with phenocurves drawn on it. The "runner," put into the longitudinal edges of the annual-trend plate, slides over this trend along the vertical axis.*

The annual temperature trend must be rectilinear the full length of n days from the starting developmental stage to the forecast stage. However, this length on the annual trend graph is twice as large as the length of n with the scale of a phenological curve. It follows that there is a simple criterion for fitness of this latest method for a particular phenological problem: the solution will be correct if the point of the intersection of the annual trend with the phenological curve has not overstepped the middle of the annual trend's rectilinear segment beginning from the starting point of the calculation (this starting point is different in different phenological problems). For example, in the problem for wheat there must be the following: $pa \leqslant ac$ (Fig. 10.2).

The evaluation of phenoforecasts for a number of regions in the USSR (Kursk and Kuybishev—center of the country, Zaporozhye—Ukraine, Samarkand—Middle Asia, Ulan-Ude—Siberia) showed that the forecasts obtained using the annual temperature trend differ by 0 to 4 days from the forecasts made on PTNs, with an average foresight period of 40 days, and do not differ at all when the foresight period is 20 days or less (provided the angle of slope of the phenocurve to the axis of abscissae is as wide as 40°). This can be seen on Figure 10.2: besides the annual temperature trend, a number of average temperature lines are drawn that concern the heat resources net. All these average temperature lines coincide well with the annual temperature trend within its rectilinear segments which correspond to 30 to 50 days, and only in the extreme sections of the annual trend do the coinciding segments diminish to 10 to 20 days (the average temperature lines are given in Fig. 10.2 only as a visual aid; they are unnecessary for solving problems). Therefore, the annual temperature trend can be a basis for solving numerous problems in forecasting the duration of the incubation period of plant illnesses, the duration of developmental periods of a number of geohelminths, and the developmental dates of medicinal plants and some agricultural crops. Here we usually deal with short interdevelomental stages. But it is hardly advisable to calculate with the simplified method such problems as, for example, the survival period of beef tapeworm's oncospheres or of the foot-and-mouth disease virus in the environment, since these periods often have a duration of more than 80 days.

On the annual temperature trend, problems of the same rank as on the PTN are solved: phenoforecasting with the use of actual temperatures over a part of the forecast period, with the use of a long-range weather forecast, the calculation of given developmental stage/temperature combinations (important for crop breeding), combined phenoforecasting for living components of biogeocenoses, and so on.

On the annual temperature trend, as well as on the PTN, it is also possible to take into account the nontemperature factors: if, for example, winter rye sproutings appear on September 1, then on a field with a moisture reserve of 33 to 55 mm in the arable layer, tillering can be expected on September 1 + 8 = September 9, where $n = 8$ days is an ordinate of the intersection of empirical curve 5 with the annual temperature trend (Fig. 10.2); on a field with moisture of 15 mm, tillering will occur on September 1 + 21 = September 22, where $n = 21$ days is an ordinate of the intersection of phenocurve 6 with the annual temperature trend.

Thus by means of the heat resources net, we have arrived at a *simple and direct* solution of two complex transcendental empirical equations. One equation is expressed by a sinusoidal type of the annual temperature trend (divided into rectilinear segments). The other equation is expressed by an exponential, parabolic, or hyperbolic phenological curve.

Thus, the first prospect: in many cases (but not in all) the temperature-phenological nomograms with heat resources nets can be replaced by a newer and simpler method, namely drawing the annual temperature trend (it only takes 20-30 minutes of manual labor) and superimposing phenocurves onto this annual trend. This entails serial production of the plastic plate with a "runner" (we have already made a pattern, as noted above). On such a plate an annual temperature trend can be drawn for a geographical region which interests the reader with the necessary phenocurves. However, it is also advisable to put the plates with PTNs in serial production (Ch. 1).

· · ·

The second prospect is the compilation of an atlas of phenological curves for plants and poikilothermic organisms, and the classification of these curves. The urgency of this problem was discussed in sufficient detail in Section 3.4.

· · ·

Remote phenological observations using spectrozonal photography from airplanes or artificial satellites are used more and more. In the American literature we can find more often the published results of such observations, such as graphs that characterize the extension of field areas under sowing, for example, winter wheat, then heading of the wheat, ripening, and harvesting, in relation to different dates and months in a given year, in a particular state. For example, an actual curve is available that represents the increase in percentage of areas with heading winter wheat on different dates in June. Then it is possible to calculate on the PTN, with high accuracy, an identical curve for heading in July. Through this it is possible to conserve flights of airplanes or to rationalize the utilization of data coming from satellites.

There are certain prospects for the use of PTNs in deciphering the sattelite's surveys, as well as in entomological and phytopathological forecasting on the basis of deciphered photos, and so on. (K. M. Stepanov and A. E. Chumakov, 1972, p. 208).

· · ·

For the time being a number of organisms remain beyond the sphere of action of the new forecasting methods: among plants, many species of weeds; among animals, fish, reptiles and some other poikilothermic organisms.

· · ·

These prospects and the above problems, solutions, and consequences do not exhaust all that will crystalize in the process of further application of the proposed methods in different fields of science and practice, including the field of inanimate nature: for example, in hydrology where the temperature totals method is used now; or in the technological world, for forecasting the aging dates of coatings such as enamel paint, polymer, and galvanic, as well as for forecasting metal corrosion, and so on.

The latter complex of problems might conceivably be solved in the following manner. In the same way as phenological curves for living objects, curves must be constructed to represent the aging dates of coatings, and corrosion dates in relation to ambient temperatures, if the key factor in this process is the temperature; or in relation to atmospheric humidity, if that is the main factor. When both these (or any other) factors are important, several aging curves must be constructed in relation to the key factor: each curve concerns a certain state of the minor argument. For example, two or three temperature curves for two or three gradations of the relative humidity or for two or three gradations of aggressiveness of the environment.

Therefore, on the graph giving such aging curves, on the horizontal axis we put the average air temperature or humidity, and on the vertical axis we put the dates (periods) when coatings or metal surfaces are worn out *to a certain extent* (when their replacement is preferable). Apparently, this way makes it possible to transform the degree of wear-and-tear (which is usually given in the literature and experiments) into a time factor. This is necessary since the phenological nomograms are constructed in coordinates of temperature (or humidity) and time. An example of such transformation has been already shown in Section 2.6.

Results of laboratory tests of materials in polythermohygrostats or in climatic chambers can be used on a larger scale for constructing the mentioned curves of aging or wearing. It is clear that for individual types and models of coatings, for each kind of metal, there will be individual curves.

The aging curves must be superimposed onto the heat or humidity resources net of the geographical region where the coatings or metals are used. The result of a forecast for the dates of aging of coating or metal corrosion (to a given degree) is read on the intersection of the curve with the temperature or humidity resources line which is designated by the date when a coating or a metal construction began to be used (i.e., it must be done in the same way as it was done above for living organisms).

Since the temperature or humidity resources net is constructed on the basis of the annual trend of ambient temperatures or humidity, the author considers it advisable, for the present, to use the entire complex of the new method only for coatings and constructions intended for work without restoration within one year. In these cases the forecasting method may be fruitful: for example, the wear resistance of enamel paint coating produced in a given region in winter, or in spring, will differ from the wear resistance of coating produced in summer. Metal corrosion, for example, in the south, will differ from that in the north, and so on.

Forecasts for the dates of aging and corrosion, which last during a number of years, can be given directly on the phenological curve presenting the duration of aging or corrosion, without a heat resources net. On the ordinate axis of such a curve will be put not days but years or months, and on the abscissa axis, for example, annual average values of temperature or of humidity.

Finally, the new method of phenological forecasting can be used in control arrangements against fungi that disturb the work of radio, electronic, and other kinds of equipment.

All this put together is *bionics*.

References and Bibliography on the New Method of Phenological Forecasting and Bioclimatic Estimations

Adzhbenov, V. K., "The Application of Nomograms in Forecasting the Development of *Hadena sordida Bkh* in Spring Wheats in the Steppe Zone of Northern Kazakhstan," in I. Ya. Polyakov, Ed., *Transactions of the All-Union Scientific Research Institute of Plant Protection* (38), VASHNIL, Leningrad, 1972, pp. 113-117.

Adzhbenov, V. K., *Recommendations on Forecasting the Development both of Spring Wheat and* Hadena sordida Bkh *in the Steppe Zone of Northern Kazakhstan,* Kainar Publishing House, Alma-Ata, 1973, pp. 1-22.

Adzhbenov, V. K., "Arrangements on Chemical Treatment Against the *Hadena sordida Bkh* on the Basis of Forecasting Methods," *Papers of the III Scientific Production Conference on Plant Protection in Kazakhstan,* Alma-Ata, 1974, pp. 3-4.

Adzhbenov, V. K., "Phenological Forecasting and Forewarning," *Methodical Instructions to Revealing, Taking into Account, and Forecasting the Development of the* Hadena sordida Bkh *and Forewarning the Dates of Pest Control,* Ministry of Agriculture of the USSR, Kolos Publishing House, Moscow, 1976, pp. 21-26.

Adzhbenov, V. K., "Combined Phenoforecasting both Spring Wheat and the *Hadena sordida Bkh,*" *J. Plant Protection,* Moscow (9), 45-47 (1979).

Agroclimatic Reference Book for the Kursk District, Chief Department of the Hydrometeorological Service of the USSR, Gydrometeoizdat Publishing House, Leningrad, 1958, pp. 1-140.

Agroclimatic Reference Book for Tadzhikistan, Chief Department of the Hydrometeorological Service of the USSR, Gydrometeoizdat Publishing House, Leningrad, 1959, pp. 1-152.

Alexandrov, M. V., *The Atmospheric Heat and Forecasting the Development of Poikilothermic Aerobes,* FAN Publishing House, Tashkent, USSR, 1974, pp. 1-164.

Almazova, V. V., "Epidemiological Significance of *Anopheles maculipennis meig* Females of Different Dates of Emergence (Different Generations)," Collected works: *Seasonal Phenomena in the Malaria Mosquito Life in the USSR,* Medgiz Publishing House, Moscow, 1957.

Almazova, V. V., "Data on the Ecology of *A. superpictus* in the Pyandzh Valley (the Western Pamirs)," in M. Ya. Rasulov, Ed., *Collected Works on the Malaria and Helminthoses (II),* Tadzhikistan Ministry of Public Health, Dushanbe, USSR, 1959, pp. 88-89.

Almazova, V. V., "Data on the Autumn Phenology of *A. superpictus* in the Valley of the Middle Reaches of the Vakhsh River," *Collected Works on the Malaria and Helminthoses* (II), Tadzhikistan Ministry of Public Health, Dushanbe, USSR, 1959, pp. 109-123.

Andrewartha, H. G., "Air Temperature Records as a Guide to the Date of the Hatching of the Nymphs of *Austrocetes cruciata Sang (Orthoptera),*" *Bull. Entomol. Res.,* **XXXV** (1) (1944).

Angus, J. F., et al. "Water Use, Growth, and Yield of Wheat in a Subtropical Environment," *Aust. J. Agric. Res.,* **31** (5), 1980, pp. 873-886.

Arapova, L. I., "The Phenology of the Colorado Potato Beetle and Forecasting its Developmental Dates in Byelorussia," in B. A. Tavrosky, Ed., *Problems of Phenological Forecasting,* Geographical Society of the USSR, Leningrad, 1970, pp. 55-56.

Arapova, L. I., "Determination of the Duration of the Colorado Potato Beetle's Development," *J. Plant Protection,* Moscow (8), 13-14 (1971).

Arapova, L. I., "Evaluation of Climatic Resources in the European Section of the USSR for Development of the Colorado Potato Beetle," in I. Ya. Polyakov, Ed., *Transactions of the All-Union Scientific Research Institute of Plant Protection* (38), VASHNIL, Leningrad, 1972, pp. 106-112.

Arapova, L. I., *Methodical Instructions to Forecasting the Development of the Colorado Potato Beetle and Forewarning the Dates for Control of this Pest in the European Section of the USSR,* Ministry of Agriculture of the USSR, Moscow, 1974, pp. 1-97.

Arapova, L. I., "The Application of Nomograms to Forecasting the Colorado Potato Beetle's Development in Order to Determine the Optimum Dates of Pest Control on Potato Crops," *Reports at the IV Interdepartmental Conference on Phenological Forecasting,* Geographical Society of the USSR, Leningrad, 1977, pp. 110-112.

Arapova, L. I., *Phenoforecasting Calendars for the Colorado Potato Beetle (Minsk District),* Byelorussian Research Institute of Scientific-Technical Information, No. 230, Minsk, 1978, pp.1-5.

Arapova, L. I., "For Forewarning the Dates of the Colorade Potato Beetle Control," *J. Plant Protection,* Moscow (2), 48-50 (1979).

Astaurov, B. L., *The Ways of Management of the Development and Vital Activity of Silkworms by Means of Thermal Conditions and Effects,* Publishing House of the Academy of Sciences of the USSR, Moscow, 1958.

Babinet, J., *Sur les Rapports de la Temperature aves le Developpement des Plantes,* Compte-Rendu des Seances de l'Akademie des Sciences, Paris, 1851.

Babushkin, L. N., "On the Methods of Agrometerological Informations and Forecasts in the Conditions of Uzbekistan," *Transactions of the Tashkent Geophysical Observatory,* Gydrometeoizdat Publishing House, Leningrad, 3 (4), 67-110 (1949).

Babushkin, L. N., *Methodical Instructions,* Central Institute of Forecasts (16), Gydrometeoizdat Publishing House, Moscow, 1951, pp. 1-47.

Barber, M. A., J. B. Rice, and A. G. Mendekoc, "The Relation of the Density of the Anopheline Mosquitoes and Transmission of Malaria," *Am. J. Hyg.,* 24 (2) (1936).

Beklemishev, V. N., *The Ecology of Malaria Mosquitoes,* Medgiz Publishing House, Moscow, 1944, pp. 1-299.

Beklemishev, V. N., et al., *Textbook of Medical Entomology,* Part I, Medgiz Publishing House, Moscow, 1949.

Beklemishev, V. N., "Determination of the Numbers of the Carrier's Population in the Malaria Breeding Ground in Connection with the Evaluation of Successfulness of

Imago Control," *Seasonal Phenomena in the Life of Malaria Mosquitoes in the USSR*, Medgiz Publishing House, Moscow (1957).

Bluck, H., "Die Entwicklung des *Dytiscus marginalis L.* vom Ei bis zur Imago," II Tl., Die Metamorphose, Zeitschr., *Wissensch. Zool.*, **121** (1923).

Bodenheimer, F. S., *Citrus Entomology in the Middle East*, Dr. W. Junk, Publisher, the Hague, Netherlands, 1951.

Borisov, G. N., *Cotton Varieties in the USSR*, Selkhozgiz Publishing House, Moscow, 1962, pp. 1-175.

Buligin, N. E., and S. M. Topper, "Comparative Effectiveness of Phenological Forecasts for the Seasonal Rhythmics of Oak and its Arbor Concomitants," Collection: *Seasonal Development of Nature in the European Section of the USSR*, Moscow Branch of the USSR Geographical Society, 1977, pp. 7-9.

Buligin, N. E., "Biological Fundamentals for the Estimation of the Influence of Thermal and Other Ecological Factors on the Seasonal Rhythmics of Arboreal Plants," Collection: *The Thermal Factor in the Development of Plants in Various Geographical Zones*, Moscow Branch of the USSR Geographical Society, 1979, pp. 7-9.

Buyanova, O. F., "Data on the Ecology of *A. superpictus* in the Valley of the Middle Reaches of the Vakhsh River," in M. Ya. Rasulov, Ed., *Collected Works on the Malaria and Helminthoses* (II), Tadzhikistan Ministry of Public Health, Dushanbe, USSR, 1959, pp. 101-107.

Chekhonadskikh, V. A., "Seminar of Forecasters," *J. Plant Protection*, Moscow (4), (1971).

Chekhonadskikh, V. A., "Forecasting the Emergence of the Hawthorn Tortricid." *J. Plant Protection*, Moscow (12), 41-42 (1973).

Cherkasov, A. F., "Comparative Analysis of Some Methods of Forecasting the Developmental Stages and the Yield of Wild Fruit-Berry Plants," Collection: *Resources of Berry Plants and Medicinal Plants and the Methods of Their Study*, Karelian Branch of the Academy of Sciences of the USSR, Petrozavodsk, 1975, pp. 8-26.

Chirkov, Yu. I., *Agricultural-Meterological Conditions and Productivity of the Corn*, Gydrometeoizdat Publishing House, Leningrad, 1969.

Chkhubianishvili, I. A., E. D. Abashidze, and G. Sh. Tatishvili, "On Working Out the Methods of Forewarning Against the Wintering Generation of the Piorcer in Kartly," *Transactions of Georgian Scientific Research Institute of Plant Protection*, Tbilisi, USSR, **XXVIII**, pp. 77-80 (1976).

Chobanov, R. E., "The Phenotemperature Nomogram as a Method of Forecasting the Developmental Dates of Geohelminth's Eggs in the Environment," *Transactions of the Scientific Research Institute of Medical Parasitology and Tropical Medicine*, Baku, Azerbaidzhan, USSR, **X** 87-93 (1978).

Chumachenko, N., "On Short-term Forecasts of Seasonal Development of *Laphygma exigua Hb*," *J. Rep. Acad. Sci. of Uzbekistan*, Tashkent (8), (1954).

Churayev, I. A., *Fall Webworm*, Selkhozizdat Publishing House, Moscow, 1962, pp. 1-103.

Danilevsky, A. S., *Photoperiodism and Seasonal Development of Insects*, Leningrad University Publishing House, 1961, pp. 1-243. (The English translation of this book was published in 1965 by Oliver and Boyd, London.)

Danilevsky, A. S., N. I. Goryshin, and V. P. Tyschenko, "Biological Rhythms in Terrestrial Arthropods," *Ann. Rev. Entomol.*, **15**, 201-244 (1970).

Davitaya, F. F., *Climatic Zones of Grapes in the USSR*, Pishchepromizdat Publishing House, Moscow, 1948, pp. 1-192.

Detinova, T. S., "Seasonal Changes in Fertility of *A. m. messeae* Females," *J. Med. Parasitol. and Parasit. Dis.*, Moscow, **V** (4), 566-568 (1936).

Detinova, T. S., "On the Biology of Mosquitoes of the *Aëdes* Genus," *J. Med. Parasitol.*, Moscow, **11** (3) (1942).

Detinova, T. S., "Trend of Numbers and the Age Structure of the Population of *Anopheles maculipennis messeae* in the Moscow Suburbs in 1946," *J. Med. Parasitol.*, Moscow, (6), (1947).

Dmitrienko, V. P., "The Ways to Determine the Length of Interdevelopmental Stage Periods Using Air Temperature and Day Length," *Transactions of the Ukrainian Scientific Research Institute of Hydrometeorology* (37), Gydrometeoizdat Publishing House, Leningrad, 1963.

Dobrovolsky, B. V., *The Phenology of Insects—Agricultural Pests*, "Higher School" Publishing House, Moscow, 1969, pp. 108-111, 120-121, 165-166, 173, 221.

Doroganevskaya, E. A., "On Temperature Totals," *Bot. J.*, Leningrad, **38** (1), (1953).

Dorozhkin, N. A., L. V. Bondar, and N. A. Shtiryonok, "The Application of the Phenotemperature Nomograms to Forecasting the Maturation of Ascospores of the Apple Scab Causal Agent," *Reports at the IV Interdepartmental Conference on Phenological Forecasting*, Geographical Society, USSR, Leningrad, 1977, pp. 122-124.

Dover, Sh., Ed., Scientific Report of Jacob Blaustein Institute for Desert Research, Ben-Gurion University of the Negev, 3rd Annual, Israel, 1980 (Contents include: 1. Podolsky, A., "Phenological Forecasting as a Method in the Study and Management of Biogeocenoses," pp. 64-65; 2. Podolsky, A., "The Forecasting for Preservation Dates of Foot-and-Mouth Disease Virus," p. 175).

Drachovska, M. D., *Forecasting in Plant Protection,"* Selkhozgiz Publishing House, Moscow, 1962, pp. 1-352 (translation from Czech into Russian).

Druzhelubova, T. S., and L. A. Makarova, "Corrective Coefficients for Forecasting the Insect Development Using the Effective Temperature Totals (on the Example of the Turnip Moth)," *Zool. J.*, Leningrad, **47** (I), 73-78 (1968).

Fastovskaya, E. I., "Results of the Three-year Work on the Sanitation of Malaria Breeding Grounds in the Mountain-River Zone of the Hissar Region," *Collected Works on the Malaria and Helminthoses* (II), Tadzhikistan Ministry of Public Health, Dushanbe, USSR, 1959, pp. 25-31.

Fedra, K., "Mathematical Modeling—a Management Tool for Aquatic Ecosystems," *Helg Meer*, **34** (2) (1980).

Galuzo, I. G., *Bloodsucking Ticks in Kazakhstan*, Vol. II, Publishing House of Kazakhstan Academy of Sciences, Alma-Ata, USSR, 1947, pp. 1-282.

Gaplevskaya, L. N., "Evaluation of the Methods Used for Determination of the Developmental Dates of Cotton Bollworm Following the Results of Testing in the Conditions of Tadzhikistan," Collected Work: *Scientific Research in Plant Protection*, Uzbekistan Ministry of Agriculture, Tashkent, 1960, pp. 193-195.

Gavrilova, T. N., "Forecasting the Flowering Dates of the Common Buckwheat," *J. Bee Farming*, Moscow (11), 19-20 (1972).

Gedikh, V. B., "Forecasting the Occurrence of Phenophases in the Species of Cowberry Family in Byelorussia," *J. Ecol.*, Academy of Sciences of the USSR, (6), 78-80 (1976).

Gnedenko, B. V., "On Some Sections of the Theory of Chances Directly Concerning the Problems of Biology and Medicine," Collected Works: *The Application of Mathematical Methods in Biology*, Leningrad State University Publishing House, 1960.

Golovin, V. V., "The New Method of Agrometeorological Forecasting," *J. Agric. in the Amur District*, USSR (3), 41-45 (1959).

Goltzberg, I. A., "Thermal Resources for Fruit Production on the Northern Slope of the Trans-Iliyan Alatau," *Proceedings of the All-Union Geographical Society*, Vol. 97, USSR, 1965.

Goltzberg, I. A., "Agroclimatic Analogs of the Regions Where the Wheat Grows Within the Limits of the Wheat's World Area," *Transactions of the Chief Geophysical Observatory*, (192), Gydrometeoizdat Publishing House, Leningrad, 1966.

Gorski, T., and Jakubczak, Z., "W Sprawie Metody Sum Temperatur w Agrometeorologii," *Roczn. Nauk rolni*, Warszawa, Ser. A, **90** (2), 215-231 (1965) (in Polish).

Goryshin, N. I., and I. A. Kuznetzova, "Duration of Insect Development at Constant and Variable Temperatures," in I. Ya. Polyakov, Ed., *Transactions of the All-Union Scientific Research Institute of Plant Protection* (38), VASHNIL, Leningrad, 1972, pp. 18-27.

Gozodova, G. E., Z. I. Martinova, and A. V. Kondrashin, "The Structure of the Extreme Section of the Dermal Leishmaniasis Area in the Western Pamirs," Interrepublican Scientific Conference on the Problem, *The Major Parasitic Diseases as well as the Prevention and Medical Treatment Against Them,* Tashkent, USSR, 1967, pp. 120-127.

Hopkins, A. D., *Bioclimatics*, Government Printing Office, Washington, D.C., 1938, pp. 1-188.

Hunger, F., and A. W. Aessman, "Effects of Constant and Fluctuating Temperatures on the Rate of Development of California Red Scale," *J. Econ. Entom.*, **41,** 425-427 (1948).

Ilyashenko, L. Ya., "Some Observations for *A. superpictus* in the Valley of the Upper Vakhsh River," *Collected Works on the Malaria and Helminthoses* (II), Tadzhikistan Ministry of Public Health, Dushanbe, USSR, 1959, pp. 119-123.

Ioff, I. G., *Problems of the Ecology of Fleas in Relation to the Epidemiological Significance of Fleas*, Pyatigorsk, USSR, 1941.

Isetov, B. I., "Some Results of the Malaria-Epidemiological Examination in the Vanch District in 1956," *Collected Works on the Malaria and Helminthoses* (II), Tadzhikistan Ministry of Public Health, Dushanbe, USSR, 1959, pp. 81-87.

Isetov, B. I., "Experiences in the Study of the Malarial Epidemiology in the Pamirs in 1952-53," *Collected Works on the Malaria and Helminthoses* (II), Tadzhikistan Ministry of Public Health, Dushanbe, USSR, 1959, pp.57-69.

Ivanov, E. N., "The Forum of Forecasters," *J. Plant Protection*, Moscow (2), (1962).

Jones, E. P., "The Bionomics and Ecology of Red Scale *Aonidiella aurantii* Haskell in Southern Rhodesia," British South Africa Publishing Co., *Ann. Rep.*, (5), 13-52 (1935).

Kalmikov, E. S., and A. Ya. Lysenko, "Data on the Rationalization of the Methods for the Application of DDT with the Antimalarial Purpose in the Zone of Distribution of *Anopheles superpictus*," *Collected Works on the Malaria and Helminthoses* (II), Tadzhikistan Ministry of Public Health, Dushanbe, 1959, USSR, pp. 163-173.

Kalmikov, E. S., "On the Season of Effective Infectibility of *Anopheles superpictus* by the Malarial Agents," *Collected Works on the Malaria and Helminthoses* (II), Tadzhikistan Ministry of Public Health, Dushanbe, 1959, USSR, pp. 163-173.

Kekukh, A. M., "Climate and Plant Protection Against Pests and Illnesses," *J. News Agric. Sci.*, Moscow, (10), 144-146 (1970).

Kharchenko, N. N., "Phenological Forecasting of the Development of Cabbage Flies," *J. Rep. Byelorussian Acad. Sci.*, Minsk, **XVII** (6), 573-575 (1973).

Kharchenko, N. N., and V. P. Bunyakin, *Phenoforecasting Calendars for the Cabbage Moth (Minsk District)*, Byelorussian Research Institute of Scientific-Technical Information, No. 102, Minsk, 1978, pp. 1-4.

Kharchenko, N. N., *Phenoforecasting Calendars for the Cabbage Fly* Hylemyia brassicae *(Minsk District)*, Byelorussian Research Institute of Scientific-Technical Information, No. 104, Minsk, 1978, pp. 1-4.

Kharchenko, N. N., and V. P. Bunyakin, "Forecasting the Control Dates Against Some Pests of Cruciferous Crops," *J. Plant Protection*, Moscow (1), 40-41 (1979).

Kharchenko, N. N., V. P. Bunyakin, and L. S. Kononuchenko, "On Forecasting the Control Dates of the Main Pest Species of Cruciferous Vegetable Crops," *Express-Information*, Agricultural Series, Byelorussian Research Institute of Scientific-Technical Information, Minsk, 1979, pp. 1-40.

Kheisin, E. M., "Observations of the Development of *Ixodes persulcatus P.ScH* and *Ixodes ricinus L.* in Laboratory Conditions," *Transactions of Karelian-Finnish State University*, V(3), Petrozavodsk, USSR, 1954, pp. 88-106.

Koloskov, P. I., *Agroclimatic Division into Regions in Kazakhstan*, Academy of Sciences of the USSR, Moscow, 1947.

Korbut, V., and V. Kovzan, "Bread: from Field to Elevator," *J. Sci. and Life*, Moscow, (6) (1976).

Korol, I. T., *Entomopathogenic Microorganisms and the Application of Them to the Pest Control in Byelorussian Fruit Plantations*, Byelorussian State University, Minsk, USSR, 1960, pp. 1-29.

Kozhanchikov, I. V., *The Methods Used for the Research in the Ecology of Insects*, "Higher School" Publishing House, Moscow, 1961, pp. 1-286.

Kozicheva, Z. V., "The All-Union Scientific Conference," *J. Plant Protection*, Moscow, (5) (1972).

Krasichkov, V. P., "A Valuable Investigation," *J. Collect. Farm—State Farm Prod. in Tadzhikistan*, Dushanbe, USSR, (9), 54 (1963).

Kuchko, A. A., "Medicinal Herbs and Shrubs in the Birch Woods in Southern Karelia," Collected Works: *Resources of Berries and Medicinal Plants and the Methods in the Study of Them*, Karelian Branch of the USSR Academy of Sciences, Petrozavodsk, 1975, pp. 113-122.

Kudina, Zh. D., "The Influence of Meterological Factors on the Occurrence of the Preimago Developmental Stages in the Corn Borer," *Informative Collected Works*, 1(87), Ukrainian Department of Hydrometeorological Service, Kiev, USSR, 1971, pp. 30-34.

Kulakova, I. T., "Forecasting the Date of the Beginning of Piorcer Caterpillars' Emergence in the Conditions of the Gorky District," Collected Works: *Problems of Indicative Phenology and of Phenological Forecasting*, The USSR Geographical Society, Leningrad, 1972, pp. 40-45.

Kunitzkaya, N. T., V. N. Kunitzky, D. M. Gauzshtein, and N. M. Savelova, "The Physiological Age of Fleas and Analysis of Age Composition of the Natural Population of *Xenopsylla gerbilly Wagn*,"*J. Parasitol.*, Academy of Sciences of the USSR, XI(3), Leningrad, 202-209 (1977).

Kunitzky, V. N., et al., "The Reproduction Rhythm of *Xenopsylla gerbilly* in the Southern Balkhash Territory," *Papers of VIII Scientific Conference of Antiplague Establishments in Middle Asia and Kazakhstan*, Middle-Asian Scientific-Research Antiplague Institute, Alma-Ata, USSR, 1974, pp. 330-333.

Kunitzky, V. N., V. M. Volkov, Z. F. Lelikova, and O. S. Agunkina, "The Number of Generations of *Xenopsylla skrjabini* in the Conditions of the Caspian Lowland," *Papers of the VIII Scientific Conference of the Antiplague Establishments in Middle Asia and Kazakhstan*, Middle-Asian Scientific Research Antiplague Institute, Alma-Ata, USSR, 1974, pp. 328-330.

Learmonth, A. T. A. (England), "On Work of the Commission of Medical Geography," *Program of Reports at the XXIII International Geographical Congress*, Vneshtorgizdat Publishing House, Moscow, 1976.

Linskens, H. F. (Netherlands), "Translocation Phenomena in the Floral Region," *Abstracts of the Papers Presented at the XII International Botanical Congress*, Part II, Nauka Publishing House, Leningrad, 1975, p. 300.

Losev, O. L., A. Ya. Lisenko, Z. I. Martinova, and A. V. Kondrashin, "Methhods Used for Revealing and for Quantitative Evaluation of the Distribution of Infectious Diseases," *Transactions of the Third Scientific Conference on Medical Geography*, Leningrad, 1968, pp. 80-85.

Loske, E. G., *Agricultural Meterology*, Yuryev University, 1913.

Lotka, A. J., *Elements of Mathematical Biology*, Dover, New York, 1956, pp. 1-465.

Luppova, E. P., "The Dynamics of Number and the Gonotrophic Cycle of the *Anopheles superpictus Grassi*—the Major Carrier of Malaria in Stalinabad," *Transactions of the Institute of Zoology and Parasitology*, Tadzhikistan Academy of Sciences, Vol. XXI, Dushanbe, USSR, 1954.

Macfadyen, A., *Animal Ecology, Aims and Methods*, 2nd ed., Sir Issac Pitman and Sons LTD, London, 1963.

Marani, A., "Growth Rate of Cotton Bolls and Their Components," *Field Crops Res.*, Amsterdam, 169-175 (1979).

Marani, A., and D. N. Baker, *Development of a Predictive Dynamic Stimulation Model of Growth and Yield in Acala Cotton*, Hebrew University of Jerusalem, 1981, pp. 1-114.

Markovich, N. Ya., "Geographical Variability of Imaginal Diapause," Collected Works: *Seasonal Phenomena in the Malaria Mosquito Life in the USSR*, Medgiz Publishing House, Moscow, 1957.

Martini, E., *Berechungen und Beobachtungen zur Epidemiologie und Bekämpfung der Malaria*, Hamburg, 1921.

Maurin, A. M., and B. N. Tardov, *Biological Forecasting*, Latvian State University, Riga, 1975, pp. 15-16, 19, 149, 186-187, 190, 199.

Ministry of Agriculture of the USSR, *Information on Arrangements, New Methods, and Equipment Used by the Countries—Members of the Council for Mutual Economic Aid for Forewarning and Forecasting the Appearance and Development of Pests and Illnesses*, Moscow, 1968, pp. 1-27 (for Podolsky's method, see pp. 18-19).

Ministry of Agriculture of the USSR, *Syllabus of the Course "Forewarning and Forecasting the Propagation of Agricultural Pests and Crop Illnesses" for Agricultural Universities*, Moscow, 1975, pp. 1-10.

Mishchenko, Z. A., *The 24-hour Temperature Trend of the Air and Its Agroclimatic Importance*, Gydrometeoizdat Publishing House, Leningrad, 1962.

Monchadsky, A. S., and A. A. Shtakelberg, *Malaria Mosquitoes in Tadzhikistan and Arrangements of Their Control*, Dushanbe, USSR, 1943.

Mordasov, P. M., "Hemosporidioses in Cattle in the Moscow District," *Transactions of the All-Union Institute of Experimental Veterinary*, Selkhozizdat Publishing House, Moscow, 1957.

Moshkovsky, Sh. D., *The Main Principles of the Malaria Epidemiology*, Publishing House of the USSR Academy of Medicinal Sciences, Moscow, 1950, pp. 1-324.

Moshkovsky, Sh. D., M. G. Rashina, N. N. Dukhanina et al., *Epidemiology and Medical Parasitology for Entomologists*, Medgiz Publishing House, Moscow, 1951, pp. 1-456.

Mustafayev, A. R., "Phenoforecasting and Bioclimatic Estimations Using the Phenotemperature Nomograms for Kura-Araxian Lowland (Azerbaidzhan)," *Data on Specialized Hydrometeorological Service for National Economy*, Ministry of Agriculture of Azerbaidzhan, Baku, USSR, 1967, pp. 5-19.

Mustafayev, A. R., "On Constants of Effective Temperature Totals," *Collected Works of the Baku Hydrometobservatory*, Hydrometeorological Service of Azerbaidzhan, Baku, 1967, pp. 128-137.

Nazarova, M. N., and S. I. Mashkin, "Phenotemperature Nomograms and Forecasting the Flowering Dates of the Species of the *Cerasus Juss* Genus," Collected Works: *Thermal Factor in the Development of Plants of Different Geographical Zones*, The Moscow Branch of the USSR Geographical Society, 1979, pp. 90-92.

Nichiporovich, A. A., "Photosynthesis and the Theory of Obtaining High Yields," *Readings Devoted to Timiryazev*, XV, Academy of Sciences of the USSR, Moscow, 1956.

Nichiporovich, A. A., and K. A. Asrorov, "On Some Principles of Optimization of Photosynthetic Activity in Crops," Collected Works: *Photosynthesis and the Application of Solar Energy*, Academy of Sciences of the USSR, Leningrad, 1971, pp. 5-17.

Nikolayev, V. P., "On Effects of Temperature on the Development of Plasmodia in the Mosquito's Body," *Transactions of the Pasteur Institute of Epidemiology and Bacteriology*, Vol. 2, Moscow, 1935.

Ostapenko, G. M., "On Results of Testing the New Method of Phenological Forecasting Proposed by A. S. Podolsky," in *The Informative Collection*, No. 4, Gydrometeoizdat Publishing House, Moscow, 1962, pp. 22-26.

Pavlovsky, E. N., *Manual of the Human Parasitology with the Teaching of Carriers of Transmissible Diseases*, Medgiz Publishing House, Moscow, 1951.

Pavlovsky, E. N., *Natural Endemicity of Vector-Borne Diseases*, Academy of Sciences of the USSR, Moscow-Leningrad, 1964.

Perry, J. N., E. D. Macaulay, and B. T. Emhett, "Phenological and Geographical Relationships between Catches of Pea Moth in Sex-Attractant Traps," *Ann. AP Biol.*, England, **97**(1), 17-26 (1981).

Piontkovsky, Yu. A., *Data on the Biology and Ecology of the Cotton Spider Mite*, SAOGIZ Publishing House, Tashkent, USSR, 1932.

Plokhinsky, N. A., Ed., *Biometrical Methods*, Publishing House of Moscow University, 1975, pp. 1-167.

Podolsky, A. S., "Effects of the Heat Factor on the Development of Crops and Pests (Experimental and Mathematical Heat Phenology)," *J. Agric. of Tadzhikistan*, Dushanbe, USSR, (4), 21-31 (1957).

Podolsky, A. S., "A New Method for Phenological Forecasting and Agroclimatic Estimations Using the Heat Factor," *J. Rep. Acad. Sci. USSR*, Moscow, **121**, (5), 932-935 (1958).

Podolsky, A. S., "A Method for Phenological Forecasting and Agroclimatic Estimations (A Detailed Report)," *J. Agric. of Tadzhikistan*, Dushanbe, USSR, (10), 17-29 (1958).

Podolsky, A. S., "Microclimatic Peculiarities of Radiation Inversions of Temperatures," *Transactions of the Agricultural Institute of Tadzhikistan*, Biological Series, Vol. 1, Ministry of Agriculture of the USSR, Dushanbe, USSR, 1958, pp. 335-354.

Podolsky, A. S., "Weather Forecasting on the Basis of Local Indications," *J. Agric. of Tadzhikistan*, Dushanbe, USSR, (10), 55-58 (1959).

Podolsky, A. S., "On the Early Fall Frosts and Late Spring Frosts," *J. Agric. of Tadzhikistan*, Dushanbe, USSR, (9), 53-55 (1959).

Podolsky, A. S., "Some Peculiarities of the Climate of Mountain Slopes," *J. Proc. Acad. Sci. USSR, Geogr. Ser.*, Moscow, (4), 85-90 (1959).

Podolsky, A. S., "A New Method of Forecasting the Developmental Dates of Agricultural Pests," Collected Works: *Scientific Research in Plant Protection*, Uzbekistan Academy of Agricultural Sciences, Tashkent, USSR, 1960, pp. 62-73.

Podolsky, A. S., *Mathematical Prediction in the Agroclimatology of Plants and Concomitant Pests (for Yavan and Dangara)*, Tadzhikgosizdat Publishing House, Dushanbe, USSR, 1963, pp. 1-37.

Podolsky, A. S., "The Application of Some Mathematical Calculations in Plant Growing," *J. Cotton Growing,* Moscow, (2), 48-52 (1965).

Podolsky, A. S., "Mathematical Forecasting of the Developmental Dates of Poicilothermic Animals and Plants," *J. Agric. Biol.,* Moscow, **1**(1), 151-159 (1966).

Podolsky, A. S., *A Novelty in Phenological Forecasting (Mathematical Forecasting in Ecology),* Kolos Publishing House, Moscow, 1967, pp. 1-232 + 4 Insets.

Podolsky, A. S., *New Principles and Methods for Making Some Ecological Forecasts,* The All-Union Institute of Plant Growing, Leningrad, 1968, pp. 1-62.

Podolsky, A. S., "New Methods of Forecasting the Developmental Dates of Plants, Their Pests, as well as of Causal Agents and Carriers of the Diseases of Warm-Blooded Organisms," in *Problems of Phenological Forecasting,* Geographical Society of the USSR, Leningrad, 1970, pp. 3-4.

Podolsky, A. S., "New Aspects in the Mathematical Method of Phenoforecasting," *Abstracts of the Papers Presented at the Interdepartmental Conference on Indicative Phenology and Phenoforecasting,* The USSR Geographical Society, Leningrad, 1972, p. 6.

Podolsky, A. S., *Phenological Forecasting (Mathematical Forecasts in Ecology),* 2nd ed., extended and revised, Kolos Publishing House, Moscow, 1974, pp. 1-287 + 5 Insets.

Podolsky, A. S., *Methodical Instructions to Forecasting the Developmental Dates of Pests, Illnesses, and Crops Hurt by Them,* Ministry of Agriculture of the USSR, Moscow, 1974, pp. 1-101.

Podolsky, A. S., *Phenoforecasting Calendars for the Colorado Potato Beetle,* The Kursk Center of Scientific-Technical Information, 1975, pp. 1-8.

Podolsky, A. S., "New Principles and Methods of Phenological Forecasting and Bioclimatic Estimations," *Abstracts of the Papers Presented by the Soviet Participators at the VIII International Congress on Plant Protection,* Publisher: Organization Committee of Congress, Moscow, 1975, pp. 40-42.

Podolsky, A. S., "The Study of Biogeocenoses and the Introduction of Plants Using the New Method of Phenological Forecasting and Bioclimatic Estimations," *Abstracts of Papers Presented at the XII International Botanical Congress,* Part I, Nauka Publishing House, Leningrad, 1975, p. 196.

Podolsky, A. S., "Phenological Forecasting as a Method for Study and Management of Biogeocenoses," in A. M. Grin and V. D. Utekhin, Eds., Collection: *Biogeophysical and Mathematical Methods for Investigations of Geosystems,* Institute of Geography of the USSR Academy of Sciences, Moscow, 1978, pp. 223-243.

Podolsky, A. S., and A. F. Brudnaya, "More Precise and Concrete Calculation of the Early Fall Frosts and Late Spring Frosts," *Collected Works of the Dushanbe Hydrometeorological Observatory,* (1), Hydrometeorological Service of Tadzhikistan, Tashkent, USSR, 1964, pp. 67-75.

Podolsky, A. S., and V. V. Denisov, "Solving the Phenological Problems by Combining Phenological Curves with the Annual Temperature Trend," *J. Agric. Biol.,* Moscow, **X**(1), 118-122 (1975).

Podolsky, A. S., and V. V. Denisov, "Solving the Phenological Problems Using the Annual Temperature Trend, without Use of Totals, Thresholds, and Nets," *Transactions of the Voronezh and Kursk Agricultural Institutes,* Vol. 76, Ministry of Agriculture of the USSR, Voronezh, USSR, 1975, pp. 73-79.

Podolsky, A. S., and I. F. Pustovoy, "The Forecast for the Developmental Dates of Larvae of the *Haemonchus* and *Bunostomum,*" Collected Works: *Natural Endemic Diseases and the Problems of Parasitology in Middle Asian Republics and in Kazakhstan* (5), Donish Publishing House, Dushanbe, 1969, pp. 175-179.

Podolsky, A. S., and E. A. Sadomov, "Mathematical Forecasting of the Developmental Dates of the Fall Webworm," *J. Plant Protection,* Moscow (9), 48-49 (1974).

Podolsky, A. S., and E. A. Sadomov, "Mathematical Forecasting of the Developmental Dates of the Fall Webworm in the Odessa District," *Plant Quarantine (Methodical Instructions)* (21), Kolos Publishing House, Moscow, 1976, pp. 34-42.

Podolsky, A. S., and E. A. Sadomov, "Mathematical Forecasting of the Developmental Dates of the Fall Webworm in Moldavia," *Quarantine Pests and Plant Illnesses*, Part I, Moscow, (3), 178-181 (1977).

Podolsky, A. S., and A. A. Slepov, "Forecasting the Inactivation Dates of the Foot-and-mouth Disease Virus in the Environment," *Urgent Problems of Veterinary Virology*, Abstracts of Papers, Part I, VASHNIL, Vladimir, USSR, 1976, pp. 179-180.

Podolsky, A. S., and O. S. Yermakov, *Phenoforecasting Calendars for Some Spring Crops (Wheats, Barley, Oats)*, The Kursk Center of Scientific-Technical Information, USSR, 1973, pp. 1-8.

Podolsky, A. S., and O. S. Yermakov, *Phenoforecasting Calendars for the Turnip Moth and Trichogramma evanescens Westw.*, The Kursk Center of Scientific-Technical Information, USSR, 1973, pp. 1-8.

Podolsky, A. S., and O. S. Yermakov, *Phenoforecasting Calendars for the Piorcer and Trichogramma cacoecia pallida Meyer.*, The Kursk Center of Scientific-Technical Information, USSR., 1973, pp. 1-8.

Podolsky, A. S., and O. S. Yermakov, "Combined Taking into Account the Temperature and Day Length Using the New Method of Phenological Forecasting," *Transactions of the Voronezh and Kursk Agricultural Institutes*, VIII(5), Ministry of Agriculture of the USSR, Voronezh, USSR, 1973, pp. 139-151.

Podolsky, A. S., and O. S. Yermakov, "The Application of the New Method of Phenological Forecasting in the Central Black-Earth Zone," *Transactions of the Kursk and Voronezh Agricultural Institutes*, VIII(5), Ministry of Agriculture of the USSR, Voronezh, USSR., 1973, pp. 152-157.

Podolsky, A. S., S. Alikhanyan, N. Amosov, A. Blokhin, Ya. Zeldovich, et al., Annual Collection: *Evrika-73*, Molodaya Gvardia Publishing House, Moscow, 1973, pp. 1-480 ("Phenological Forecast," pp. 357-360).

Podolsky, A. S., R. I. Babayeva, and A. N. Brudastov, "Forecasting the Survival Dates of Oncospheres of the Beef Tapeworm for Various Districts of Uzbekistan," in *Urgent Problems of Medical Parasitology*, (1), Uzbekistan Institute for Scientific Research in Medical Parasitology, Samarkand, USSR, 1973, pp. 65-71.

Podolsky, A. S., L. V. Shatunova, and V. D. Tokareva, "Mathematical Forecasting of the Developmental Dates of Medicinal Plants for the Purposes of their Rational Procurement (on the Example of the Lily-of-the-Valley)," *J. Veg. Resour.*, Academy of Sciences of the USSR, Leningrad, X(3), 320-328 (1974).

Podolsky, A. S., V. D. Tokareva, and L. V. Shatunova, "Mathematical Forecasting of the Developmental Dates of Medical Herbaceous Plants," Collection: *Seasonal Development of Nature*, Moscow Branch of the USSR Geographical Society, 1975, pp. 71-75.

Podolsky, A. S., L. V. Shatunova, and V. D. Tokareva, "Phenoforecasting Calendars for the Lily-of-the-Valley," Collection: *Seasonal Development of Nature (the Phenology of Medicinal Plants)*, Moscow Branch of the USSR Geographical Society, 1977, pp. 7-9.

Polyakov, I. Ya., *Forecasting the Distribution of Agricultural Pests*, Kolos Publishing House, Leningrad, 1964, pp. 1-326.

Polyakov, I. Ya., "Scientific Bases for Using the Agroclimatic Units in Plant Protection," *Trans. VIZR* Leningrad, (38), 5-10 (1972).

Polyakov, I. Ya., et al., *Forecasting the Development of Agricultural Pests*, Kolos Publishing House, Leningrad, 1975, pp. 1-238.

Porodenko, V. V., "Methods for the Calculation of the Incubation Period of the Stem Rust

Causal Agent on Wheats," *J. Mycol. and Phytopathol.*, Academy of Sciences of the USSR, Leningrad, **6**(3), 277-283 (1972).

Pravikov, G. A., and L. V. Popov, "Observations on the Phenology of Malaria Mosquitoes in Turkmenistan," Collection: *Seasonal Phenomena in the Malaria Mosquito Life in the Soviet Union*, Medgiz Publishing House, Moscow, 1957.

Protserov, A. V., and E. S. Ulanova, *Applicability of the Methods of Agrometeorological Forecasting in Different Regions of the USSR (Results of Testing in Production)*, Gydrometeoizdat Publishing House, Moscow, 1961.

Psaryova, V. V., and L. S. Kononuchenko, "The Phytophthora Disease of Tomatoes and Control Arrangements Against It," *Express-Information*, Byelorussian Research Institute of Scientific-Technical Information, Minsk, USSR, 1978, pp. 1-21.

Pustovoy, I. F., "The Method of Phenological Forecasting in Helminthology, on the Basis of the Heat Factor," *Collected Works on Helminthology Published in Commemoration of the 90th Anniversary of Academician K. I. Skryabin*, Nauka Publishing House, Moscow, 1968, pp. 290-297.

Rashevsky, N., *Some Medical Aspects of Mathematical Biology*, Charles C. Thomas Publisher, Springfield, Ill., 1964.

Rivnay, E., and J. Perzelan, "Insects Associated with *Pseudococcus spp.* (Homoptera) in Palestine, with Notes on their Biology and Economic Status," *J. Ent. Soc. S. Africa*, **6,** 9-28 (September 1943).

Rodd, A., "Summarizing the Experiences in Cotton Protection against the Cotton Bollworm," *A Brief Report of the Middle-Asian Station of Plant Protection for 1951*, Tashkent, USSR, 1952.

Ross, R., *The Prevention of Malaria*, London, 1911.

Round, F. E., "The Ecology of Benthic Algae," in D. F. Jackson, Ed., *Algae and Man*, Plenum Press, New York, 1964, p. 179.

Rudenko, A. I., *Determining the Development Stages in Crops*, Publisher: Moscow Society of Naturalists, Moscow, 1950.

Rudnev, G. V., *Agricultural Meteorology*, Gydrometeoizdat Publishing House, Leningrad, 1964, pp. 256-258.

Rudnev, G. V., *Agricultural Meteorology*, 2nd ed., revised and extended, Gydrometeoizdat Publishing House, Leningrad, 1973, pp. 300-305.

Ryabov, V. A., "The Relationship of the Development of Arboreal Plants to the Air Temperature, and Phenological Forecasting," Collection: *Seasonal Development of Nature in the European Section of the USSR*, Moscow Branch of the USSR Geographical Society, 1974, pp. 8-9.

Ryabov, V. A., "Experience of Phytophenological Forecasting in the Central Black-Earth Reservation," *Abstracts of the Papers at the Conference on Theoretical Problems of the Reservation Affairs in the USSR*, Chief Department of Hunt and Reservations, Kursk, 1975, pp. 26-27.

Ryabov, V. A., "Arboreal Plants and the Heat Factor," *Papers on the Study of Natural Ecosystems in the Central Forest-Steppe Region of the Russian Plain*, Transactions of the Reservation (XIII), Central Black-Earth Publishing House, Voronezh, 1977, pp. 33-41.

Ryabov, V. A., "Phenological Forecasting of the Common Valerian," Collection: *Seasonal Development of Nature*, Moscow Branch of the USSR Geographical Society, 1977, pp. 26-31.

Ryabov, V. A., "Forecasting the Flowering of Melliferous Plants," *J. Bee Farming*, Moscow, (4), 11-13 (1978).

Ryabov, V. A., "Experience of Phytophenological Forecasting in the Central Black-Earth Reservation," *Bot. J.*, Academy of Sciences of the USSR, Leningrad, **63** (11), 1656-1663 (1978).

Ryabov, V. A., "Plant Sensitivity to Heat at the Period Before the Beginning of Flowering," Collection: *Thermal Factor in Plant Development*, Moscow Branch of the USSR Geographical Society, 1979, pp. 104-105.

Ryabtseva, N. I., et al., *Pests and Crop Illnesses in Tadzhikistan*, Tadzhikgosizdat Publishing House, Dushanbe, 1964, pp. 1-399.

Sadomov, E. A., "On the Problem of Chances of the Fall Webworm's Acclimatization in the European Section of the USSR," *Abstracts of Papers of the Soviet Participators at the VIII International Congress on Plant Protection*, Publisher: Organization Committee of Congress, Moscow, 1975, pp. 298-299.

Samurov, M. A., "Forecasting the Numbers of *Xenopsylla conformis* fleas in the Volga-Urals Sand Territory," *Zoo. J.* USSR Academy of Sciences, LVI(11), 1649-1653 (1977).

Savelyev, O. L., and M. A. Kuznetzova, "Importance of Phenological Observations for the Optimization of Dates of Gathering the Medicinal Plant Raw Materials," Collection: *Seasonal Development of Nature (the Phenology of Medicinal Plants)*, Moscow Branch of the USSR Geographical Society, 1977, pp. 3-6.

Schrödinger, Erwin, *What Is Life? The Physical Aspect of the Living Cell*, 1955.

Selyaninov, G. T., "On the Agricultural Estimation of Climate," *Transactions on Agricultural Meteorology* (XX), AGMI, Leningrad, 1928.

Selyaninov, G. T., Ed., *World Agroclimatic Reference Book*, AGMI, Leningrad-Moscow 1937.

Selyaninov, G. T., "The Afterword to L. N. Babushkin's Article," *Transactions on Agricultural Meteorology* (XXV), AGMI, Moscow-Leningrad, 1938, pp. 110-111.

Sergiev, P. G., and N. N. Dukhanina, *System of Arrangements on Preventing the Malaria Occurrence in the USSR, and Epidemiological Bases of this System*, Moscow, 1961.

Sergiev, P. G., and A. I. Yakusheva, *Malaria and Control of It in the USSR*, Medgiz Publishing House, Moscow, 1956, pp. 1-307.

Sharpe, Peter J. H., and Don W. DeMichele, "Reaction Kinetics of Poikilotherm Development," *J. Theor. Biol.*, **64**, 649-670 (1977).

Shashko, D. I., *Climatic Conditions of the Agriculture in the Central Yakutia*, Publishing House of the USSR Academy of Sciences, 1961.

Shatilov, I. S., "Ecological, Biological, and Agrotechnical Conditions for Obtaining the Planed Yields," *Proc. Timiryazev Agric. Acad.*, Moscow, (1), 60-68 (1970).

Shatsky, A. L., "On the Temperature Total as an Agricultural-Climatic Index," *Transactions on Agricultural Meteorology* (XXI), AGMI, Leningrad, 1930.

Sheludko, A. D., "Zones of Harmfulness of the Weevil *Tanymecus dilaticollis*, Forecasting Its Numbers and Forewarning the Dates of Emergence after Wintering," *Informative Collection*, 1(87), Hydrometeorological Department of the Ukraine, Kiev, 1971, pp. 26-29.

Shchotkin, N. L., *Higher Scaly-Winged Organisms in the Vakhsh Valley (Tadzhikistan)*, Transactions of E. N. Pavlovsky Institute of Zoology and Parasitology, Vol. XIX, Dushanbe, USSR, 1960.

Shigolev, A. A., *A Guide in the Treatment of Phenological Observations and Making Phenological Forecasts*, Moscow, 1941.

Shigolev, A. A., *A Guide in Making Phenological Forecasts* (Methodical Instructions of Central Forecast Institute) (15), (25), Gydrometeoizdat Publishing House, Leningrad, 1951, 1954, pp. 1-44.

Shigolev, A. A., *A Guide in the Control and Treatment of Observations of Crop Developmental Stages*, Gydrometeoizdat Publishing House, Leningrad, 1955.

Shipitzina, N. K., "Characteristic Features of the Autumn Phenological Development of *Anopheles maculipennis* in the Soviet Union," Collected Works: *Seasonal Phenomena in the Malaria Mosquito Life*, Medgiz Publishing House, Moscow, 1957.

Shipitzina, N. K., "Seasonal Periodicity in the Life of *A. maculipennis* Malaria Mosquito and the Importance of the Study of this Periodicity for the Malaria Control in the Soviet Union," Collected Works: *Seasonal Phenomena in the Malaria Mosquito Life,* Medgiz Publishing House, Moscow, 1957.

Shipitzina, N. K., "Phenological Bases of the Rationalization of Antimalarial Arrangements and Rational Arrangement Dates in the Soviet Union," Collected Works: *Seasonal Phenomena in the Malaria Mosquito Life,* Medgiz Publishing House, Mosow, 1957.

Shultz, G. E., "Intrazonal Phenological Parallels," *Bot. J.,* Leningrad, (8), (1967).

Shultz, G. E., "Phytophenology at the XII International Botanical Congress," *Bot. J.,* Leningrad, **63,** 928-932 (1978).

Shultz, G. E., Thermal Regime as a Factor of Seasonal Development of Higher Plants," Collection: *Thermal Factor in the Development of Plants in Different Geographical Zones,* Moscow Branch of the USSR Geographical Society, 1979, pp. 4-5.

Sibiryakova, O. A., and M. A. Gobova, "Duration and Intensity of the *A. m. messeae's* Diapause in Irkutsk," Collection: *Seasonal Phenomena in the Malaria Mosquito Life in the Soviet Union,* Medgiz Publishing House, Moscow, 1957.

Sinitsina, N. I., I. A. Goltzberg, and E. A. Strunnikov, *Agricultural Climatology,* Gydrometeoizdat Publishing House, Leningrad, 1973, pp. 1-344.

Smirnov, V. A., *Crops Sown after Harvest and Climate,* Gydrometeoizdat Publishing House, Leningrad, 1960, pp. 1-98.

Stativkin, V. G., Ed., *Survey of the Development and Distribution of Pests and Crop Illnesses in Tadzhikistan in 1959 and Forecast of their Occurrence in 1960,* Tadzhikistan Ministry of Agriculture, Dushanbe, USSR, 1959, pp. 7, 21.

Stepanov, K. M., and A. E. Chumakov, *Forecasting the Crop Illnesses,* Kolos Publishing House, Leningrad, 1972, pp. 1-271.

Stinner, R. E., et al., "Simulation of Temperature-Dependent Development in Population Dynamics Models," *J. Can. Entomol.,* **107**(11), 1167-1174 (1975).

Storozheva, A. Ya., "The Biology of *Anopheles maculipennis mg.* and the Significance of this Insect in the Malaria Epidemiology in the Malo-Kabardian District," Collection: *Problems of the Malaria Mosquito Physiology and Ecology,* Moscow, 1948.

Storozheva, A. Ya., "Digestion Rate at Different Temperatures and the Duration of the Diapause of *A. m. sacharovi* Females in Tashauz (Turkmenia)," Collection: *Seasonal Phenomena in the Malaria Mosquito Life,* Medgiz Publishing House, Moscow, 1957.

Sukhov, K. S., and G. M. Razvyazina, *The Biology of Viruses and Viral Diseases of Plants,* Sovetskaya Nauka Publishing House, Moscow, 1955.

Sumanov, E. Ya., "The Influence of the Ear Coloration on the Developmental Dates and Other Biological Indices of Wheats," *Problems of Phenological Forecasting,* Geographical Society of the USSR, Leningrad, 1970, pp. 19-20.

Suslov, I. M., "On Methods of Forecasting the Ascaridiasis," *J. Med. Parasitol. and Parasit. Dise.,* Moscow, (2), 19-29 (1973).

Tarshis, M. G., and V. M. Konstantinov, *Mathematical Methods in Epizootology,* Kolos Publishing House, Moscow, 1975.

Tavrovsky, V. A., "Conference on the Phenological Forecasting Problem," *Zoo. J.,* Leningrad, **XLIX**(10), (1970).

Ter-Avanesyan, D. V., "On the Genetics of the Cotton Vegetation Period," *J. Proc. Acad. Sci. Uzbekistan,* Tashkent, (4) (1949).

Ter-Avanesyan, D. V., "On Methods for the Evaluation of the Ripening Rate of Cotton Varieties," *J. Cotton Growing,* Moscow, (8) (1954).

Ter-Avanesyan, D. V., "Intravarietal Crossing of the Cotton Crops Sown in Different Times," *J. Agric. Biol.,* **6**(90) (1954).

Ter-Avanesyan, D. V., and N. G. Protsenko, "Geographical Variability of Both Patterns of the World Cotton Collection and their Hybrids," *Transactions on Applied Botany, Genetics, and Crop Breeding*, **XXX**(3), VIR, Leningrad, 1957.

Terentyev, P. V., Ed., *The Application of Mathematical Methods in Biology*, Leningrad University Publishing House, 1960; 1963, pp. 1-227; 1-239.

Terentyev, P. V., "On the History of Biometry," *Biometric Methods*, Moscow-Leningrad, 5-22 (1977).

The All-Union Institute of Plant Protection (VIZR), in I. Ya. Polyakov, Ed., *Review of the Distribution of the Main Mass Pests and Crop Illnesses in 1959 and the Forecast of their Occurrence in 1960*, Leningrad, 1960, pp. 66-68.

Ulanova, E. S., *Methods of Agrometeorological Forecasting*, Gydrometeoizdat Publishing House, Leningrad, 1959, pp. 1-280.

Ushatinskaya, R. S., "Physiological Peculiarities of Insect Diapause," *Transactions of the 3rd Ecological Conference*, Part 1, Kiev, 1954.

Vannovsky, T. Ya., "Epizootological Peculiarities of the Occurrence and Development of Hemoparasitical Diseases in Cattle, and the Efficiency of Preventive Arrangements against these Diseases in Tadzhikistan," *Transactions of Tadzhikistan Agricultural Institute*, Ministry of Agriculture of the USSR, Vol. 5, Dushanbe, USSR, 1965.

Vavilov, N. I., *Tadzhikistan's Cultivated Flora in its Past and Future, Selected Works*, Vol. V, Nauka Publishing House, Moscow-Leningrad, 1965.

Vavilov, N. I., *The World Experience in the Agricultural Development of Highlands, Selected Works*, Vol. V, Nauka Publishing House, Moscow-Leningrad, 1965.

Vavilov, N. I., *World Centers of Varieties Resources, Selected Works*, Vol. V, Nauka Publishing House, Moscow-Leningrad, 1965.

Vavilov, N. I., *Scientific Foundations of the Introduction and Search of New Crops, Selected Works*, Vol. V, Nauka Publishing House, Moscow-Leningrad, 1965.

Ventzkevich, G. Z., "From the Experience of Critical Treatment of Phenological Data Given in the Agroclimatical Annuals," *Transactions of Central Institute of Forecasts* (98), Gydrometeoizdat Publishing House, Moscow, 1960.

Verderevsky, D. D., et al., *Methods of Forewarning the Dates of Chemical Treatment against the Mildew in Vineyards*, Kishinev, Moldavia, USSR, 1961.

Vinogradskaya, O. N., "Drought Resistance and Hydrophilia of Anopheles Species," *J. Med. Parasitol. and Parasit. Dis.*, Moscow, (2) (1945).

Vitkevich, V. I., *Agricultural Meterology*, Selkhozgiz Publishing House, Moscow, 1960, pp. 284-290, and 1965 (in Russian).

Vitkevich, V. I., *Agricultural Meteorology*, Program for Scientific Translations, Jerusalem, 1963, pp. 194-198 (translated from Russian).

Volvach, V. V., *Methodical Instructions for Hydrometeorological Stations and Posts on Forecasting the Dates of the Colorado Potato Beetle's Development*, Chief Department of Hydrometeorological Service, Institute of Experimental Meteorology, Gydrometeoizdat Publishing House, Moscow, 1975, pp. 1-16.

Wang, J. Yu., *Agricultural Meteorology*, Madison Pacemaker Press, Milwaukee, 1963, pp. 124-126, 128-129, 271, 399, University of Wisconsin.

Wang, J. Yu., *Agricultural Meteorology*, 2nd ed., Milieu Information Service, 1967, California State University.

Wang, J. Yu., *Agricultural Meteorology*, 3rd ed., revised, Milieu Information Service, 1972, pp. 290, 353-354, 369, California State University.

Waterman, T. H., and H. T. Morowitz, *Theoretical and Mathematical Biology*, Blaisdell pub., New York-Toronto-London, 1965.

Wellings, P. W., "The Effect of Temperature on the Growth and Reproduction of Two Closely Related Aphid Species on Sycamore," *Econ. Entomol.*, **6**(2) (1981).

Wiggles-Worth, B. B., *The Principles of Insect Physiology*, Methuen, London, 1965.

Willard, J. R., "Studies on Rates of Development and Reproduction of California Red Scale *Aonidiella aurantii (Mask)* on Citrus," *J. Zool.*, **20**, 37-47 (1972).

Yakunin, B. M., N. A. Chernova, and N. T. Kunitzkaya, "On the Number of Generations of *Xenopsylla skrjabini* Fleas (Aphaniptera) in Mangyshlak," *J. Parasitol.*, Academy of Sciences of the USSR, Leningrad, **XIII**(5), 510-515 (1979).

Yermakov, O. S., "The Application of the Mathematical Phenoforecasting Method in the Central Black-Earth Zone's Districts," *Abstracts of the Papers Presented at the Conference on the Indicatory Phenology and Phenoforecasting*, the USSR Geographical Society, Leningrad, 1972, p. 12.

Author Index

Subject Index